MW00417111

MATHEMATICAL STATISTICS WITH RESAMPLING AND R

MATHEMATICAL STATISTICS WITH RESAMPLING AND R

LAURA CHIHARA
Carleton College

TIM HESTERBERG
Google

Published by John Wiley & Sons, Inc., Hoboken, New Jersey
Published simultaneously in Canada

For general information on our other products and services or for technical support, please contact our Customer Care Department within the United States at (800) 762-2974, outside the United States at (317) 572-3993 or fax (317) 572-4002.

Wiley also publishes its books in a variety of electronic formats. Some content that appears in print may not be available in electronic formats. For more information about Wiley products, visit our web site at www.wiley.com.

Library of Congress Cataloging-in-Publication Data:

ISBN: 978-1-118-02985-5

Printed in Singapore.

10 9 8 7 6 5 4 3 2 1

The world seldom notices who teachers are;
but civilization depends on what they do.
— Lindley Stiles[*]

To
Theodore S. Chihara

To
Bev Hesterberg

[*]Stiles founded The Best Should Teach Initiative http://thebestshouldteach.org

CONTENTS

Preface **xiii**

1 Data and Case Studies **1**

1.1 Case Study: Flight Delays, 1
1.2 Case Study: Birth Weights of Babies, 2
1.3 Case Study: Verizon Repair Times, 3
1.4 Sampling, 3
1.5 Parameters and Statistics, 5
1.6 Case Study: General Social Survey, 5
1.7 Sample Surveys, 6
1.8 Case Study: Beer and Hot Wings, 8
1.9 Case Study: Black Spruce Seedlings, 8
1.10 Studies, 8
1.11 Exercises, 10

2 Exploratory Data Analysis **13**

2.1 Basic Plots, 13
2.2 Numeric Summaries, 16
2.2.1 Center, 17
2.2.2 Spread, 18
2.2.3 Shape, 19
2.3 Boxplots, 19
2.4 Quantiles and Normal Quantile Plots, 20
2.5 Empirical Cumulative Distribution Functions, 24
2.6 Scatter Plots, 26

2.7 Skewness and Kurtosis, 28
2.8 Exercises, 30

3 **Hypothesis Testing** **35**

3.1 Introduction to Hypothesis Testing, 35
3.2 Hypotheses, 36
3.3 Permutation Tests, 38
3.3.1 Implementation Issues, 42
3.3.2 One-Sided and Two-Sided Tests, 47
3.3.3 Other Statistics, 48
3.3.4 Assumptions, 51
3.4 Contingency Tables, 52
3.4.1 Permutation Test for Independence, 54
3.4.2 Chi-Square Reference Distribution, 57
3.5 Chi-Square Test of Independence, 58
3.6 Test of Homogeneity, 61
3.7 Goodness-of-Fit: All Parameters Known, 63
3.8 Goodness-of-Fit: Some Parameters Estimated, 66
3.9 Exercises, 68

4 **Sampling Distributions** **77**

4.1 Sampling Distributions, 77
4.2 Calculating Sampling Distributions, 82
4.3 The Central Limit Theorem, 84
4.3.1 CLT for Binomial Data, 87
4.3.2 Continuity Correction for Discrete Random Variables, 89
4.3.3 Accuracy of the Central Limit Theorem, 90
4.3.4 CLT for Sampling Without Replacement, 91
4.4 Exercises, 92

5 **The Bootstrap** **99**

5.1 Introduction to the Bootstrap, 99
5.2 The Plug-In Principle, 106
5.2.1 Estimating the Population Distribution, 107
5.2.2 How Useful Is the Bootstrap Distribution?, 109
5.3 Bootstrap Percentile Intervals, 113
5.4 Two Sample Bootstrap, 114
5.4.1 The Two Independent Populations Assumption, 119
5.5 Other Statistics, 120
5.6 Bias, 122
5.7 Monte Carlo Sampling: The "Second Bootstrap Principle", 125
5.8 Accuracy of Bootstrap Distributions, 125
5.8.1 Sample Mean: Large Sample Size, 126

5.8.2 Sample Mean: Small Sample Size, 127
5.8.3 Sample Median, 127
5.9 How Many Bootstrap Samples are Needed?, 129
5.10 Exercises, 129

6 Estimation **135**

6.1 Maximum Likelihood Estimation, 135
6.1.1 Maximum Likelihood for Discrete Distributions, 136
6.1.2 Maximum Likelihood for Continuous Distributions, 139
6.1.3 Maximum Likelihood for Multiple Parameters, 143
6.2 Method of Moments, 146
6.3 Properties of Estimators, 148
6.3.1 Unbiasedness, 148
6.3.2 Efficiency, 151
6.3.3 Mean Square Error, 155
6.3.4 Consistency, 157
6.3.5 Transformation Invariance, 160
6.4 Exercises, 161

7 Classical Inference: Confidence Intervals **167**

7.1 Confidence Intervals for Means, 167
7.1.1 Confidence Intervals for a Mean, σ Known, 167
7.1.2 Confidence Intervals for a Mean, σ Unknown, 172
7.1.3 Confidence Intervals for a Difference in Means, 178
7.2 Confidence Intervals in General, 183
7.2.1 Location and Scale Parameters, 186
7.3 One-Sided Confidence Intervals, 189
7.4 Confidence Intervals for Proportions, 191
7.4.1 The Agresti–Coull Interval for a Proportion, 193
7.4.2 Confidence Interval for the Difference of Proportions, 194
7.5 Bootstrap t Confidence Intervals, 195
7.5.1 Comparing Bootstrap t and Formula t Confidence Intervals, 200
7.6 Exercises, 200

8 Classical Inference: Hypothesis Testing **211**

8.1 Hypothesis Tests for Means and Proportions, 211
8.1.1 One Population, 211
8.1.2 Comparing Two Populations, 215
8.2 Type I and Type II Errors, 221
8.2.1 Type I Errors, 221
8.2.2 Type II Errors and Power, 226
8.3 More on Testing, 231

8.3.1 On Significance, 231
8.3.2 Adjustments for Multiple Testing, 232
8.3.3 P-values Versus Critical Regions, 233
8.4 Likelihood Ratio Tests, 234
8.4.1 Simple Hypotheses and the Neyman–Pearson Lemma, 234
8.4.2 Generalized Likelihood Ratio Tests, 237
8.5 Exercises, 239

9 Regression **247**

9.1 Covariance, 247
9.2 Correlation, 251
9.3 Least-Squares Regression, 254
9.3.1 Regression Toward the Mean, 258
9.3.2 Variation, 259
9.3.3 Diagnostics, 261
9.3.4 Multiple Regression, 265
9.4 The Simple Linear Model, 266
9.4.1 Inference for α and β, 270
9.4.2 Inference for the Response, 273
9.4.3 Comments About Assumptions for the Linear Model, 277
9.5 Resampling Correlation and Regression, 279
9.5.1 Permutation Tests, 282
9.5.2 Bootstrap Case Study: Bushmeat, 283
9.6 Logistic Regression, 286
9.6.1 Inference for Logistic Regression, 291
9.7 Exercises, 294

10 Bayesian Methods **301**

10.1 Bayes' Theorem, 302
10.2 Binomial Data, Discrete Prior Distributions, 302
10.3 Binomial Data, Continuous Prior Distributions, 309
10.4 Continuous Data, 316
10.5 Sequential Data, 319
10.6 Exercises, 322

11 Additional Topics **327**

11.1 Smoothed Bootstrap, 327
11.1.1 Kernel Density Estimate, 328
11.2 Parametric Bootstrap, 331
11.3 The Delta Method, 335
11.4 Stratified Sampling, 339
11.5 Computational Issues in Bayesian Analysis, 340
11.6 Monte Carlo Integration, 341

11.7 Importance Sampling, 346
11.7.1 Ratio Estimate for Importance Sampling, 352
11.7.2 Importance Sampling in Bayesian Applications, 355
11.8 Exercises, 359

Appendix A Review of Probability **363**

A.1 Basic Probability, 363
A.2 Mean and Variance, 364
A.3 The Mean of a Sample of Random Variables, 366
A.4 The Law of Averages, 367
A.5 The Normal Distribution, 368
A.6 Sums of Normal Random Variables, 369
A.7 Higher Moments and the Moment Generating Function, 370

Appendix B Probability Distributions **373**

B.1 The Bernoulli and Binomial Distributions, 373
B.2 The Multinomial Distribution, 374
B.3 The Geometric Distribution, 376
B.4 The Negative Binomial Distribution, 377
B.5 The Hypergeometric Distribution, 378
B.6 The Poisson Distribution, 379
B.7 The Uniform Distribution, 381
B.8 The Exponential Distribution, 381
B.9 The Gamma Distribution, 382
B.10 The Chi-Square Distribution, 385
B.11 The Student's t Distribution, 388
B.12 The Beta Distribution, 390
B.13 The F Distribution, 391
B.14 Exercises, 393

Appendix C Distributions Quick Reference **395**

Solutions to Odd-Numbered Exercises **399**

Bibliography **407**

Index **413**

PREFACE

Mathematical Statistics with Resampling and R is a one-term undergraduate statistics textbook for sophomores or juniors who have taken a course in probability (at the level of, for instance, Ross (2009), Ghahramani (2004), and Scheaffer and Young (2010)) but may not have had any previous exposure to statistics.

What sets this book apart from other mathematical statistics texts is the use of modern resampling techniques—permutation tests and bootstrapping. We begin with permutation tests and bootstrap methods before introducing classical inference methods. Resampling helps students understand the meaning of sampling distributions, sampling variability, P-values, hypothesis tests, and confidence intervals. We are inspired by the textbooks of Waldrop (1995) and Chance and Rossman (2005), two innovative introductory statistics books that also take a nontraditional approach in the sequencing of topics.

We believe the time is ripe for this book. Many faculty have learned resampling and simulation-based methods in graduate school and use them in their own work and are eager to incorporate these ideas into a mathematical statistics course. Students and faculty today have access to computers that are powerful enough to perform resampling quickly.

A major topic of debate about the mathematical statistics course is how much theory to introduce. We want mathematically talented students to get excited about statistics, so we try to strike a balance between theory, computing, and applications. We feel that it is important to demonstrate some rigor in developing some of the statistical ideas presented here, but that mathematical theory should not dominate the text. And of course, if additions are made to a syllabus, then deletions must also be made. Thus, some topics such as sufficiency, Fisher information, and ANOVA have been omitted in order to make room for permutation testing, bootstrap; and other modern computing methods (though we plan to make some of these omitted topics available as supplements on the text web page `https://sites.google.com/site/ChiharaHesterberg`). This site will also contain `R` scripts for the text, and errata.

We have compiled the definitions and theorems of the important probability distributions in Appendix B. Instructors who want to prove results on distributional theory can refer to this appendix. Instructors who wish to skip the theory can continue without interrupting the flow of the statistical discussion.

Incorporating resampling and bootstrapping methods requires that students use statistical software. We use `R` because it is freely available (`http://www.r-project.org/`), powerful, flexible, and a valuable tool in future careers. One of us works at Google where there is an explosion in the use of `R`, with more and more nonstatisticians learning `R` (the statisticians already know it). We realize that the learning curve for `R` is high, but believe that the time invested in mastering `R` is worth the effort. We have written some basic materials on `R` that are available on the web site for this text. We recommend that instructors work through the introductory worksheet with the students on the first or second day of the term, in a computer lab if possible. We also provide `R` script files with code found in the text and additional examples.

Statistical computing is necessary in statistical practice and for people working with data in a wide variety of fields. There is an explosion of data—more and more data—and new computational methods are continuously being developed to handle this explosion. Statistics is an exciting field, dare we even say sexy?[1]

ACKNOWLEDGMENTS

This textbook could not have been completed without the assistance of many colleagues and students. In particular, we would like to thank Professor Katherine St. Clair of Carleton College who bravely tested an early rough draft in her Introduction to Statistical Inference class during Winter 2010. In addition, Professor Julie Legler of St. Olaf College adopted the manuscript in her Statistical Theory class during Fall 2010. Both instructors and students provided valuable feedback that improved the exposition and content of this textbook.

We would also like to thank Siyuan (Ernest) Liu and Chen (Daisy) Sun, two Carleton College students, for solving many of the exercises and writing up the solutions with LaTeX.

Finally, the staff at Wiley, including Steve Quigley, Sanchari Sil, Dean Gonzalez, and Jackie Palmieri, provided valuable assistance in preparing this book.

LAURA CHIHARA
Northfield, MN
TIM HESTERBERG
Seattle, WA

[1] Try googling "statistics sexy profession."

1

DATA AND CASE STUDIES

Statistics is the art and science of collecting and analyzing data and understanding the nature of variability. Mathematics, especially probability, governs the underlying theory, but statistics is driven by applications to real problems.

In this chapter, we introduce several data sets that we will encounter throughout the text in the examples and exercises.

1.1 CASE STUDY: FLIGHT DELAYS

If you have ever traveled by air, you probably have experienced the frustration of flight delays. The Bureau of Transportation Statistics maintains data on all aspects of air travel, including flight delays at departure and arrival (`http://www.bts.gov/xml/ontimesummarystatistics/src/index.xml`).

LaGuardia Airport (LGA) is one of three major airports that serves the New York City metropolitan area. In 2008, over 23 million passengers and over 375,000 planes flew in or out of LGA. United Airlines and American Airlines are two major airlines that schedule services at LGA. The data set `FlightDelays` contains information on all 4029 departures of these two airlines from LGA during May and June 2009 (Tables 1.1 and 1.2).

Each row of the data set is an *observation*. Each column represents a *variable*—some characteristic that is obtained for each observation. For instance, on the first observation listed, the flight was a United Airlines plane, flight number 403, destined for

Mathematical Statistics with Resampling and R, First Edition. By Laura Chihara and Tim Hesterberg.

TABLE 1.1 Partial View of `FlightDelays` Data

Flight	Carrier	FlightNo	Destination	DepartTime	Day
1	UA	403	DEN	4–8 a.m.	Fri
2	UA	405	DEN	8–noon	Fri
3	UA	409	DEN	4–8 p.m.	Fri
4	UA	511	ORD	8–noon	Fri
		⋮			

Denver, and departing on Friday between 4 a.m. and 8 a.m. This data set consists of 4029 observations and 9 variables.

Questions we might ask include the following: Are flight delay times different between the two airlines? Are flight delay times different depending on the day of the week? Are flights scheduled in the morning less likely to be delayed by more than 15 min?

1.2 CASE STUDY: BIRTH WEIGHTS OF BABIES

The birth weight of a baby is of interest to health officials since many studies have shown possible links between this weight and conditions in later life, such as obesity or diabetes. Researchers look for possible relationships between the birth weight of a baby and the age of the mother or whether or not she smoked cigarettes or drank alcohol during her pregnancy. The Centers for Disease Control and Prevention (CDC), using data provided by the U.S. Department of Health and Human Services, National Center for Health Statistics, the Division of Vital Statistics as well as the CDC, maintain a database on all babies born in a given year (`http://wonder.cdc.gov/natality-current.html`). We will investigate different samples taken from the CDC's database of births.

One data set we will investigate consists of a random sample of 1009 babies born in North Carolina during 2004 (Table 1.3). The babies in the sample had a gestation period of at least 37 weeks and were single births (i.e., not a twin or triplet).

TABLE 1.2 Variables in Data Set `FlightDelays`

Variable	Description
Carrier	UA=United Airlines, AA=American Airlines
FlightNo	Flight number
Destination	Airport code
DepartTime	Scheduled departure time in 4 h intervals
Day	Day of week
Month	September or October
Delay	Minutes flight delayed (negative indicates early departure)
Delayed30	Departure delayed more than 30 min?
FlightLength	Length of time of flight (minutes)

TABLE 1.3 Variables in Data Set `NCBirths2004`

Variable	Description
Age	Mother's age
Tobacco	Mother used tobacco?
Gender	Gender of baby
Weight	Weight at birth (grams)
Gestation	Gestation time (weeks)

In addition, we will also investigate a data set, `Girls2004`, consisting of a random sample of 40 baby girls born in Alaska and 40 baby girls born in Wyoming. These babies also had a gestation period of at least 37 weeks and were single births.

The data set `TXBirths2004` contains a random sample of 1587 babies born in Texas in 2004. In this case, the sample was not restricted to single births, nor to a gestation period of at least 37 weeks. The numeric variable `Number` indicates whether the baby was a single birth, or one of a twin, triplet, and so on. The variable `Multiple` is a factor variable indicating whether or not the baby was a multiple birth.

1.3 CASE STUDY: VERIZON REPAIR TIMES

Verizon is the primary local telephone company (incumbent local exchange carrier, ILEC) for a large area of the eastern United States. As such, it is responsible for providing repair service for the customers of other telephone companies known as competing local exchange carriers (CLECs) in this region. Verizon is subject to fines if the repair times (the time it takes to fix a problem) for CLEC customers are substantially worse than those for Verizon customers.

The data set `Verizon` contains a random sample of repair times for 1664 ILEC and 23 CLEC customers (Table 1.4). The mean repair time for ILEC customers is 8.4 hours, while that for CLEC customers is 16.5 h. Could a difference this large be easily explained by chance?

1.4 SAMPLING

In analyzing data, we need to determine whether the data represent a *population* or a *sample*. A *population* represents all the individual cases, whether they are babies,

TABLE 1.4 Variables in Data Set `Verizon`

Variable	Description
Time	Repair times (in hours)
Group	ILEC or CLEC

fish, cars, or coin flips. The data from Flight Delays Case Study in Section 1.1 are *all* the flight departures of United Airlines and American Airlines out of LaGuardia Airport in May and June 2009; thus, this data set represents the population of all such flights. On the other hand, the North Carolina data set contains only a subset of 1009 births from over 100,000 births in North Carolina in 2004. In this case, we will want to know how representative statistics computed from this sample are of the entire population of North Carolina babies born in 2004.

Populations may be finite, such as births in 2004, or infinite, such as coin flips or births next year.

Throughout this chapter, we will talk about drawing random samples from a population. We will use capital letters (e.g., X, Y, Z, and so on) to denote random variables and lowercase letters (e.g., x_1, x_2, x_3, and so on) to denote actual values or data.

There are many kinds of random samples. Strictly speaking, a "random sample" is any sample obtained using a random procedure. However, in this book we use *random sample* to mean a sample of independent and identically distributed (i.i.d.) observations from the population, if the population is infinite.

For instance, suppose you toss a fair coin 20 times and consider each head a "success." Then your sample consists of the random variables $X_1, X_2, \ldots, X_{20}$, each a Bernoulli random variable with success probability 1/2. We use the notation $X_i \sim \text{Bern}(1/2)$, $i = 1, 2, \ldots, 20$.

If the population of interest is finite $\{x_1, x_2, \ldots, x_N\}$, we can choose a random sample as follows: label N balls with the numbers $1, 2, \ldots, N$ and place them in an urn. Draw a ball at random, record its value $X_1 = x_{i_1}$, and then replace the ball. Draw another ball at random, record its value, $X_2 = x_{i_2}$, and then replace. Continue until you have a sample $x_{i_1}, x_{i_2}, \ldots, x_{i_n}$. This is *sampling with replacement*. For instance, if $N = 5$ and $n = 2$, then there are $5 \times 5 = 25$ different samples of size 2 (where order matters). (Note: By "order matters" we do not imply that order matters in practice, rather we mean that we keep track of the order of the elements when enumerating samples. For instance, the set $\{a, b\}$ is different from $\{b, a\}$.)

However, in most real situations, for example, in conducting surveys, we do not want to have the same person polled twice. So we would sample *without replacement*, in which case, we will not have independence. For instance, if you wish to draw a sample of size $n = 2$ from a population of $N = 10$ people, then the probability of any one person being selected is $1/10$. However, after having chosen that first person, the probability of any one of the remaining people being chosen is now $1/9$.

In cases where populations are very large compared to the sample size, calculations under sampling without replacement are reasonably approximated by calculations under sampling with replacement.

Example 1.1 Consider a population of 1000 people, 350 of whom are smokers and the rest are nonsmokers. If you select 10 people at random but with replacement, then the probability that 4 are smokers is $\binom{10}{4}(350/1000)^4(650/1000)^6 \approx 0.2377$. If you select without replacement, then the probability is $\binom{350}{4}\binom{650}{6}/\binom{1000}{10} \approx 0.2388$. □

1.5 PARAMETERS AND STATISTICS

When discussing numeric information, we will want to distinguish between populations and samples.

Definition 1.1 A *parameter* is a (numerical) characteristic of a population or of a probability distribution.

A *statistic* is a (numerical) characteristic of data. ||

Any function of a parameter is also a parameter; any function of a statistic is also a statistic. When the statistic is computed from a random sample, it is itself random, and hence is a random variable.

Example 1.2 μ and σ are parameters of the normal distribution with pdf $f(x) = (1/\sqrt{2\pi}\sigma)e^{-(x-\mu)^2/(2\sigma^2)}$.

The variance σ^2 and *signal-to-noise ratio* μ/σ are also parameters. □

Example 1.3 If $X_1, X_2, \ldots, X_n$ are a random sample, then the mean $\bar{X} = 1/n \sum_{i=1}^{n} X_i$ is a statistic. □

Example 1.4 Consider the population of all babies born in the United States in 2004. Let μ denote the average weight of all these babies. Then μ is a parameter. The average weight of a sample of 2000 babies born in that year is a statistic. □

Example 1.5 If we consider the population of all adults in the United States today, the proportion p who approve of the president's job performance is a parameter. The fraction $\hat{p}$ who approve in any given sample is a statistic. □

Example 1.6 The average weight of 1009 babies in the North Carolina Case Study in Section 1.2 is 3448.26 g. This average is a statistic. □

Example 1.7 If we survey 1000 adults and find that 60% intend to vote in the next presidential election, then $\hat{p} = 0.60$ is a statistic: it estimates the parameter p, the proportion of all adults who intend to vote in the next election. □

1.6 CASE STUDY: GENERAL SOCIAL SURVEY

The General Social Survey (GSS) is a major survey that has tracked American demographics, characteristics, and views on social and cultural issues since the 1970s. It is conducted by the National Opinion Research Center (NORC) at the University of Chicago. Trained interviewers meet face-to-face with the adults chosen for the survey and question them for about 90 minutes in their homes.

TABLE 1.5 Variables in Data Set `GSS2002`

Variable	Description
Region	Interview location
Gender	Gender of respondent
Race	Race of respondent: White, Black, other
Marital	Marital status
Education	Highest level of education
Happy	General happiness
Income	Respondent's income
PolParty	Political party
Politics	Political views
Marijuana	Legalize marijuana?
DeathPenalty	Death penalty for murder?
OwnGun	Have gun at home?
GunLaw	Require permit to buy a gun?
SpendMilitary	Amount government spends on military
SpendEduc	Amount government spends on education
SpendEnv	Amount government spends on the environment
SpendSci	Amount government spends on science
Pres00	Whom did you vote for in the 2000 presidential election?
Postlife	Believe in life after death?

The GSS Case Study includes the responses of 2765 participants selected in 2002 to about a dozen questions, listed in Table 1.5. For example, one of the questions (`SpendEduc`) asked whether the respondent believed that the amount of money being spent on the nation's education system was too little, too much, or the right amount.

We will analyze the GSS data to investigate questions such as the following: Is there a relationship between the gender of an individual and whom they voted for in the 2000 presidential election, are people who live in certain regions happier, are there educational differences in support for the death penalty? These data are archived at the Computer-Assisted Survey Methods Program at the University of California (`http://sda.berkeley.edu/`).

1.7 SAMPLE SURVEYS

"When do you plan to vote for in the next presidential election?" "Would you purchase our product again in the future?" "Do you smoke cigarettes? If yes, how old were you when you first started?" Questions such as these are typical of sample surveys. Researchers want to know something about a population of individuals, whether they are registered voters, online shoppers, or American teenagers, but to poll every individual in the population—that is, to take a *census*—is impractical and costly.

Thus, researchers will settle for a sample from the target population. But if, say, 60% of those in your sample of 1000 adults intend to vote for candidate Smith in the next election, how close is this to the actual percentage who will vote for Smith? How can we be sure that this sample is truly representative of the population of all voters? We will learn techniques for *statistical inference*, drawing a conclusion about a population based on information about a sample.

When conducting a survey, researchers will start with a *sampling frame*—a list from which the researchers will choose their sample. For example, to survey all students at a college, the campus directory listing could be a sampling frame. For pre-election surveys, many polling organizations use a sampling frame of registered voters. Note that the choice of sampling frame could introduce the problem of *undercoverage*: omitting people from the target population in the survey. For instance, young people were missed in many election surveys during the 2008 Obama–McCain presidential race because they had not yet registered to vote.

Once the researchers have a sampling frame, they will then draw a random sample from this frame. Researchers will use some type of *probability (scientific) sampling scheme*, that is, a scheme that gives everybody in the population a positive chance of being selected. For example, to obtain a sample of size 10 from a population of 100 individuals, write each person's name on a slip of paper, put the slips of paper into a basket, and then draw out 10 slips of paper. Nowadays, statistical software is used to draw out random samples from a sampling frame.

Another basic survey design uses *stratified sampling*: the population is divided into nonoverlapping strata and then random samples are drawn from each stratum. The idea is to group individuals who are similar in some characteristic into homogeneous groups, thus reducing variability. For instance, in a survey of university students, a researcher might divide the students by class: first year, sophomores, juniors, seniors, and graduate students. A market analyst for an electronics store might choose to stratify customers based on income levels.

In *cluster sampling*, the population is divided into nonoverlapping clusters and then a random sample of clusters is drawn. Every person in a chosen cluster is then interviewed for the survey. An airport wanting to conduct a customer satisfaction survey might use a sampling frame of all flights scheduled to depart from the airport on a certain day. A random sample of flights (clusters) is chosen and then all passengers on these flights are surveyed. A modification of this design might involve sampling in stages: for instance, the analysts might first choose a random sample of flights, and then from each flight choose a random sample of passengers.

The General Social Survey uses a more complex sampling scheme in which the sampling frame is a list of counties and county equivalents (standard metropolitan statistical areas) in the United States. These counties are stratified by region, age, and race. Once a sample of counties is obtained, a sample of block groups and enumeration districts is selected, stratifying these by race and income. The next stage is to randomly select blocks and then interview a specific number of men and women who live within these blocks.

Indeed, all major polling organizations such as Gallup or Roper as well as the GSS use a multistage sampling design. In this book, we use the GSS data or polling results

TABLE 1.6 Variables in Data Set `Beerwings`

Variable	Description
Gender	Male or female
Beer	Ounces of beer consumed
Hotwings	Number of hot wings eaten

for examples as if the survey design used simple random sampling. Calculations for more complex sampling scheme are beyond the scope of this book and we refer the interested reader to Lohr (1991) for details.

1.8 CASE STUDY: BEER AND HOT WINGS

Carleton student Nicki Catchpole conducted a study of hot wings and beer consumption at the Williams Bar in the Uptown area of Minneapolis (Catchpole (2004)). She asked patrons at the bar to record their consumption of hot wings and beer over the course of several hours. She wanted to know if people who ate more hot wings would then drink more beer. In addition, she investigated whether or not gender had an impact on hot wings or beer consumption.

The data for this study are in `Beerwings` (Table 1.6). There are 30 observations and 3 variables.

1.9 CASE STUDY: BLACK SPRUCE SEEDLINGS

Black spruce (*Picea mariana*) is a species of a slow-growing coniferous tree found across the northern part of North America. It is commonly found on wet organic soils. In a study conducted in the 1990s, a biologist interested in factors affecting the growth of the black spruce planted its seedlings on sites located in boreal peatlands in northern Manitoba, Canada (Camill et al. (2010)).

The data set `Spruce` contains a part of the data from the study (Table 1.7). Seventy-two black spruce seedlings were planted in four plots under varying conditions (fertilizer–no fertilizer, competition–no competition) and their heights and diameters were measured over the course of 5 years.

The researcher wanted to see whether the addition of fertilizer or the removal of competition from other plants (by weeding) affected the growth of these seedlings.

1.10 STUDIES

Researchers carry out studies to understand the conditions and causes of certain outcomes: Does smoking cause lung cancer? Do teenagers who smoke marijuana

TABLE 1.7 Variables in Data Set `Spruce`

Variable	Description
Tree	Tree number
Competition	C (competition), CR (competition removed)
Fertilizer	F (fertilized), NF (not fertilized)
Height0	Height (cm) of seedling at planting
Height5	Height (cm) of seedling at year 5
Diameter0	Diameter (cm) of seedling at planting
Diameter5	Diameter (cm) of seedling at year 5
Ht.change	Change (cm) in height
Di.change	Change (cm) in diameter

tend to move on to harder drugs? Do males eat more hot wings than females? Do black spruce seedlings grow taller in fertilized plots?

The Beer and Hot Wings Case Study in Section 1.8 is an example of an *observational study*, a study in which researchers observe participants but do not influence the outcome. In this case, the student just recorded the number of hot wings eaten and beer consumed by the patrons of Williams Bar.

Example 1.8 The first Nurse's Health Study is a major observational study funded by the National Institutes of Health. Over 120,000 registered female nurses who, in 1976, were married, between the ages of 33 and 55 years, and who lived in the 11 most populous states, have been responding every 2 years to written questions about their health and lifestyle, including smoking habits, hormone use, and menopause status. Many results on women's health have come out of this study, such as finding an association between taking estrogen after menopause and lowering the risk of heart disease, and determining that for nonsmokers there is no link between taking birth control pills and developing heart disease.

Because this is an observational study, no *cause and effect* conclusions can be drawn. For instance, we cannot state that taking estrogen after menopause will *cause* a lowering of the risk for heart disease. In an observational study, there may be many unrecorded or hidden factors that impact the outcomes. Also, because the participants in this study are registered nurses, we need to be careful about making inferences about the general female population. Nurses are more educated and more aware of health issues than the average person. □

On the other hand, the Black Spruce Case Study in Section 1.9 was an *experiment*. In an experiment, researchers will manipulate the environment in some way to observe the response of the objects of interest (people, mice, ball bearings, etc.). When the objects of interest in an experiment are people, we refer to them as *subjects*; otherwise, we call them *experimental units*. In this case, the biologist randomly assigned the experimental units—the seedlings—to plots subject to four *treatments*: fertilization with competition, fertilization without competition, and no fertilization with competition, and no fertilization with no competition. He then recorded their height over a period of several years.

A key feature in this experiment was the *random assignment* of the seedlings to the treatments. The idea is to spread out the effects of unknown or uncontrollable factors that might introduce unwanted variability into the results. For instance, if the biologist had planted all the seedlings obtained from one particular nursery in the fertilized, no competition plot and subsequently recorded that these seedlings grew the least, then he would not be able to discern whether this was due to this particular treatment or due to some possible problem with seedlings from this nursery. With random assignment of treatments, the seedlings from this particular nursery would usually be spread out over the four treatments. Thus, the differences between the treatment groups should be due to the treatments (or chance).

Example 1.9 Knee osteoarthritis (OA) that results in deterioration of cartilage in the joint is a common source of pain and disability for the elderly population. In a 2008 paper, "Tai Chi is effective in treating knee osteoarthritis: A randomized controlled trial," Wang et al. (2009) at Tufts University Medical School describe an experiment they conducted to see whether practicing Tai Chi, a style of Chinese martial arts, could alleviate pain from OA. Forty patients over the age of 65 with confirmed knee OA but otherwise in good health were recruited from the Boston area. Twenty were randomly assigned to attend twice weekly 60 min sessions of Tai Chi for 12 weeks. The remaining 20 participants, the *control group*, attended twice weekly 60 min sessions of instructions on health and nutrition, as well as some stretching exercises.

At the end of the 12 weeks, those in the Tai Chi group reported a significant decrease in knee pain. Because the subjects were randomly assigned to the two treatments, the researchers can assert that the Tai Chi sessions lead to decrease in knee pain due to OA. Note that because the subjects were recruited, we need to be careful about making an inference about the general elderly population: Somebody who voluntarily signs up to be in an experiment may be inherently different from the average person. □

1.11 EXERCISES

1. For each of the following, describe the population and, if relevant, the sample. For each number presented, determine if it is a parameter or a statistic (or something else).
   (a) A survey of 2000 high school students finds that 47% watch the television show "Glee."
   (b) The 2000 U.S. Census reports that 13.9% of the U.S. population was between the ages of 15 and 24 years.
   (c) Based on the rosters of all National Basketball Association teams for the 2006–2007 season, the average height of the players was 78.93 in.
   (d) A December 2009 Gallup poll of 1025 national adults, aged 18 and older, shows that 47% would advise their member of Congress to vote for health care legislation (or lean toward doing so).

2. Researchers reported that moderate drinking of alcohol was associated with a lower risk of dementia ((Mukamal et al. (2003)). Their sample consisted of 373 people with dementia and 373 people without dementia. Participants were asked how much beer, wine, or shot of liquor they consumed. Thus, participants who consumed 1–6 drinks a week had a lower risk of dementia than those who abstained from alcohol.
   (a) Was this study an observational study or an experiment?
   (b) Can the researchers conclude that drinking alcohol causes a lower risk of dementia?
3. Researchers surveyed 959 ninth graders who attended three large U.S. urban high schools and found that those who listened to music that had references to marijuana were almost twice as likely to have used marijuana as those who did not listen to music with references to marijuana (Primack et al. (2010)).
   (a) Was this an observational study or an experiment?
   (b) Can the researchers conclude that listening to music with references to marijuana causes students to use drugs?
   (c) Can the researchers extend their results to all urban American adolescents?
4. Duke University researchers found that diets low in carbohydrates are effective in controlling blood sugar levels (Westman et al. (2008)). Eighty-four volunteers with obesity and type 2 diabetes were randomly assigned to either a diet of less than 20 g of carbohydrates/day or a low-glycemic, reduced calorie diet (500 calories/day). Ninety-five percent of those on the low-carbohydrate diet were able to reduce or eliminate their diabetes medications compared to 62% on the low-glycemic diet.
   (a) Was this study an observational study or an experiment?
   (b) Can researchers conclude that a low-carbohydrate diet causes an improvement in type 2 diabetes?
   (c) Can researchers extend their results to a more general population? Explain.
5. In a population of size N, the probability of any subset of size n being chosen is $1/\binom{N}{n}$. Show this implies that any one person in the population has a n/N probability of being chosen in a sample. Then, in particular, every person in the population has the same probability of being chosen.
6. A typical Gallup poll surveys about $n = 1000$ adults. Suppose the sampling frame contains 100 million adults (including you). Now, select a random sample of 1000 adults.
   (a) What is the probability that you will be in this sample?
   (b) Now suppose that 2000 such samples are selected, each independent of the others. What is the probability that you will *not* be in any of the samples?
   (c) How many samples must be selected for you to have a 0.5 probability of being in at least one sample?

2

EXPLORATORY DATA ANALYSIS

Exploratory data analysis (EDA) is an approach to examining and describing data to gain insight, discover structure, and detect anomalies and outliers. John Tukey (1915–2000), an American mathematician and statistician who pioneered many of the techniques now used in EDA, stated in his 1977 book *Exploratory Data Analysis* (Tukey (1977)) that "Exploratory data analysis is detective work—numerical detective work—counting detective work—or graphical detective work." In this chapter, we will learn many of the basic techniques and tools for gaining insight into data.

Statistical software packages can easily do the calculations needed for the basic plots and numeric summaries of data. We will use the software package `R`. We will assume that you have gone through the introduction to `R` available at the web site `https://sites.google.com/site/ChiharaHesterberg`.

2.1 BASIC PLOTS

In Chapter 1, we described data on the lengths of flight delays of two major airlines flying from LaGuardia Airport in New York City in 2009. Some basic questions we might ask include how many of these flights were flown by United Airlines and how many by American Airlines? How many flights flown by each of these airlines were delayed more than 30 min?

A *categorical* variable is one that places the observations into groups. For instance, in the `FlightDelays` data set, `Carrier` is a categorical variable (we will also

Mathematical Statistics with Resampling and R, First Edition. By Laura Chihara and Tim Hesterberg.

TABLE 2.1 Counts for the `Carrier` Variable

	Carrier	
	American Airlines	United Airways
Number of flights	2906	1123

call this a *factor* variable) with two *levels*, UA and AA. Other data sets might have categorical variables such as `gender` (with two levels, Male or Female) or `size` (with levels Small, Medium, and Large).

A *bar chart* is used to describe the distribution of a categorical (factor) variable. Bars are drawn for each level of the factor variable and the height of the bar is the number of observations in that level. For the `FlightDelays` data, there were 2906 American Airlines flights and 1123 United Airlines flights (Table 2.1). The corresponding bar chart is shown in Figure 2.1.

We might also be interested in investigating the relationship between a carrier and whether or not a flight was delayed more than 30 min. A *contingency table* summarizes the counts in the different categories.

From Table 2.2, we can compute that 13.5% of American Airlines flights were delayed more than 30 min compared to 18.2% of United Airlines flights. Is this difference in percentages *statistically significant*? Could the difference in percentages be due to *natural variability*, or is there a systematic difference between the two airlines? We will address this question in the following chapters.

With a numeric variable, we will be interested in its distribution: What is the range of values? What values are taken on most often? Where is the *center*? What is the *spread*?

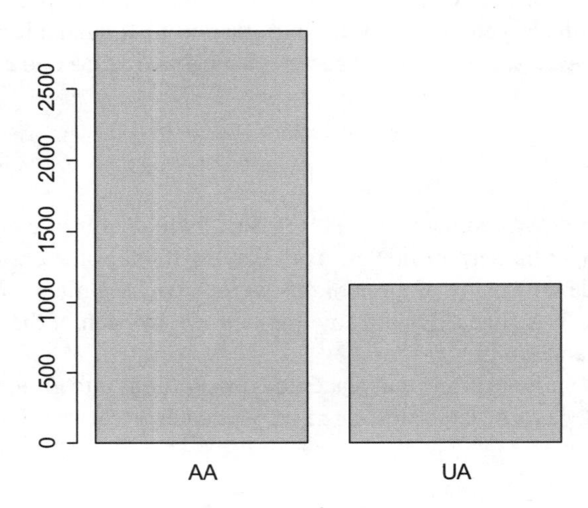

FIGURE 2.1 Bar chart of `Carrier` variable.

TABLE 2.2 Counts of Delayed Flights Grouped by Carrier

Carrier	Delayed More Than 30 Min?		
	No	Yes	Total
American Airlines	2513	393	2906
United Airlines	919	204	1123

For the flight delays data set, although we can inspect the distribution of the lengths of the delays with a table by partitioning the values into nonoverlapping intervals (Table 2.3), a visual representation is often more informative.

A *histogram* corresponding to Table 2.3 is shown in Figure 2.2. Note that the height of each bar reflects the frequency of flights whose delays fall in the corresponding interval. For example, 722 flights departed on time or earlier than scheduled, while 249 flights were delayed by at most 50 min. Some software will give users the option to create bar heights equal to proportions or percentages.

We describe this distribution as *right skewed.* Most of the flights departed on time (or were early) and the counts of late departures decrease as time increases.

Average January temperatures in the state of Washington follow a *left-skewed distribution* (Figure 2.3): in most years, average temperate fell in the 30–35 °F interval, and the number of years in which temperatures were less than 30 °F decreases as temperature decreases.

Remark The exact choice of subintervals to use is discretionary. Different software packages utilize various algorithms for determining the length of the subintervals; also, some software packages may use subintervals of the form $[a, b)$ instead of $(a, b]$. ||

For small data sets, a *dot plot* is an easy graph to create by hand. A dot represents one observation and is placed above the value it represents. The number of dots in a column represents the frequency of that value.

The dot plot for the data 4, 4.5, 4.5, 5, 5, 5, 6, 6, 6.5, 7, 7, 7 is shown in Figure 2.4.

TABLE 2.3 Distribution of Length of Flight Delays for United Airlines

Time Interval	Number of Flights
(−50, 0]	722
(0, 50]	249
(50, 100]	86
(100, 150]	39
(150, 200]	14
(200, 250]	7
(250, 300]	3
(350, 400]	2
(400, 450]	1

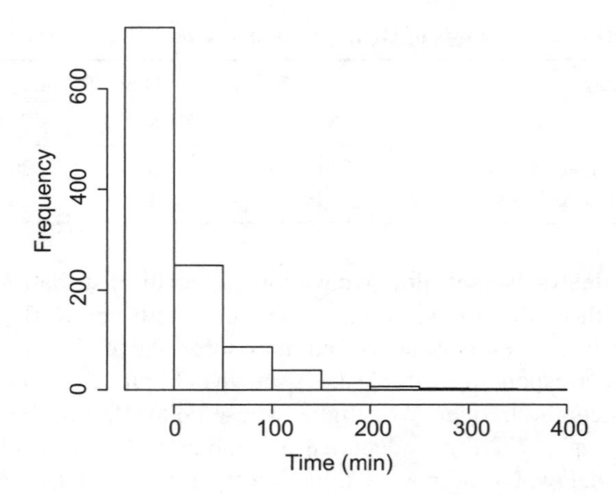

FIGURE 2.2 Histogram of lengths of flight delays for United Airlines. The distribution is right skewed.

2.2 NUMERIC SUMMARIES

It is often useful to have numerical descriptions of variables. Unfortunately, the old adage "a picture is worth a thousand words" cuts both ways—doing without a picture limits what we can say without thousands of words. So we focus on key characteristics—center, spread, and sometimes shape.

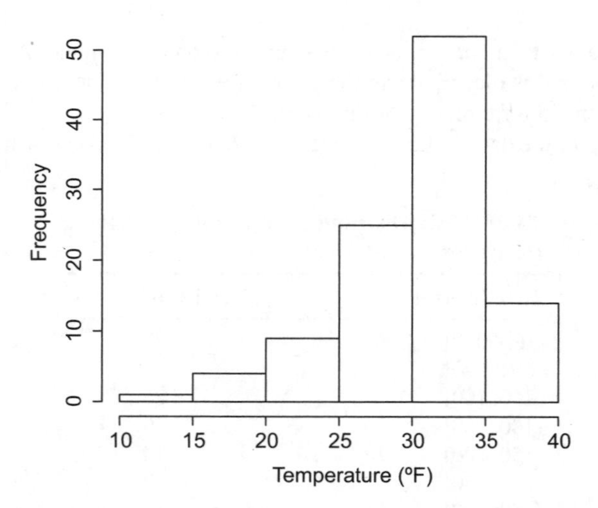

FIGURE 2.3 Histogram of average January temperatures in Washington state (1895–1999). The distribution is left skewed.

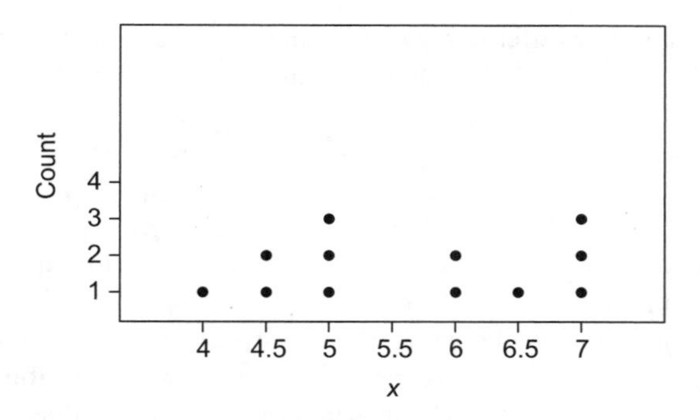

FIGURE 2.4 Example of a dot plot.

2.2.1 Center

First consider *center*. By eyeballing the histogram (Figure 2.2) of flight delay times, we might put the center at around 0. Two statistics commonly used to describe the *center* of a variable include the *mean* and *median*.

If $x_1, x_2, \ldots, x_n$ are n data values, then the mean is

$$\bar{x} = \frac{1}{n} \sum_{i=1}^{n} x_i.$$

The median is the middle value in a sorted arrangement of the values; that is, half the values are less than or equal to the median and half are greater. If $y_1 \leq y_2 \leq \cdots \leq y_n$ denotes a sorted list of values and n is odd, the median is the middle value $y_{(n+1)/2}$. If n is even, then the median is the average of the two middle values, $(1/2)(y_{n/2} + y_{(n/2+1)})$.

A compromise between the mean and the median is a *trimmed mean*. The mean is the average of all observations, while the median is the average of the middle one or two observations. For a 25% trimmed mean, for example, you sort the data, omit 25% of the observations on each side, and take the mean of the remaining middle 50% of the observations. The 25% trimmed mean is also known as *midmean*.

Example 2.1 The mean of the 12 values $1, 3, 3, 4, 4, 7, 8, 10, 14, 21, 24, 26$ is 10.42, the median is the average of the sixth and seventh values, $(7 + 8)/2 = 7.5$, and the midmean is the average of fourth through ninth values, 7.83.

The mean of the 15 values $1, 3, 3, 4, 4, 7, 8, 10, 14, 21, 24, 28, 30, 30, 34$ is 14.73, the median is the 8th value, 10, and the midmean is the average of the 4th through 12th values, 13.33. □

Example 2.2 The mean length of a departure delay for United Airlines was 15.9831 min. The median length of a departure delay was −1.00 min; that is, half of the flights left more than 1 min earlier than their scheduled departure time. □

Remark Software may differ in how it calculates trimmed means. In `R`, `mean(x, trim = 0.25)` rounds $0.25n$ down; thus, for $n = 15$, three observations are omitted. ||

2.2.2 Spread

To describe spread, three common choices are the range, the interquartile range, and the standard deviation.

The *range* is the difference between the largest and smallest values.

The *interquartile range* (IQR) is the difference between the third and the first quartiles. It gives a better measure of the center of the data than does the range and is not sensitive to outliers.

The *sample standard deviation*, or *standard deviation*, is

$$s = \sqrt{\frac{1}{n-1}\sum_{i=1}^{n}(x_i - \bar{x})^2}. \tag{2.1}$$

To motivate the standard deviation, we begin with a less common measure of spread, the *mean absolution deviation* (MAD), $(1/n)\sum_{i=1}^{n}|x_i - \bar{x}|$. This is the average distance from the mean and is a natural measure of spread. In contrast, the standard deviation is roughly the average squared distance from the mean, followed by a square root; the combination is roughly equal to the MAD, though usually a bit larger. The standard deviation tends to have better statistical properties.

There are a couple of versions of standard deviation. The *population standard deviation* is the square root of the *population variance*, which is the average of the squared distances from the mean, $(1/n)\sum_{i=1}^{n}(x_i - \bar{x})^2$. The *sample variance* is similar but with a divisor of $n-1$,

$$s^2 = \frac{1}{n-1}\sum_{i=1}^{n}(x_i - \bar{x})^2, \tag{2.2}$$

and the sample standard deviation is its square root. The population versions are appropriate when the data are the whole population. When the data are a sample from a larger population, we use the sample versions; in this case, the population versions tend to be too small—they are *biased*; we will return to this point in Section 6.3.1. For n large, there is little practical difference between using $n - 1$ or n.

Example 2.3 The standard deviation of the departure delay times for United Airlines flights is 45.119 min. Since the observations represent a population (we compiled *all* United Airlines flights for the months of May and June), we use the definition with the $1/n$ term rather than the $1/(n-1)$ term. Using Equation 2.1 gives 45.139. □

2.2.3 Shape

To describe the shape of a data set, we may use skewness and kurtosis (see page 28.) However, more common and intuitive is to use the *five-number summary*: the minimum, first quartile, median, third quartile, and maximum value.

Example 2.4 Consider the 15 numbers 9, 10, 11, 11, 12, 14, 16, 17, 19, 21, 25, 31, 32, 41, 61.

The median is 17. Now, find the median of the numbers less than or equal to 17. This will be the first quartile $Q_1 = 11.5$. The median of the numbers greater than or equal to 17 is the third quartile $Q_3 = 28$. Thus, the five-number summary is 9, 11.5, 17, 28, 31. □

Remark Different software packages use different algorithms for computing quartiles, so do not be alarmed if your results do not match exactly. ||

2.3 BOXPLOTS

A boxplot is a type of graph that can be used to visualize the five-number summary of a set of numeric values.

Example 2.5 Consider the following set of 21 values (Table 2.4).

TABLE 2.4 A Set of 21 Data Values

5	6	6	8	9	11	11
14	17	17	19	20	21	21
22	23	24	32	40	43	49

The five-number summary for these data is 5, 11, 19, 23, 48 and the interquartile range is $23-11 = 12$. The corresponding boxplot is shown in Figure 2.5. □

To create a boxplot

- Draw a box with the bottom placed at the first quartile and the top placed at the third quartile. Draw a line through the box at the median.
- Compute the number $Q_3 + 1.5 \times \text{IQR}$, called the *upper fence*, and then place a cap at the largest observation that is less than or equal to this number.
- Similarly, compute the *lower fence*, $Q_1 - 1.5 \times \text{IQR}$, and place a cap at the smallest observation that is greater than or equal to this number.
- Extend *whiskers* from the edge of the box to the caps.
- The observations that fall outside the caps will be considered *outliers* and separate points are drawn to indicate these values.

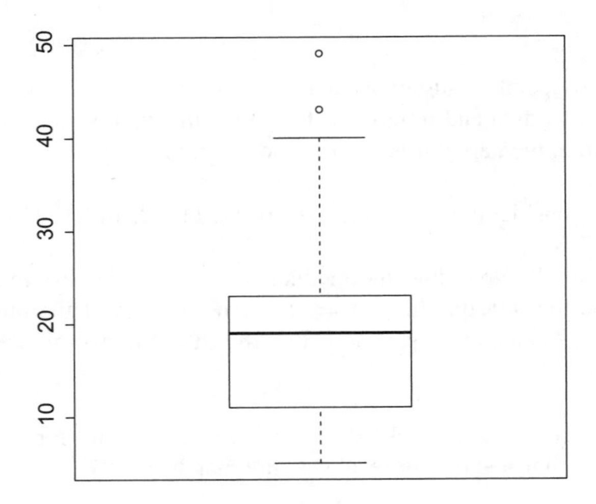

FIGURE 2.5 Boxplot for Table 2.4.

In the above example, the upper fence is $23 + 1.5 \times 12 = 41$. The largest observation that falls below this fence is 40, so a cap is drawn at 40. The lower fence is $11 - 1.5 \times 12 = -7$. The smallest observation that falls above this fence is 5, so a cap is drawn at 5. The outliers are 43 and 49.

Example 2.6 For the length of United Airlines flight delays, the five-number summary is -17.00, -5.00, -1.00, 12.50, 377.00. Thus, the interquartile range is $12.50 - (-5.00) = 17.50$ and half of the 1123 values are contained in an interval of length 17.50. □

Boxplots are especially useful in comparing the distribution of a numeric variable across levels of a factor variable.

Example 2.7 We can compare the lengths of the flight delays for United Airlines across the days of the week for which the departure was scheduled.

For instance, we can see that the most variability in delays seems to occur on Thursdays and Fridays (Figure 2.6). □

2.4 QUANTILES AND NORMAL QUANTILE PLOTS

For the random sample of 1009 babies born in North Carolina in 2004, the distribution of their weights is unimodal and roughly symmetric (Figure 2.7). We introduce another graph that allows us to compare this distribution with the normal distribution.

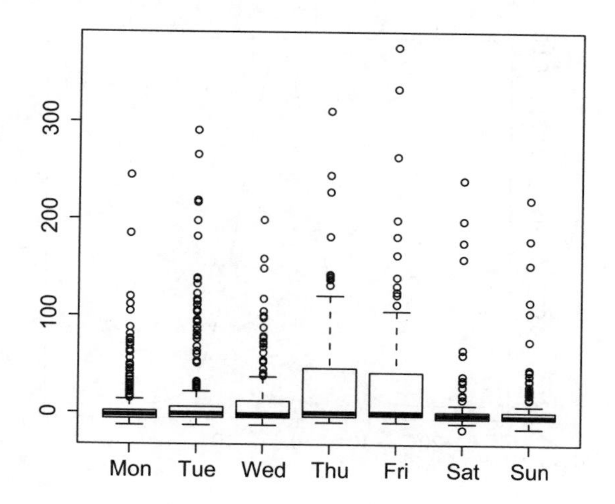

FIGURE 2.6 Distribution of lengths of the flight delays for United Airlines across the days of the week.

Definition 2.1 Let X denote a random variable. For $0 \leq p \leq 1$, the pth *quantile* of X is the value q_p such that $P(X \leq q_p) = p$. That is, q_p is the value at which the amount of area under the density curve (of X) to the left of q_p is p, or $p \times 100\%$ of the area under the curve is to the left of q_p.

Some books may use the term *percentile* rather than quantile. For instance, the 0.3 quantile is the same as the 30th percentile. ||

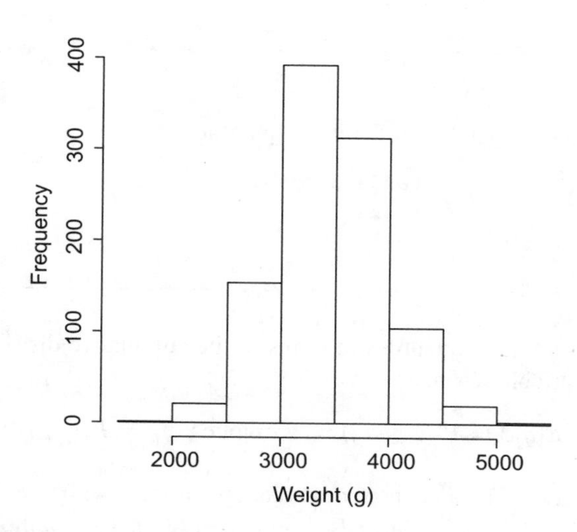

FIGURE 2.7 Distribution of birth weights for North Carolina babies.

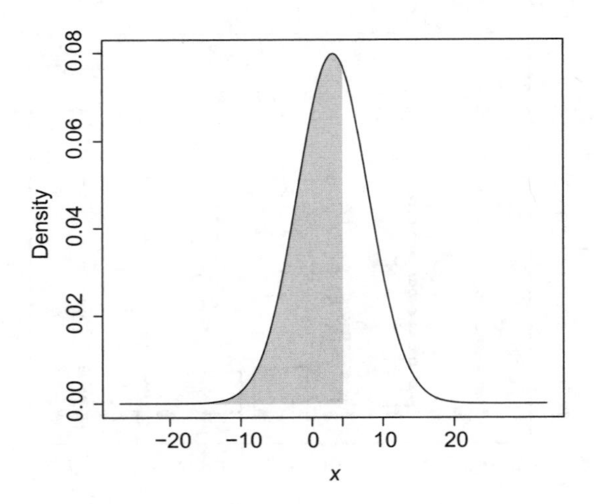

FIGURE 2.8 Density for $N(3, 5^2)$ with $P(X \leq 4.27) = 0.6$.

Example 2.8 Let Z denote the standard normal distribution. Let $p = 0.5$. Then, the 0.5 quantile of Z is 0 since $P(Z \leq 0) = 0.5$. That is, 0 is the 50th percentile of the standard normal distribution.

Let $p = 0.25$. Then, $q_{0.25} = -0.6744$ since $P(Z \leq -0.6744) = 0.25$. That is, -0.6744 is the 25th percentile of the standard normal distribution. □

Example 2.9 Let X be a normal random variable, $N(3, 5^2)$. Find the $p = 0.6$ quantile.

We want q_p such that $P(X \leq q_p) = 0.6$. The desired value is $q_p = 4.3$ (see Figure 2.8). □

R Note:

Use the qnorm command to find normal quantiles:

```
> qnorm(.25)          # standard normal
[1] -0.6744898
> qnorm(.6, 3, 5)     # N(3,5^2)
[1] 4.266736
```

We can also formulate quantiles in terms of the cumulative distribution function F of a random variable X since

$$F(q_p) = P(X \leq q_p) = p \text{ implies } q_p = F^{-1}(p).$$

Example 2.10 Let X be an exponential random variable with $\lambda = 3$. The cdf of X is given by $F(x) = 1 - e^{-3x}$. Since $F^{-1}(y) = (-1/3)\ln(1 - y)$, the pth quantile is given by $q_p = (-1/3)\ln(1 - p)$.

Alternatively, since we know the pdf of X is $f(x) = e^{-3x}$, $x \geq 0$, we could also solve for q_p in

$$p = P(X \leq q_p) = \int_0^{q_p} e^{-3t}\, dt. \qquad \square$$

Suppose the (sorted) data are $x_1 \leq x_2 \leq \cdots \leq x_n$ and we wish to see if these data come from the normal distribution, $N(0, 1)$. The *normal quantile plot* for comparing distributions is a plot of the x's against $(q_1, x_1), (q_2, x_2), \ldots, (q_n, x_n)$, where q_k is the $k/(n+1)$ quantile of the standard normal distribution. If these points fall (roughly) on a straight line, then we conclude that the data follow an approximate normal distribution. This is one type of *quantile–quantile plot*, or *qq plot* for short, in which quantiles of a data set are plotted against quantiles of a distribution or of another data set.

Example 2.11 Here, there are $n = 10$ points. We will look at the $i/(n+1) = i/(11\text{th})$ quantiles, $i = 1, \ldots, 10$, of the standard normal.

x	17.7	22.6	26.1	28.3	30.0	31.2	31.5	33.5	34.7	36.0
p_i	1/11	2/11	3/11	4/11	5/11	6/11	7/11	8/11	9/11	10/11
q_p	−1.34	−0.91	−0.60	−0.35	−0.11	0.11	0.35	0.60	0.91	1.34

For instance, the q_p entry corresponding to $p_5 = 5/11 = 0.455$ (the 45.5th percentile) is

$$q_{0.455} = -0.11 \text{because} P(Z \leq -0.11) = 0.455.$$

To create a normal quantile plot, we graph the pairs (q_p, x). A straight line is often drawn through the points corresponding to the first and third quartiles of each variable (see Figure 2.9).

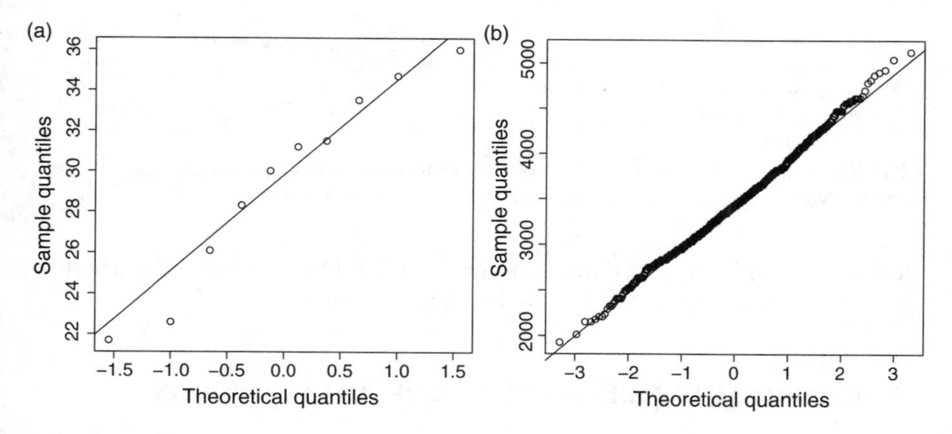

FIGURE 2.9 (a) Example of normal quantile plot for data in Example 2.11. (b) Normal quantile plot for weights of NC babies.

R Note:

The commands qqnorm and qqline can be used to create normal quantile plots:

```
x <- c(21.7, 22.6, 26.1, 28.3, 30, 31.2, 31.5, 33.5, 34.7, 36)
qqnorm(x)       # plot points
qqline(x)       # add straight line

qqnorm(NCBirths$Weight)
qqline(NCBirths$Weight)
```

The qqnorm command plots the quantiles of the standard normal on the x axis. The qqline command adds a straight line through the first and third quartiles of the data.

□

Recall that the distribution of the flight delay times for United Airlines is strongly right skewed (Figure 2.2). The normal quantile plots for these data and for the left-skewed distribution of average January temperatures in Washington state (Figure 2.3) are shown in Figure 2.10.

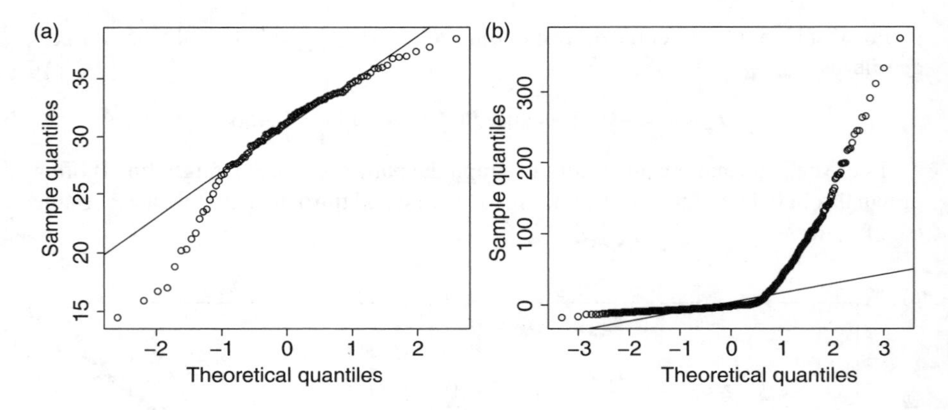

FIGURE 2.10 (a) Normal quantile plot for average January temperatures in Washington state. (b) Normal quantile plot for flight delay times for United Airlines.

Remark Even for samples drawn from a normal distribution, the points on a normal quantile plot do not lie *exactly* on a straight line. See Exercise 14. ||

2.5 EMPIRICAL CUMULATIVE DISTRIBUTION FUNCTIONS

The *empirical cumulative distribution function* (ecdf) is an estimate of the underlying cumulative distribution function (page 363) for a sample. The empirical cdf, denoted by $\hat{F}$, is a step function

$$\hat{F}(x) = \frac{1}{n}\big(\text{number of values} \leq x\big),$$

where n is the sample size.

For instance, consider the set of values 3, 6, 15, 15, 17, 19, 24. Then, $\hat{F}(18) = 5/7$ since there are five data values less than or equal to 18.

More generally,

$$\hat{F}(x) = \begin{cases} 0, & x < 3, \\ 1/7, & 3 \leq x < 6, \\ 2/7, & 6 \leq x < 15, \\ 4/7, & 15 \leq x < 17, \\ 5/7, & 17 \leq x < 19, \\ 6/7, & 19 \leq x < 24, \\ 1, & x \geq 18. \end{cases}$$

Figure 2.11 displays the empirical cdf for this example as well as the ecdf for a random sample of size 25 from the standard normal distribution. The graph of the cdf for the standard normal $\Phi(t)$ is added for comparison.

The empirical cumulative distribution function is useful for comparing two distributions. Figure 2.12 shows the ecdf's of beer consumption for males and females

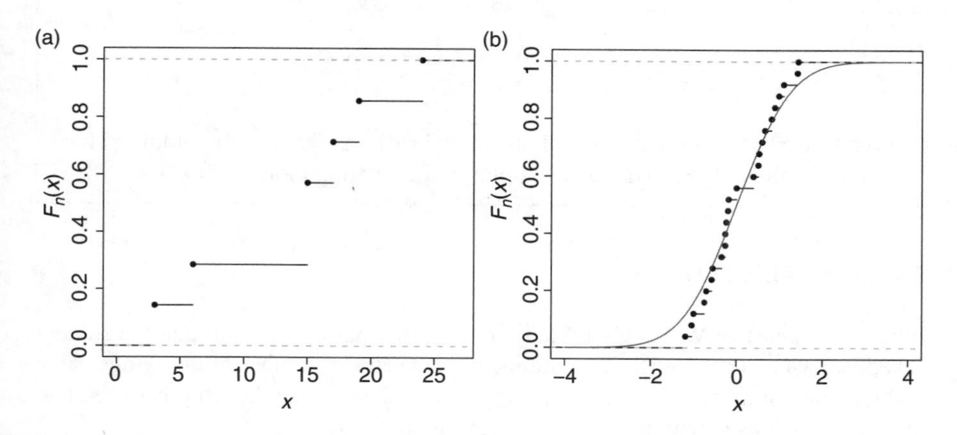

FIGURE 2.11 (a) Empirical cumulative distribution function for the data 3, 6, 15, 15, 17, 19, 24. (b) Ecdf for a random sample from $N(0, 1)$ with the cdf for the standard normal.

from the Beer and Hot Wings Case Study in Section 1.8. With the vertical line at 25 ounces, we can see that about 30% of the males and nearly 70% of the females have consumed 25 or fewer ounces of beer.

R Note:

The command `plot.ecdf` plots the empirical cumulative distribution function.

```
x <- c(3, 6, 15, 15, 17, 19, 24)
plot.ecdf(x)
x <- rnorm(25)                   # random sample of size 25 from N(0,1)
plot.ecdf(x, xlim = c(-4, 4))                 # adjust x range
curve(pnorm(x), col = "blue", add = TRUE) # impose normal cdf
abline(v = 25, col = "red")                   # add vertical line
```

For the Beer and Hot Wings Case Study, we first create vectors that hold the data for the men and women separately.

```
beerM <- subset(Beerwings, select = Beer, subset = Gender == "M",
           drop = T)
beerF <- subset(beerwings, select = Beer, subset = Gender == "F",
          drop = T)
```

The `subset` command creates a new vector from the data set `Beerwings` by selecting the column `Beer` and extracting those rows corresponding to the males (`subset=Gender=="M"`) or females (`subset=Gender=="F"`). The `drop=T` argument ensures that we have a vector object (as opposed to a data frame).

```
plot.ecdf(beerM, xlab = "ounces")
plot.ecdf(beerF, col = "blue", pch = 2, add = TRUE)
abline(v = 25, lty = 2)
legend(c(5, .8), legend = c("Males", "Females"),
   col = c("black", "blue"), pch = c(19, 2))
```

In the last `plot.ecdf` command above, the `pch=2` changes the plotting character while the `add=TRUE` adds this plot to the existing plot.

2.6 SCATTER PLOTS

In the Beer and Hot Wings Case Study in Section 1.8, one question that the student asked was whether there was a relationship between the number of hot wings eaten and the amount of beer consumed. A way to visualize the relationship between two numeric variables is with a scatter plot (see Figure 2.13).

Each point in the scatter plot represents a single observation—that is, a single person who took part in the study. From the graph, we note that there is a positive,

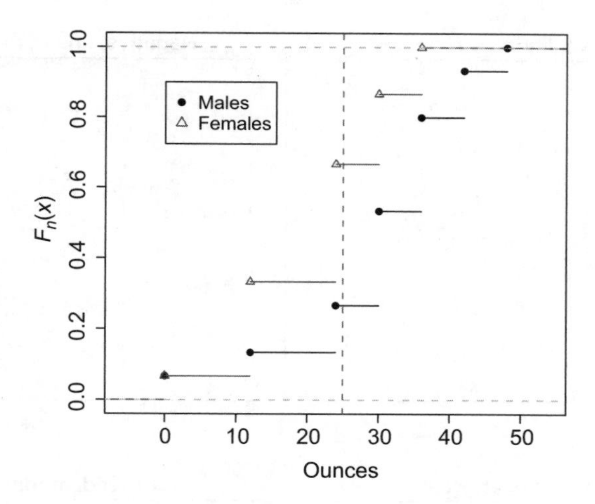

FIGURE 2.12 Ecdf's for male and female beer consumption. The vertical line is at 25 ounces.

roughly linear, association between hot wings and beer: as the number of hot wings eaten increases, the amount of beer consumed also increases.

Remark In statistics, the convention is to put the variable of primary interest on the y-axis, and the variable that may help predict or explain that variable as x, and to "plot y against x." ||

Further examples are shown in Figure 2.14. In general, when describing the relationship between two numeric variables, we will look for *direction*, *form*, and *strength*.

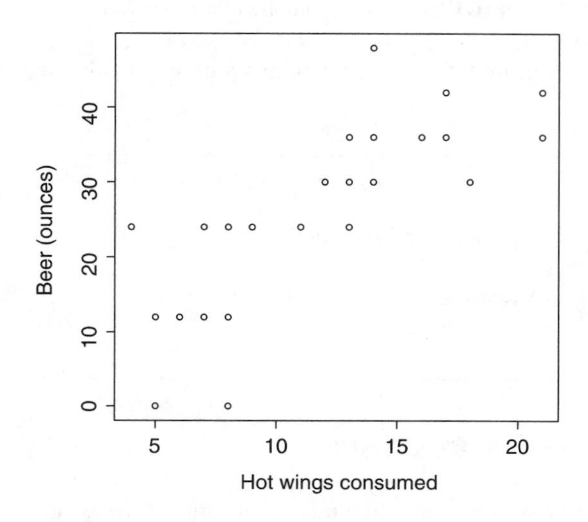

FIGURE 2.13 A scatter plot of `Beer` against `Hotwings`.

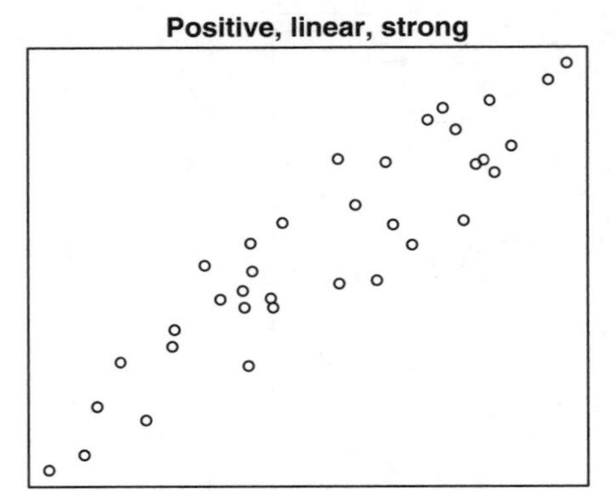

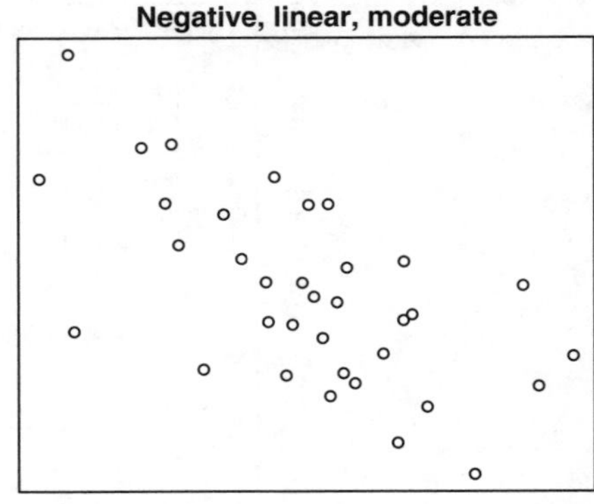

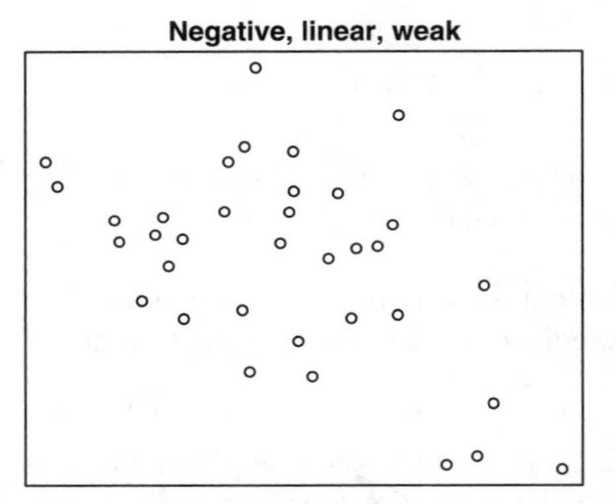

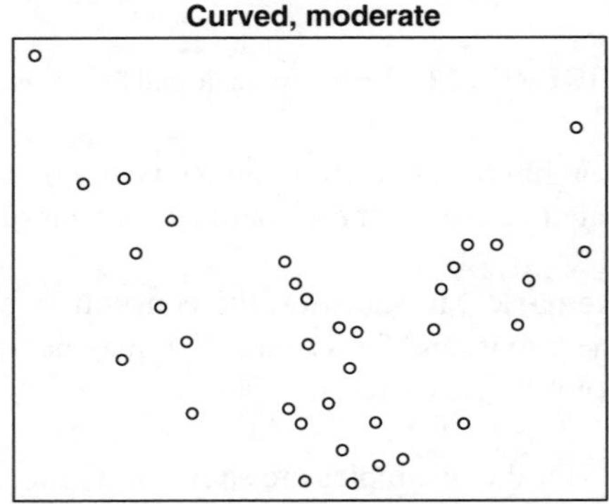

FIGURE 2.14 Examples of scatter plots.

In Chapter 9, we will investigate the relationship between two numeric variables in more detail.

R Note:

For scatter plots, use the `plot` command:

```
plot(Beerwings$Hotwings, Beerwings$Beer, xlab = "Hot wings eaten",
  ylab = "Beer consumed")
```

2.7 SKEWNESS AND KURTOSIS

Asymmetry and peakedness are often measured using *skewness* and *kurtosis*, which are defined using third and fourth central moments (Section A.7).

Definition 2.2 Let X be a random variable with mean μ and standard deviation σ. The *skewness* of X is

$$\gamma_1 = \mathrm{E}\left[\left(\frac{X-\mu}{\sigma}\right)^3\right] = \frac{\mu_3}{\sigma^3} \tag{2.3}$$

and the *kurtosis* of X is

$$\gamma_2 = \frac{\mu_4}{\sigma^4} - 3. \tag{2.4}$$

||

A variable with positive skewness typically has a longer or heavier tail on the right than on the left; for negative skewness, the opposite holds. A variable with positive kurtosis typically has a higher central peak and a longer or heavier tail on at least one side than a normal distribution, while a variable with negative skewness is flatter in the middle and has shorter tails. Figure 2.15 shows some examples.

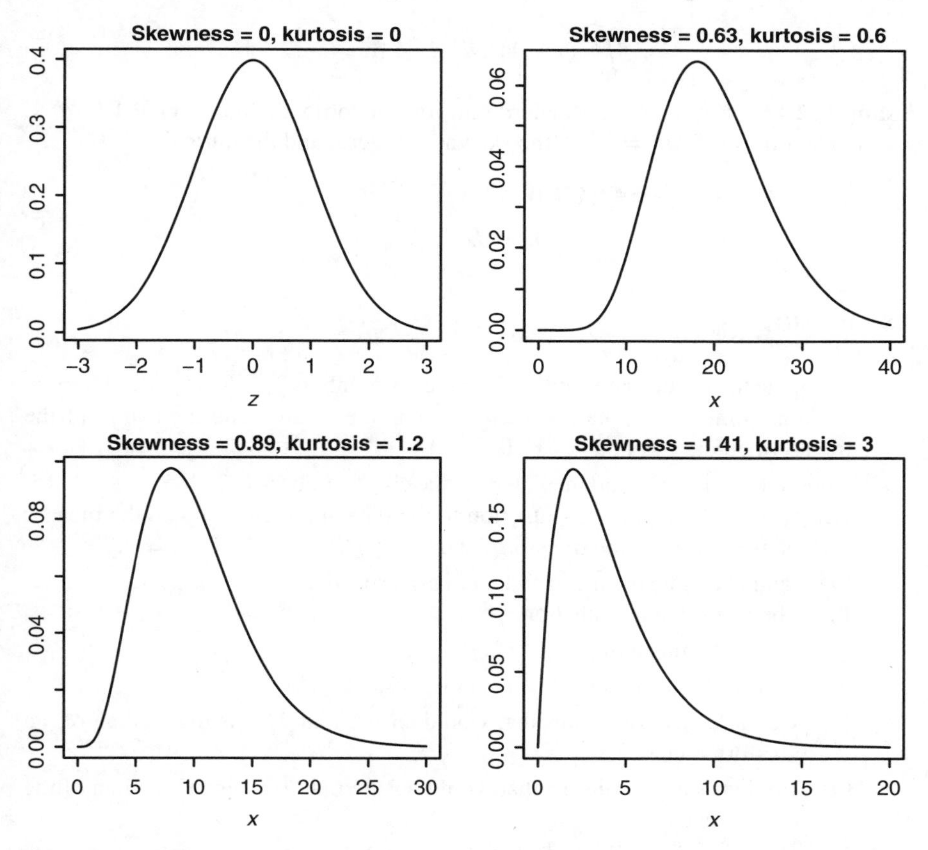

FIGURE 2.15 Examples of skewness and kurtosis for four distributions, including the standard normal (top left).

Example 2.12 Let Z be the standard normal variable with $\mu = 0$ and $\sigma = 1$. Then the skewness of Z is

$$\frac{1}{\sqrt{2\pi}} \int_{-\infty}^{\infty} z^3 e^{-z^2/2}\, dz = 0$$

and the kurtosis is

$$\frac{1}{\sqrt{2\pi}} \int_{-\infty}^{\infty} z^4 e^{-z^2/2}\, dz - 3 = 0.$$

□

Example 2.13 Let X be an exponential random variable with parameter $\lambda = 1$. Then $\mu = 1 = \sigma$, the skewness of X is

$$\int_0^{\infty} (x-1)^3 e^{-x}\, dx = 2,$$

and the kurtosis is

$$\int_0^{\infty} (x-1)^4 e^{-x}\, dx - 3 = 6.$$

□

Example 2.14 Let X be the standard uniform random variable, $f(x) = 1$ for $0 < x < 1$. Then $\mu = 0.5$, $\sigma^2 = 1/12$, the skewness is zero, and the kurtosis is

$$\frac{\int_0^1 (x-0.5)^4\, dx}{(\int_0^1 (x-0.5)^2\, dx)^2} - 3 = -1.2.$$

□

2.8 EXERCISES

1. Compute the mean $\bar{x}$ and median m of the six numbers 3, 5, 8, 15, 20, 21. Apply the logarithm to the data and then compute the mean $\tilde{x}$ and median $\tilde{m}$ of the transformed data. Is $\ln(\bar{x}) = \tilde{x}$? Is $\ln(m) = \tilde{m}$?
2. Compute the mean $\bar{x}$ and median m of the eight numbers 1, 2, 4, 5, 6, 8, 11, 15. Let $f(x) = \sqrt{x}$. Apply this function to the data and then compute the mean $\tilde{x}$ and the median $\tilde{m}$ of the transformed data. Is $f(\bar{x}) = \tilde{x}$? Is $f(m) = \tilde{m}$?
3. Let $\bar{x}$ and m denote the mean and median, respectively, of $x_1 < x_2 < \cdots < x_n$. Let f be a real-valued function.
   (a) Is $f(\bar{x})$ the mean of $f(x_1), f(x_2), \ldots, f(x_n)$?
   (b) Is $f(m)$ the median of $f(x_1), f(x_2), \ldots, f(x_n)$?
   (c) Are there any conditions that would ensure that $f(\bar{x})$ is the median of the transformed data?
   (d) Are there any conditions that would ensure that $f(m)$ is the median of the transformed data?
4. Import data from the Flight Delays Case Study in Section 1.1 data into `R`.
   (a) Create a table and a bar chart of the departure times (`DepartTime`).

(b) Create a contingency table of the variables `Day` and `Delayed30`. For each day, what is the proportion of flights delayed at least 30 min?

(c) Create side-by-side boxplots of the lengths of the flights, grouped by whether or not the flight was delayed at least 30 min.

(d) Do you think that there is a relationship between the length of a flight and whether or not the departure is delayed by at least 30 min?

5. Import data from the General Social Survey Case Study in Section 1.6 into R.

(a) Create a table and a bar chart of the responses to the question about the death penalty.

(b) Use the `table` command and the `summary` command in R on the gun ownership variable. What additional information does the `summary` command give that the `table` command does not?

(c) Create a contingency table comparing responses to the death penalty to the question about gun ownership.

(d) What proportion of gun owners favor the death penalty? Does it appear to be different from the proportion among those who do not own guns?

6. Import data from the Black Spruce Case Study in Section 1.9 into R.

(a) Compute the numeric summaries for the height changes (`Ht.Change`) of the seedlings.

(b) Create a histogram and normal quantile plot for the height changes of the seedlings. Is the distribution approximately normal?

(c) Create a boxplot to compare the distribution of the change in diameters of the seedlings (`Di.change`) grouped by whether or not they were in fertilized plots.

(d) Use the `tapply` command to find the numeric summaries of the diameter changes for the two levels of fertilization.

(e) Create a scatter plot of the height changes against the diameter changes and describe the relationship.

7. Let $x_1 < x_2 < \cdots < x_n$ and $y_1 < y_2 < \cdots < y_n$ be two sets of data with means $\bar{x}, \bar{y}$ and medians m_x, m_y, respectively. Let $w_i = x_i + y_i$ for $i = 1, 2, \ldots, n$.

(a) Prove or give a counterexample: $\bar{x} + \bar{y}$ is the mean of $w_1, w_2, \ldots, w_n$.

(b) Prove or give a counterexample: $m_x + m_y$ is the median of $w_1, w_2, \ldots, w_n$.

8. Find the median m and the first and third quartiles for the random variable X having

(a) the exponential distribution with pdf $f(x) = \lambda e^{-\lambda x}$.

(b) the Pareto distribution with parameter $\alpha > 0$ with pdf $f(x) = \alpha/x^{\alpha+1}$ for $x \geq 1$.

9. Let the random variable X have a Cauchy distribution with pdf $f(x) = 1/(\pi(1 + (x - \theta)^2))$ for $-\infty < x < \infty$. Show that θ is the median of the distribution.

10. Find
    (a) the 30th and 60th percentiles for $N(10, 17^2)$.
    (b) the 0.10 and 0.90 quantile for $N(25, 32^2)$.
11. Let X be a random variable with cdf $F(x) = x^2/a^2$ for $0 \leq x \leq a$. Find an expression for the $\alpha/2$ and $(1 - \alpha/2)$ quantiles, where $0 < \alpha < 1$.
12. Let X be a random variable with cdf $F(x) = 1 - 9/x^2$ for $x \geq 3$. Find an expression for the pth quantile of X.
13. Let $X \sim \text{Binom}(20, 0.3)$ and let F denote its cdf. Does there exist a q such that $F(q) = 0.05$?
14. In this exercise, we investigate normal quantile plots using R.
    (a) Draw a random sample of size $n = 15$ from $N(0, 1)$ and plot both the normal quantile plot and the histogram. Do the points on the quantile plot appear to fall on a straight line? Is the histogram symmetric, unimodal, and mound shaped? Do this several times.

R Note:

```
x <- rnorm(15)        # draw random sample of size 15 from N(0,1)
par(mfrow=c(2,1)) # set up plot area to place 2 graphs on one sheet
qqnorm(x)
qqline(x)
hist(x)
```

    (b) Repeat part (a) for samples of size $n = 30$, $n = 60$, and $n = 100$.
    (c) What lesson do you draw about using graphs to assess whether or not a data set follows lesson a normal distribution?

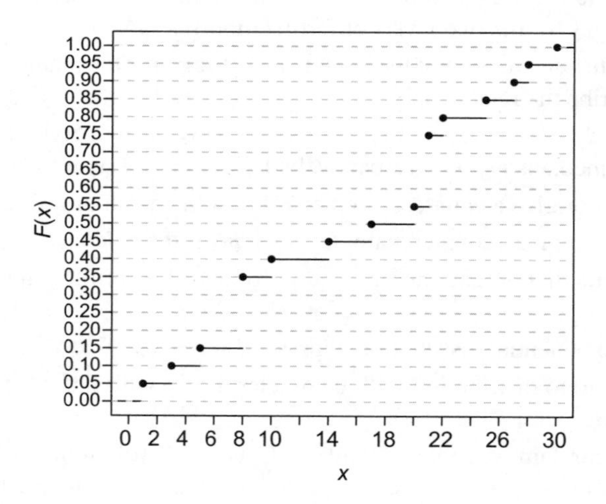

FIGURE 2.16 Empirical cdf for a data set, $n = 20$.

15. Plot by hand the empirical cumulative distribution function for the set of values 4, 7, 8, 9, 9, 13, 18, 18, 18, 21.
16. The ecdf for a data set with $n = 20$ values is given in Figure 2.16.
    (a) How many values are less than or equal to 7?
    (b) How many times does the value 8 occur?
    (c) In a histogram of these values, how many values fall in the bin (20, 25]?
17. Compare the ecdf's for United Airlines and American Airlines lengths of flight delays in the Flight Delays Case Study in Section 1.1.

3

HYPOTHESIS TESTING

3.1 INTRODUCTION TO HYPOTHESIS TESTING

Suppose scientists invent a new drug that supposedly will inhibit a mouse's ability to run through a maze. The scientists design an experiment in which three mice are randomly chosen to receive the drug and another three mice serve as controls by ingesting a placebo. The time each mouse takes to go through a maze is measured in seconds. Suppose the results of the experiment are as follows:

Drug			Control		
30	**25**	**20**	18	21	22

The average time for the drug group is 25 s and the average time for the control group is 20.33 s. The mean difference in times is $25 - 20.33 = 4.67$ s.

The average time for the mice given the drug is greater than the average time for the control group, but this could be due to random variability rather than a real drug effect. We cannot, however, tell for sure whether there is a real effect. What we do instead is that we estimate how easily pure random chance would produce a difference this large. If that probability is small, then we conclude there is something other than pure random chance at work, and hence there is a real effect.

If the drug does not really influence times, then split of the six observations into two groups was essentially random. The outcomes could just as easily be distributed:

Mathematical Statistics with Resampling and R, First Edition. By Laura Chihara and Tim Hesterberg.

Drug			Control		
30	**25**	18	**20**	21	22

In this case, the mean difference is $((30+25+18)/3) - ((20+21+22)/3) = 3.33$.

There are $\binom{6}{3} = 20$ ways to distribute six numbers into two sets of size 3, ignoring any ordering with each set. Of the 20 possible differences in means, 3 are as large or larger than the observed 4.67; so the probability that pure chance would give a difference this large is $3/20 = 0.15$.

The value of 15% is small, but not small enough to be remarkable. It is plausible that chance alone is the reason the mice in the drug group ran slower (had larger times) through the maze.

For comparison, suppose a friend claims that she can control the flip of a coin, producing a head at will. You are skeptical; you give her a coin, and she indeed flips a head, three times. Are you convinced? I hope not; that could easily occur by chance, with a 12.5% probability.

This is the core idea of *statistical significance* or classical *hypothesis testing*—to calculate how often pure random chance would give an effect as large as that observed in the data, in the absence of any real effect. If that probability is small enough, we conclude that the data provide convincing evidence of a real effect.

If the probability is not small, we do not make that conclusion. This is not the same as concluding that there is no effect; it is only that the data available do not provide convincing evidence that there is an effect. In practice, there may be just too little data to provide convincing evidence. If the drug effect is small, it may be possible to distinguish the effect from random noise with 60 mice, but not 6. More flips might make your friend's claim convincing, though it would be prudent to check for a two-headed coin. (One of the authors had such a coin, and also had a professor who could reliably flip a coin as desired; he had earlier been a professional magician; see `http://news-service.stanford.edu/news/2004/june9/diaconis-69.html`.)

3.2 HYPOTHESES

We formalize the core idea using the language of *statistical significance testing*, also known as *hypothesis testing*.

Definition 3.1 The *null hypothesis*, denoted H_0, is a statement that corresponds to no real effect. This is the status quo, in the absence of the data providing convincing evidence to the contrary.

The *alternative hypothesis*, denoted H_A, is a statement that there is a real effect. The data may provide convincing evidence that this hypothesis is true.

A hypothesis should involve a statement about a population parameter or parameters, commonly referred to as θ; the null hypothesis is H_0: $\theta = \theta_0$ for some θ_0.

A *one-sided alternative hypothesis* is of the form $H_A\colon \theta > \theta_0$ or $H_A\colon \theta < \theta_0$; a *two-sided alternative hypothesis* is $H_A\colon \theta \neq \theta_0$. ||

Example 3.1 Consider the mice example in Section 3.1. Let μ_d denote the true mean time that a randomly selected mouse that received the drug takes to run through the maze; let μ_c denote the true mean time for a control mouse. Then, $H_0\colon \mu_d = \mu_c$. That is, on average, there is no difference in the mean times between mice who receive the drug and mice in the control group.

The alternative hypothesis is $H_A\colon \mu_d > \mu_c$. That is, on average, mice who receive the drug have slower times (larger values) than the mice in the control group.

The hypotheses may be rewritten as $H_0\colon \mu_d - \mu_c = 0$ and $H_A\colon \mu_d - \mu_c > 0$; thus, $\theta = \mu_d - \mu_c$ (any function of parameters is itself a parameter). □

The next two ingredients in hypothesis testing are a numerical measure of the effect and the probability that chance alone could produce that measured effect.

Definition 3.2 A *test statistic* is a numerical function of the data whose value determines the result of the test. The function itself is generally denoted $T = T(\mathbf{X})$, where $\mathbf{X}$ represents the data, for example, $T = T(X_1, X_2, \ldots, X_n)$ in a one-sample problem, or $T = T(X_1, X_2, \ldots, X_m, Y_1, \ldots, Y_n)$ in a two-sample problem. After being evaluated for the sample data $\mathbf{x}$, the result is called an *observed test statistic* and is written in lowercase, $t = T(\mathbf{x})$. ||

Definition 3.3 The *P-value* is the probability that chance alone would produce a test statistic as extreme as the observed test statistic. For example, if large values of the test statistic support the alternative hypothesis, the P-value is $P(T \geq t)$. ||

Definition 3.4 A result is *statistically significant* if it would rarely occur by chance. How rarely? It depends on the context, most common is a 5% threshold but 1% or 10% are also common. It is more informative to give the P-value, for example, a result is *statistically significant* ($p = c$) if the P-value is c. Other common terminology is to declare a result *statistically significant at the 5% level* if the P-value is less than 0.05; however, this is less informative than giving the P-value, as it does not distinguish between, say, $p = 0.049$ (barely significant) and $p = 0.0001$ (extremely unlikely to occur by chance alone).

The symbol α is used to denote the significance level. ||

Example 3.2 In the mice example (Section 3.1), we let the test statistic be the difference in means, $T = T(X_1, X_2, X_3, Y_1, Y_2, Y_3) = \bar{X} - \bar{Y}$ with observed value $t = \bar{x} - \bar{y} = 4.67$. Large values of the test statistic support the alternative hypothesis, so the P-value is $P(T \geq 4.67) = 3/20$. □

Rather than just calculating the probability, we often begin by answering a larger question—What is the distribution of the test statistic when there is no real effect?

TABLE 3.1 All Possible Distributions of {30, 25, 20, 18, 21, 22} into Two Sets

Drug			Control			$\bar{X}_D$	$\bar{X}_C$	Difference in means
18	20	21	22	25	30	19.67	25.67	−6.00
18	20	22	21	25	30	20	25.33	−5.33
18	20	25	21	22	30	21	24.33	−3.33
18	20	30	21	22	25	22.67	22.67	0.00
18	21	22	20	25	30	20.33	25	−4.67
18	21	25	20	22	30	21.33	24	−2.67
18	21	30	20	22	25	23	22.33	0.67
18	22	25	20	21	30	21.67	23.67	−2.00
18	22	30	20	21	25	23.33	22	1.33
18	25	30	20	21	22	24.33	21	3.33
20	21	22	18	25	30	21	24.33	−3.33
20	21	25	18	22	30	22	23.33	−1.33
20	21	30	18	22	25	23.67	21.67	2.00
20	22	25	18	21	30	22.33	23	−0.67
20	22	30	18	21	25	24	21.33	2.67
20	25	30	18	21	22	**25**	**20.33**	**4.67** *
21	22	25	18	20	30	22.67	22.67	0.00
21	22	30	18	20	25	24.33	21	3.33
21	25	30	18	20	22	**25.33**	**20**	**5.33** *
22	25	30	18	20	21	**25.67**	**19.67**	**6.00** *

Rows where the difference in means exceeds the original value are highlighted.

For example, Table 3.1 gives all values of the test statistic in the mice example; each value has the same probability if there is no drug effect.

Definition 3.5 The *null distribution* is the distribution of the test statistic if the null hypothesis is true. ||

You can think of the null distribution as a reference distribution; we compare the observed test statistic with this reference to determine how unusual the observed test statistic is. Figure 3.1 shows the cumulative distribution function of the null distribution in the mice example.

There are different ways to calculate exact or approximate null distributions and P-values. For now, we focus on one method—permutation tests.

3.3 PERMUTATION TESTS

In the mice example in Section 3.1, we compared the test statistic to a reference distribution using permutations of the observed data. We investigate this approach in more detail.

Recall the Beer and Hot Wings Case Study in Section 1.8. The mean number of wings consumed by females and males were 9.33 and 14.53, respectively, while the standard deviations were 3.56 and 4.50, respectively. See Figure 3.2 and Table 3.2.

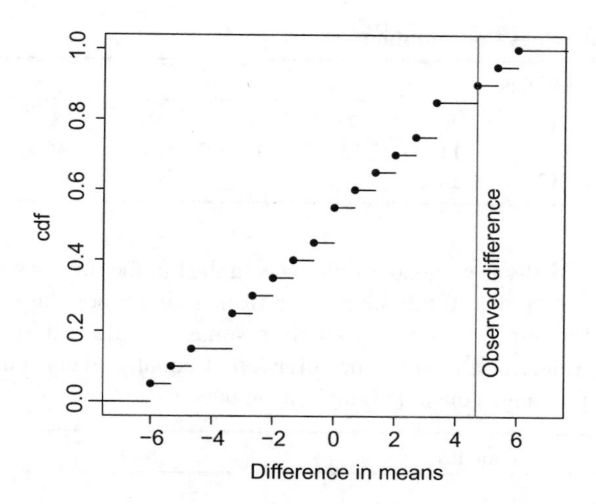

FIGURE 3.1 Empirical cumulative distribution function of the null distribution for difference in means for mice.

The sample means for the males and females are clearly different, but the difference (14.33 − 9.33 = 5.2) could have arisen by chance. Can the difference *easily* be explained by chance alone? If not, we will conclude that there are genuine gender differences in hot wings consumption.

For a hypothesis test, let μ_M denote the mean number of hot wings consumed by males and μ_F denote the mean number of hot wings consumed by females. We test

$$H_0\colon \mu_M = \mu_F \quad \text{versus} \quad H_A\colon \mu_M > \mu_F$$

or equivalently

$$H_0\colon \mu_M - \mu_F = 0 \quad \text{versus} \quad H_A\colon \mu_M - \mu_F > 0.$$

We use $T = \bar{X}_M - \bar{X}_F$ as a test statistic, with observed value $t = 5.2$.

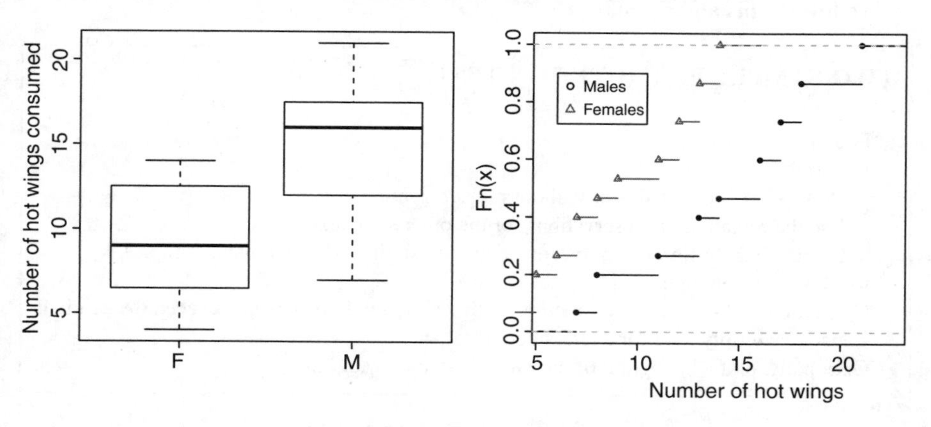

FIGURE 3.2 Number of hot wings consumed, by gender.

TABLE 3.2 Hot Wings Consumption

Females					Males				
4	**5**	**5**	**6**	**7**	7	8	8	11	13
7	**8**	**9**	**11**	**12**	13	14	16	16	17
12	**13**	**13**	**14**	**14**	17	18	18	21	21

Suppose there really is no gender influence on the number of hot wings consumed by bar patrons. Then, the 30 numbers come from a single population, and the way they were divided into 2 groups (by labeling some as male and others as female) was essentially random, and any other division is equally likely. For instance, the distribution of hot wings consumed might have been as below:

Females					Males				
5	**6**	**7**	**7**	**8**	**4**	**5**	7	8	**9**
8	**11**	**12**	**13**	**14**	11	**12**	13	13	**13**
14	14	16	16	21	17	17	18	18	21

In this case, the difference in means is $12.4 - 11.47 = 0.93$.

We could proceed, as in the mice example, calculating the difference in means for *every* possible way to split the data into two samples of size 15 each. This would result in $\binom{30}{15} = 155117520$ differences! In practice, such exhaustive calculations are impractical unless the sample sizes are small, so we resort to sampling instead.

We create a *permutation resample*, or *resample* for short, by drawing $m = 15$ observations *without* replacement from the pooled data to be one sample (the males), leaving the remaining $n = 15$ observations to be the second sample (the females). We calculate the statistic of interest, for example, difference in means of the two samples. We repeat this many times (1000 or more). The P-value is then the fraction of times the random statistic exceeds[1] the original statistic.

We follow this algorithm:

TWO-SAMPLE PERMUTATION TEST

Pool the $m + n$ values.
repeat
Draw a resample of size m without replacement.
Use the remaining n observations for the other sample.
Calculate the difference in means or another statistic that compares samples.
until we have enough samples
Calculate the P-value as the fraction of times the random statistics exceed the original statistic. Multiply by 2 for a two-sided test.
Optionally, plot a histogram of the random statistic values.

[1] In this chapter, "exceeds" generally means $\geq$ rather than $>$.

The distribution of this difference across all permutation resamples is the *permutation distribution*. This may be exact (calculated exhaustively) or approximate (implemented by sampling). In either case, we usually use statistical software for the computations. Here is the code that will perform the test in R.

R Note:

We first compute the observed mean difference in the number of hot wings consumed by males and females.

```
> tapply(Beerwings$Hotwings, Beerwings$Gender, mean)
       F         M
  9.333333 14.533333
> observed <- 14.5333 - 9.3333  # store observed mean difference
> observed
[1] 5.2
```

Since we will be working with the hot wings variable, we will create a vector holding these values. Then, we will draw a random sample of size 15 from the numbers 1 through 30 (there are 30 observations is total). The hot wing values corresponding to these positions will be values for the males and the remaining ones for the females. The mean difference of this permutation will be stored in `result`. This will be repeated many times.

```
hotwings <- Beerwings$Hotwings
# Another way:
# hotwings <- subset(Beerwings, select = Hotwings, drop = T)

N <- 10^5 - 1         # number of times to repeat this process
result <- numeric(N) # space to save the random differences
for (i in 1:N)
{ # sample of size 15, from 1 to 30, without replacement
  index <- sample(30, size=15, replace=FALSE)
  result[i] <- mean(hotwings[index]) - mean(hotwings[-index])
}
```

We first create a histogram of the permutation distribution and add a vertical line at the observed mean difference.

```
hist(result, xlab = "xbar1 - xbar2",
     main = "Permutation Distribution for hot wings")
abline(v = observed, col = "blue")
# add line at observed mean diff.
```

We determine how likely it is to obtain an outcome as large or larger than what we observed.

```
> (sum(result >= observed) + 1)/(N + 1)  # P-value
[1] 0.000831                           # results will vary
```

The code snippet `result >=observed` results in a vector of `TRUE`'s and `FALSE`'s depending on whether or not the mean difference computed for a resample is greater than the observed mean difference.

`sum(result >= observed)` counts the number of `TRUE`'s. Thus, the computed P-value is just the proportion of statistics (including the original) that are as large or larger than the original mean difference.

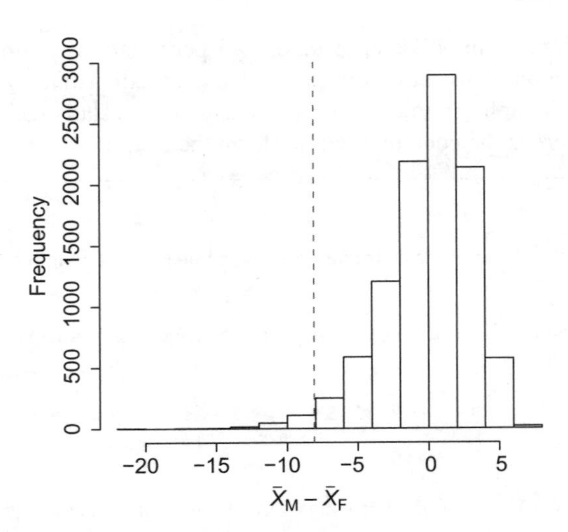

FIGURE 3.3 Permutation distribution of the difference in means, male − female, in the beer and hot wings example.

From the output, we see that the observed difference in means is 5.2. The P-value is 0.00083. Of the $10^5 - 1$ resamples computed by R, less than 0.1% of the resampled difference in means was as large or larger than 5.2. There are two possibilities—either there is a real difference or there is no real effect, but a miracle occurred giving a difference well beyond the range of normal chance variation. We cannot rule out the miracle, but the evidence does support the hypothesis that females in this study consume fewer hot wings than males (Figure 3.3).

The participants in this study were a convenience sample: they were chosen because they happened to be at the bar when the study was conducted. Thus, we cannot make any inference about a population.

3.3.1 Implementation Issues

We note here some implementation issues for permutation tests. The first (choice of test statistic) applies to both the exhaustive and sampling implementations, while the final three (add one to both numerator and denominator, sample with replacement from null distribution, and more samples for better accuracy) are specific to sampling.

Choice of Test Statistic In the examples above, we used the difference in means. We could have equally well used $\bar{X}$ (the mean of the first sample), $m\bar{X}$ (the sum of the observations in the first sample), or a variety of other test statistics. For example, in Table 3.1, the same three rows have test statistics that exceed the observed test statistic, whether the test statistic is difference in means or $\bar{X}_\text{D}$ (the mean of the sample in the drug group).

Here is the result that states this more formally:

Theorem 3.1 *In permutation testing, if two test statistics T_1 and T_2 are related by a strictly increasing function, $T_1(\mathbf{X}^*) = f(T_2(\mathbf{X}^*))$ where $\mathbf{X}^*$ is any permutation resample of the original data $\mathbf{x}$, they yield exactly the same P-values, for both the exhaustive and resampling versions of permutation testing.*

Proof For simplicity, we consider only a one-sided (greater) test. Let $\mathbf{X}^*$ be any permutation resample. Then,

$$\begin{aligned} p_1 &= P(T_2(\mathbf{X}^*) \geq T_2(\mathbf{x})) \\ &= P(f(T_2(\mathbf{X})) \geq f(T_2(\mathbf{x}))) && \text{since } f \text{ is strictly increasing} \\ &= P(T_1(\mathbf{X}^*) \geq T_1(\mathbf{x}) && \text{by hypothesis.} \end{aligned}$$

Furthermore, in the sample implementation, exactly the same permutation resamples have $T_2(\mathbf{X}) \geq T_2(\mathbf{x})$ as have $T_1(\mathbf{X}) \geq T_1(\mathbf{x})$, so counting the number or fraction of samples that exceed the observed statistic yields the same results. □

Remark One subtle point is that the transformation need to be strictly monotone only *for the observed data*, not for all possible sets of data. For example, in the mice example, we used $p = P(\bar{X}_1 - \bar{X}_2 \geq \bar{X}_1 - \bar{X}_2)$. Let $T_1 = \bar{X} = \bar{X}_1 - \bar{X}_2$ denote the mean difference and let $T_2 = \bar{X}_1$ denote the mean of just the treatment group. Let $S_1 = 3\bar{X}_1$ and $S_2 = 3\bar{X}_2$ be the sums in the two samples, and $S = S_1 + S_2 = 136$ the overall sum; this is the same for every resample (it is the sum of the same data, albeit in a different order), so we can rewrite

$$\bar{X}_2 = \frac{S_2}{3} = \frac{S - S_1}{3} = \frac{136}{3} - \bar{X}_1$$

and

$$\bar{X}_1 - \bar{X}_2 = 2\bar{X}_1 - \frac{136}{3}.$$

Hence, the transformation is $f(T_2) = 2T_2 - 136/3$. This is linear in T_2 and hence monotone (increasing). For these data, it is true that $\bar{X}_1 - \bar{X}_2 \geq 4.67$ if and only if $\bar{X}_1 \geq 25$, but that is not true for every possible set of data.

In other words, the transformation may depend on the original data: $T_1(\mathbf{X}^*) = f(T_2(\mathbf{X}^*); \mathbf{x})$. ||

Add One to both Numerator and Denominator When computing the P-value in the sampling implementation, we add one to both numerator and denominator. This corresponds to including the original data as an extra resample. This is a bit conservative, and avoids reporting an impossible P-value of 0.0—since there is always at least one resample that is as extreme as the original data, namely, the original data itself.

Sample with Replacement from the Null Distribution In the sampling implementation, we do not attempt to ensure that the resamples are unique. In effect, we draw resamples *with replacement* from the population of $\binom{m+n}{m}$ possible resamples, and hence obtain a sample with replacement from the $\binom{m+n}{m}$ test statistics that make up the exhaustive null distribution. Sampling without replacement would be more accurate, but it is not feasible, requiring too much time and memory to check that a new sample does not match any previous sample.

More Samples for Better Accuracy In the hot wings example, we resampled 99,999 times. In general, the more the resamples, the better the accuracy. If the true P-value is p, the estimated P-value has variance approximately equal to $p(1-p)/N$, where N is the number of resamples.

Remark Just as the original n data values are a sample from the population, so are the N resampled statistics a sample from a population (in this case, the null distribution). ||

The next example features highly skewed distributions and unbalanced sample sizes, as well as the need for high accuracy.

Example 3.3 Recall the Verizon Case Study in Section 1.3. We wish to compare the repair times for ILEC and CLEC customers. This is determined using hypothesis tests under an agreement with the local Public Utilities Commission (PUC). There are thousands of significance tests performed, to compare the speed of different types of repairs, over different time periods for different competitors. The tests are performed at a 1% significance level; if substantially more than 1% of the tests come up positive, then Verizon is deemed to be discriminating.

Figure 3.4 shows the raw data for one of these tests. The mean of 1664 repairs for ILEC customers is 8.4 h, while the mean for 23 repairs for CLEC customers is 16.5 h. Could a difference that large be easily explained by chance? There appears to be one outlier in the smaller data set; perhaps that explains the difference in means. However, it would not be reasonable to throw out that observation as faulty—it is clear from the larger data set that large repair times do occur fairly frequently. Furthermore, even in the middle of both distributions, the CLEC times do appear to be longer (this is apparent in the right panel). One curious aspect of these data is the bends in the normal quantile plot due to 24 h cycles.

Let μ_1 denote the mean repair time for the ILEC customers and μ_2 the mean repair time for the CLEC customers. We test

$$H_0\colon \mu_1 = \mu_2 \quad \text{versus} \quad H_A\colon \mu_1 < \mu_2.$$

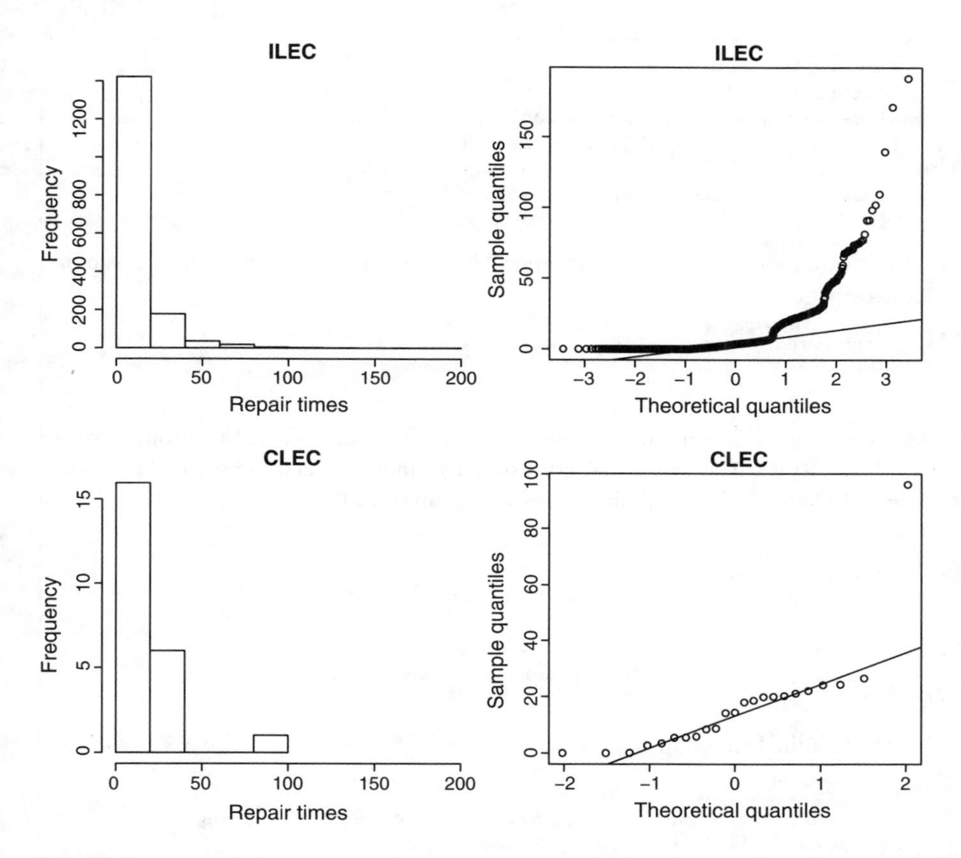

FIGURE 3.4 Distribution of repair times for Verizon (ILEC) and competitor (CLEC) customers. Note that the Y-axis scales are different.

We use a one-sided test because the alternative of interest to the PUC is that the CLEC customers are receiving worse service (longer repair times) than the ILEC customers.

R Note:

```
> tapply(Verizon$Time, Verizon$Group, mean)
    CLEC     ILEC
16.50913 8.411611
```

We will create three vectors, one containing the times for all the customers, one with the times for just the ILEC customers, and one for just the CLEC customers.

```
Time <- Verizon$Time
#Alternatively
#Time <- subset(Verizon, select = Time, drop = TRUE)
Time.ILEC <- subset(Verizon, select = Time,
                    subset = Group == "ILEC", drop = T)
Time.CLEC <- subset(Verizon, select = Time,
                    subset = Group == "CLEC", drop = T)
```

Now we compute the mean difference in repair times and store in the vector `observed`

```
> observed <-  mean(Time.ILEC) - mean(Time.CLEC)
> observed
[1] -8.09752
```

We will draw a random sample of size 1664 (size of ILEC group) from $1, 2, \ldots, 1687$. The times that correspond to these observations will be put in the ILEC group; the remaining times will go into the CLEC group.

```
N <- 10^4-1
result <- numeric(N)
for (i in 1:N)
{
  index <- sample(1687, size = 1664, replace = FALSE)
  result[i] <- mean(Time[index]) - mean(Time[-index])
}
```

First, plot the histogram

```
hist(result, xlab = "xbar1-xbar2",
     main = "Permutation distribution for Verizon times")
abline(v = observed, ,lty = 2, col = "blue")
```

Note here that we will want to find the proportion of times the resampled mean difference is less than or equal to the observed mean difference.

```
(sum(result <= observed) + 1)/(N + 1)  # P-value
[1] 0.0165
```

The P-value of 0.0165 indicates that the observed difference in means is not significant at the 1% level (though it is at the 5% level).

In the above simulation, we used $10^4 - 1$ resamples to speed up the calculations. For higher accuracy, we should use a half-million resamples; this was negotiated between Verizon and the PUC. The goal is to have only a small chance of a test wrongly being declared significant or not, due to random sampling.

The permutation distribution is shown in Figure 3.5. The P-value is the fraction of the distribution that falls to the left of the observed value.

This test works fine even with unbalanced sample sizes of 1664 and 23, and even for very skewed data. The permutation distribution is skewed to the left, but that does

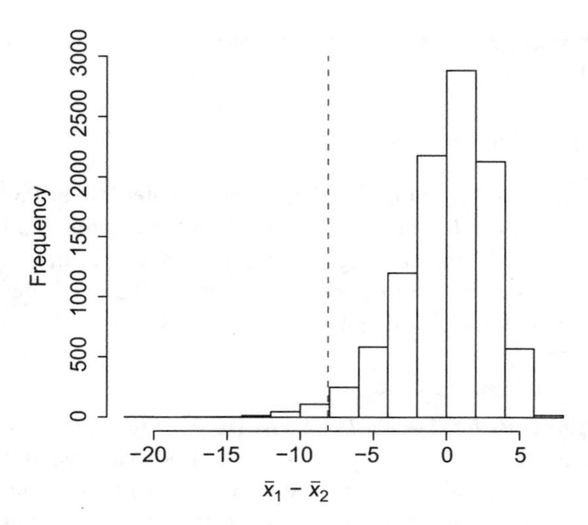

FIGURE 3.5 Permutation distribution of $\bar{x}_1 - \bar{x}_2$ for the Verizon repair time data.

not matter; both the observed statistic and the permutation resamples are affected by imbalance and skewness in the same way.

In contrast, t-tests for comparing two means (we discuss these in Chapter 8) assume normal populations and are not accurate with skewed populations and imbalanced sample sizes; the pooled-variance t-test claims a P-value of 0.0045, about four times too small. This test would claim a result is significant at the 1% level about 4% of the time. □

3.3.2 One-Sided and Two-Sided Tests

For the hypothesis test with alternative H_A: $\mu_1 - \mu_2 < 0$, we compute a P-value by finding the fraction of resample statistics that are less than or equal to the observed test statistic (or greater than or equal to for the alternative $\mu_1 - \mu_2 > 0$.)

For a two-sided test, we calculate both one-sided P-values, multiply the smaller by 2, and finally (if necessary) round down to 1.0 (because probabilities can never be larger than 1.0).

In the mice example with observed test statistic $t = 4.67$, the one-sided P-values are $3/20$ for H_A: $\mu_d - \mu_c > 0$ and 18/20 for H_A: $\mu_d - \mu_c < 0$. Hence, the two-sided P-value is $6/20 = 0.30$ (recall Table 3.1).

Two-sided P-values are the default in statistical practice—you should perform a two-sided test unless there is a clear reason to pick a one-sided alternative hypothesis. It is not fair to look at the data before deciding to use a one-sided hypothesis.

Example 3.4 We return to the `Beerwings` data set and the comparison of the mean number of hot wings consumed by males and females. Suppose prior to this study, we had no preconceived idea of which gender would consume more hot wings. Then, our hypotheses would be

$$H_0\colon \mu_M = \mu_F \quad \text{versus} \quad H_A\colon \mu_M \neq \mu_F.$$

We found the one-sided P-value (for alternative "greater") to be 0.00111 (page 42), so for a two-sided test, we double 0.00111 to obtain the P-value 0.00222.

If gender does not influence average hot wings consumption, a difference as extreme as what we observed would occur only about 0.2% of the time. We conclude that males and females do not consume, on average, the same number of hot wings. □

To Obtain P-Values in the Two-Sided Case We Multiply by 2 We multiply the smaller of the one-sided P-values by 2, using the observed test statistic. Multiplying by 2 has a deeper meaning. Because we are open to more than one alternative to the null hypothesis, it takes stronger evidence for any one of these particular alternatives to provide convincing evidence that the null hypothesis is incorrect. With two possibilities, the evidence must be stronger by a factor of 2, measured on the probability scale.

3.3.3 Other Statistics

We noted in Section 3.3.1 the possibility of using a variety of statistics and getting equivalent results, provided the statistics are related by a monotone transformation.

Permutation testing actually offers considerably more freedom than that; the basic procedure works with any test statistic. We compute the observed test statistic, resample, compute the test statistics for each resample, and compute the P-value (see the algorithm on page 40.) Nothing in the process requires that the statistic be a mean or equivalent to a mean.

This provides the flexibility to choose a test statistic that is more suitable to the problem at hand. Rather than using means, for example, we might base the test statistic on *robust statistics*, that is, statistics that are not sensitive to outliers. Two examples of robust statistics are the median and the trimmed mean. We have already encountered the median. The trimmed mean is just a variant of the mean: we sort the data, omit a certain fraction of the low and high values, and calculate the mean of the remaining values. In addition, permutation tests could also compare proportions or variances. We give examples of each of these cases next, then turn in the next section to what appears at first glance to be a completely different setup but is in fact just another application of this idea.

Example 3.5 In the Verizon example we observed that the data have a long tail—there are some very large repair times (Figure 3.4). We may wish to use a test statistic that is less sensitive to these observations. There are a number of reasons we might do this. One is to get a better measure of what is important in practice, how inconvenienced customers are by the repairs. After a while, each additional hour

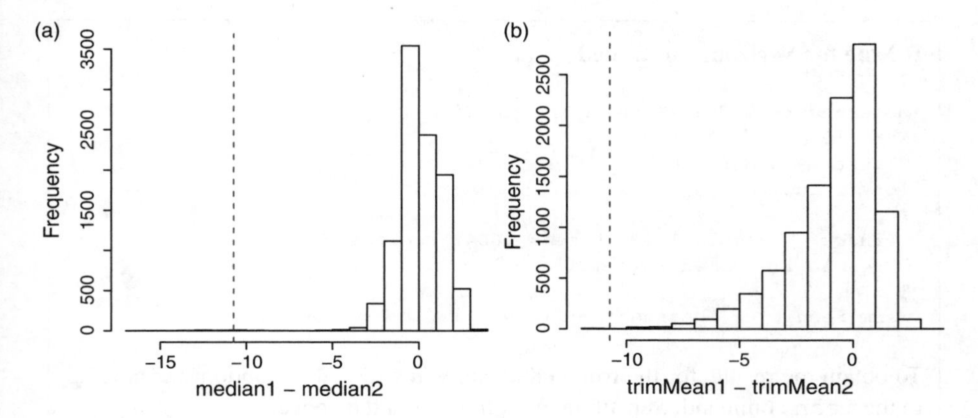

FIGURE 3.6 Repair times for Verizon data. (a) Permutation distribution for difference in medians. (b) Permutation distribution for difference in 25% trimmed means.

probably does not matter as much, yet a sample mean treats an extra 10 h on a repair time of 100 h the same as an extra 10 h on a repair time of 1 h. Second, a large recorded repair time might just be a blunder; for example, a repair time of 10^6 h must be a mistake. Third, a more robust statistic could be more sensitive at detecting real differences in the distributions—the mean is so sensitive to large observations that it pays less attention to moderate observations, whereas a statistic more sensitive to moderate observations could detect differences between populations that show up in the moderate observations.

Here is the R code for permutation tests using medians and trimmed means.

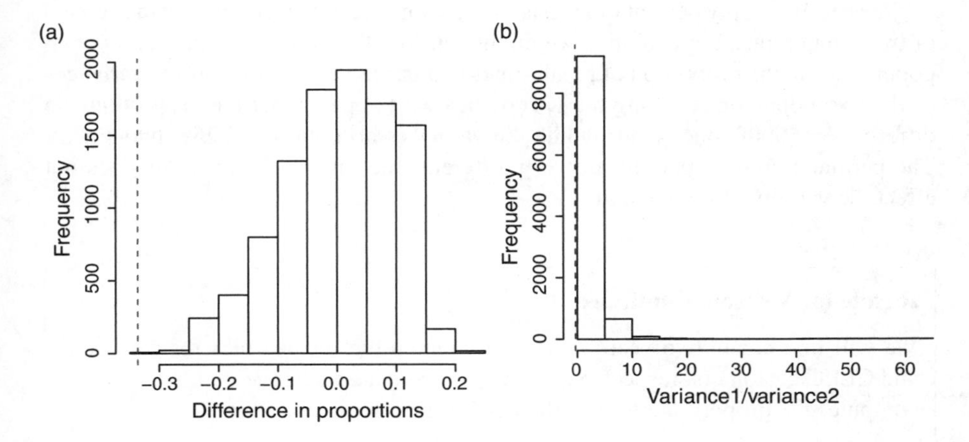

FIGURE 3.7 Repair times for Verizon data. (a) Difference in proportion of repairs exceeding 10 h. (b) Ratio of variances (ILEC/CLEC).

R Note for Verizon, Continued

```
observed <- median(Time.ILEC) - median(Time.CLEC)
N <- 10^4-1
result <- numeric(N)
for (i in 1:N)
{
  index <- sample(1687, size = 1664, replace = FALSE)
  result[i] <- median(Time[index]) - median(Time[-index])
}
(sum(result <= observed) + 1)/(N + 1)  # P-value
```

To obtain the results for the trimmed mean, you will add the option `trim=.25` to the `mean` command. Substitute the following in the above:

```
observed <- (mean(Time.ILEC, trim = .25) -
             mean(Time.CLEC, trim = .25))
result[i] <- (mean(Time[index], trim = .25) -
              mean(Time[-index], trim = .25))
```

It seems apparent that these more robust statistics are more sensitive to a possible difference between the two populations; the tests are significant with estimated *P*-values of 0.002 and 0.001, respectively. The figures also suggest that the observed statistics are well outside the range of normal chance variation.

One caveat is in order—it is wrong to try many different tests, possibly with minor variations, until you obtain a statistically significant outcome. If you try enough different things, eventually one will come out significant, whether or not there is a real difference. To guard against this, you can apply a *Sidak correction* or *Bonferroni correction* (see Section 8.3.2.)

One can also apply permutation tests to questions other than comparing the centers of two populations. For example, we might consider the difference between the two populations in the proportion of repair times that exceed 10 h or the ratio of variances of the two populations. Using the `R` code below, it appears that the proportions do differ ($p = 0.0008$, one-sided), while the variances do not ($p = 0.258$, two-sided). The permutation distributions are very different (see Figure 3.7), but this does not affect the validity of the method.

R Note for Verizon, Continued

We will first create two vectors that will contain the repair times for the ILEC and CLEC customers, respectively. The command `mean(Time.ILEC > 10)` computes the proportion of times the ILEC times are greater than 10.

```
> observed <- mean(Time.ILEC > 10) - mean(Time.CLEC > 10)
```

```
> observed
[1] -0.336852
```

Thus, about 33.7% fewer ILEC customers had repair times exceeding 10 h.

To run the simulation, reuse the code at the top of page 50 but with the following modification:

```
result[i] <- mean(Time[index]>10) - mean(Time[-index] > 10)
```

To perform the test for the variance, substitute

```
observed <- var(Time.ILEC)/var(Time.CLEC)
result[i] <- var(Time[index])/var(Time[-index])
```

□

3.3.4 Assumptions

Under what conditions can we use the permutation test? First, the permutation test makes no distributional assumption on the two populations under consideration. That is, there is no requirement that samples are drawn from a normal distribution, for example.

In fact, permutation testing does not even require that the data be drawn by random sampling from two populations. A study for the treatment of a rare disease could include all patients with the disease in the world. In this case, it does require that subjects be assigned to the two groups randomly.

In the usual case that the two groups are samples from two populations, pooling the data does require that the two *populations* have the same distribution when the null hypothesis is true. They must have the same mean, spread, and shape. This does not mean that the two *samples* must have the same mean, spread, and shape. There will always be some chance variation in the data.

In practice, the permutation test is usually robust when the two populations have different distributions. The major exception is when the two populations have different spreads and the sample sizes are dissimilar.

This exception is rarely a concern in practice, unless you have other information (besides the data) that the spreads are different. For example, one of us consulted for a large pharmaceutical company testing a new procedure for measuring a certain quantity; the new procedure was substantially cheaper, but not as accurate. The loss of accuracy was acceptable, provided that the mean measurements matched. This is a case where permutation testing would be doubtful, because it would pool data from different distributions. Even then, it would usually work fine if the sample sizes were equal.

Example 3.6 We investigate the extreme case in more detail. Suppose population A is normal with mean 0 and variance $\sigma_{\mathrm{A}}^2 = 10^6$, and population B is normal with mean 0 and variance $\sigma_{\mathrm{B}}^2 = 1$. Draw a sample of size $n_{\mathrm{A}} = 10^2$ from population A and a sample of size $n_{\mathrm{B}} = 10^6$ from population B. Thus, we have that the null hypothesis is true with both populations having mean 0. Let the test statistic be $T = \bar{X}_{\mathrm{A}}$. When drawing the original sample, T has variance $\sigma_{\mathrm{A}}^2/n_{\mathrm{A}} = 10^4$ (by Theorem A.5). What is the probability that this statistic T is greater than, say, 5? By standardizing, we find

$$P(T \geq 5) = P\left(\frac{T}{100} \geq \frac{5}{100}\right) = P(Z \geq 0.05) = 0.48.$$

Thus, with its huge variance of 10^4, there is nearly a 50% chance of T being greater than 5.

When we pool the two samples, it turns out that the variance of the permutation distribution of T is around $(n_{\mathrm{A}}\sigma_{\mathrm{A}}^2 + n_{\mathrm{b}}\sigma_{\mathrm{B}}^2)/(n_{\mathrm{A}} + n_{\mathrm{B}}) \approx 101$ (plus or minus random variation). Thus, when we perform the permutation test, the resampled T's have variance around $101/n_{\mathrm{A}} \approx 1.01$, or equivalently, a standard deviation about 1.005 (again, by Theorem A.5). So almost none of the permutation T's will be larger than 5:

$$P(T \geq 5) = P\left(\frac{T}{1.005} \geq \frac{5}{1.005}\right) = P(T \geq 4.975) = 0.$$

Thus, there is nearly a 50% chance of reporting a P-value near 0 and erroneously concluding that the means are not the same. □

3.4 CONTINGENCY TABLES

We now turn to a class of problems that appears at first glance to be very different. Yet the core ideas are the same.

As a start, we address the question of why the two-sample permutation test above is called *permutation* testing. It seems like all we are doing is splitting the data into two samples, with no hint of a permutation. Well, imagine storing the data in a table with two columns and $m + n$ rows; the first column contains labels, for example, m copies of "M" and n copies of "F", while the second contains the numerical data. We may permute the rows of either column, randomly; this is equivalent to splitting the data into two groups randomly.

Table 3.3 illustrates one such permutation of one of the columns in the `beerwings` data.

The idea of permuting the rows of one column generalizes to other situations, including, in this section, the analysis of contingency tables.

In the General Social Survey Case Study in Section 1.6, one question asked to the participant was whether he or she favored or opposed the death penalty for murder. A 5×2 contingency table summarizing the responses by education is given in Table 3.4.[2]

We see that a larger percentage of those who had a degree from a junior college favor the death penalty compared to those with a bachelors degree, 81.6% versus 65%. There are other large differences between education groups. Can these differences

[2]There are only 1307 respondents here compared to the total 2765 originally interviewed because many people chose not to respond to the death penalty question. They have been removed from this analysis.

TABLE 3.3 Partial View of `Beerwings` Data Set

	Gender	Hot Wings			Gender	Hot Wings
1	F	4		11	F	9
2	F	5		26	F	17
3	F	5		25	F	17
4	F	6		2	F	5
5	F	7		4	F	6
6	F	7	$\Longrightarrow$	8	F	8
7	M	7		3	M	5
8	F	8		20	F	14
9	M	8		10	M	8
10	M	8		18	M	13
	⋮				⋮	

The `Gender` column is held fixed and the rows of the `Hotwings` variable are permuted. The first column indicates which rows of the hot wing values were permuted.

easily be explained by chance variation, or do these data suggest that support for the death penalty depends on education?

In order to test this question, we need a test statistic and a reference distribution. We begin with the test statistic. First, what kind of results would we expect if opinions and education were independent? Of the 113 people with graduate degrees, we would expect 68.7% to favor the death penalty and 31.3% to oppose, the same proportions as the whole sample. Similarly for the other cells. Let N_{ij} be the number of people in row i and column j, let R_i be the total for row i, C_j the total for column j, and n the overall total. The *expected count* E_{ij} for any cell is the row total times the column proportion, which is equivalent to the column total times the row proportion:

$$E_{ij} = R_i(C_j/n) = C_j(R_i/n) = R_iC_j/n.$$

For instance, the expected count for the (4, 2)-cell is $87 \times 409/1307 = 222.4935$ (see Table 3.5).

It seems intuitive that the observed and expected counts should be similar if education and support for the death penalty are independent. Thus, it seems plausible

TABLE 3.4 Counts of Death Penalty Opinions Grouped by Education

	Death penalty for murder?			
Education	Favor	Oppose	Row Sum	% Favor
Bachelors	135	71	206	65.6%
Graduate	64	50	113	56.6%
HS	511	200	711	71.9%
JrColl	71	16	87	81.6%
Left HS	117	72	189	61.9%
Column sum	898	409	n = 1307	68.7%

TABLE 3.5 Expected Counts of Death Penalty Opinions Grouped by Degree

	Death Penalty?	
Education	Favor	Oppose
Bachelors	141.5363	64.4637
Graduate	78.3259	35.6741
HS	488.5065	39.1675
JrColl	59.7751	222.4935
Left HS	129.8562	59.1438

to look for a test statistic that takes into account the differences *observed count − expected count* for all cells. But just adding the differences does not work; the differences always add to zero (check this!). And both positive and negative differences should contribute to the test statistic in the same way—both large positive and large negative differences suggest dependence. So it would be reasonable to let the test statistic be the sum of absolute values of the differences or the sum of squared differences. These are legitimate test statistics, but are not ideal—they do not take size into account. A difference of say 20 in a cell with an expected count of 10 is a bigger deal than a difference of 20 in a cell with an expected count of 2000.

There is a standard test statistic in this setting that does take cell size into account. In the 1900s, Karl Pearson proposed the statistic

$$C = \sum_{\text{all cells}} \frac{(\text{observed} - \text{expected})^2}{\text{expected}}. \tag{3.1}$$

This turns out to be effective, and to have some other nice properties we will see later; we will use this statistic. This statistic is common enough to merit a name, the *chi-square test statistic*.

For the GSS example, the value of the statistic is

$$\begin{aligned} c = &\frac{(135 - 141.5363)^2}{141.5363} + \frac{(71 - 64.4637)^2}{64.4637} + \frac{(64 - 78.3259)^2}{78.3259} + \frac{(50 - 35.3611)^2}{35.3611} \\ &+ \frac{(511 - 488.5065)^2}{488.5065} + \frac{(200 - 222.4935)^2}{222.4935} + \frac{(71 - 59.7751)^2}{59.7751} + \frac{(16 - 27.2249)^2}{27.2249} \\ &+ \frac{(117 - 129.8562)^2}{129.8562} + \frac{(72 - 59.1438)^2}{59.1438} \\ = &\ 23.450. \end{aligned}$$

Next, we need a reference distribution to which we compare the observed test statistic. We will consider two approaches—permutation testing and the use of chi-square distributions.

3.4.1 Permutation Test for Independence

To do permutation testing, we create a table with two columns, education and opinion, and $n = 1307$ rows. If the null hypothesis that education and opinion are independent is correct, then we could permute the opinion (or education) values, and any other permutation would be equally likely. For each such permutation resample, we can cross-tabulate to obtain a table of observed values like Table 3.4 and compute the chi-square test statistic for the resample. Note that for every resample, the row and column totals in the contingency table are the same; only the counts in the table change. By forming many such resamples, we obtain the permutation distribution of the chi-square test statistic. We follow this algorithm:

PERMUTATION TEST FOR INDEPENDENCE OF TWO VARIABLES

Store the data in a table with one row per observation and one column per variable.
Calculate a test statistic for the original data. Normally large values of the test statistic suggest dependence.
repeat
  Randomly permute the rows in one of the columns.
  Calculate the test statistic for the permuted data.
until we have enough samples
Calculate the P-value as the fraction of times the random statistics exceed the original statistic.
Optionally, plot a histogram of the resampled statistic values.

For instance, in the `GSS2002` data set, one permutation of the values in the `DeathPenalty` column while leaving the `Education` column fixed results in the contingency table below.

	Death Penalty?	
Education	Favor	Oppose
Bachelors	142	59
Graduate	77	39
Jr Col	69	36
HS	484	219
Left HS	127	56

The corresponding chi-square statistic (from Equation 3.1) is $c = 1.128825$. Repeating this permutation many times and computing the chi-square statistic each time gives the permutation distribution of this statistic.

The permutation distribution for this example is shown in Figure 3.8. The estimated P-value, based on $10^5 - 1$ replications, is 0.00012. Thus, we conclude that there is an association between the education level of a person and his/her support for the death penalty.

The General Social Survey (GSS) draws a random sample of participants for its surveys, so we can draw an inference about a population. However, in this example, we did exclude all people who did not provide a response to the education question

(5 people) or to the death penalty question (1457 people). Do you think this affects the results in any way?

R Note:

Here is code to perform the permutation test in R.
We first define a function that computes the test statistic C.

```
chisq <- function(Obs){ # Obs is the observed contingency table
  Expected <- outer(rowSums(Obs), colSums(Obs)) / sum(Obs)
  sum((Obs - Expected)^2 / Expected)
}
```

We use this function on the contingency table for `Education` and `DeathPenalty`

```
> Education <- GSS2002$Education
> DeathPenalty <- GSS2002$DeathPenalty
> observed <- chisq(table(Education, DeathPenalty))
> observed
[1] 23.45093
```

Now, there were over 1000 people who declined to respond to either the education or death penalty question, so we will remove them from our analysis.

We first find the rows where a missing value (NA) occurs and store those row numbers in `index` and then create new objects that contain the complete responses.

```
index <- which(is.na(Education) | is.na(DeathPenalty))
Educ2 <- Education[-index]
DeathPenalty2 <- DeathPenalty[-index]
```

The `sample(DeathPenalty2)` command below will permute the values in `DeathPenalty2`.

```
N <- 10^4-1
result <- numeric(N)
for (i in 1:N)
{
  DP.permuted <- sample(DeathPenalty2)
  GSS.table <- table(Educ2, DP.permuted)
  result[i] <- chisq(GSS.table)
}

hist(result, xlab = "chi-square statistic",
    main = "Distribution of chi-square statistic")
abline(v = observed, col = "blue")
```

Check the distribution of the test statistics to help in determining the direction of the inequality when computing the P-value

```
(sum(result >= observed) + 1)/(N+1)
```

If you have a slow computer, change the number of replications N to $10^3 - 1$.

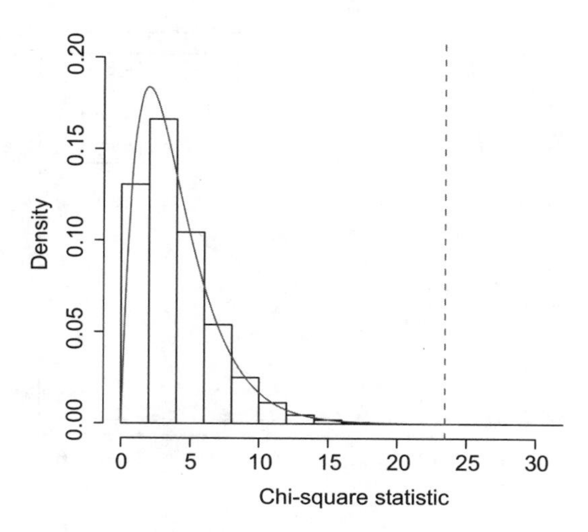

FIGURE 3.8 Null distribution for chi-square statistic for death penalty opinions; the overlaid density is a chi-square distribution with 4 degrees of freedom.

3.4.2 Chi-Square Reference Distribution

The permutation test approach is easy now, with fast computers, but was not easy in Pearson's day. Work in the 1920s by Pearson (1900, 1922, 1923) and Fisher (1922) led to a shortcut—if the expected counts are all reasonably large, then the null distribution is approximately equal to a chi-square distribution, with $(I-1)(J-1) = 4$ degrees of freedom, where I and J are the number of rows and columns, respectively. This is apparent in Figure 3.8 where the overlaid chi-square distribution is a close match to the permutation distribution.

If the chi-square statistic has a chi-square distribution with 4 degrees of freedom, then the P-value, the probability of exceeding $c = 23.45$, is 0.00010. We conclude that education and support for the death penalty are not independent.

The chi-square distribution approximation is half of the value 0.0002 estimated from simulation. The difference could be random variation in the simulation estimate or it could be that Fisher's approximation is not accurate here—while the cell sizes are large enough for the chi-square distribution to be accurate in the center of the null distribution, it may not be as accurate in the tail. In any case, the P-value is small enough for the result to be statistically significant.

The reference distribution—the chi-square distribution—that we have been discussing is described more fully in Section B.10. The pdf for a chi-square random variable is given by

$$f(x) = \frac{x^{m/2-1}e^{-x/2}}{2^{m/2}\Gamma(m/2)}, \quad 0 \le x < \infty.$$

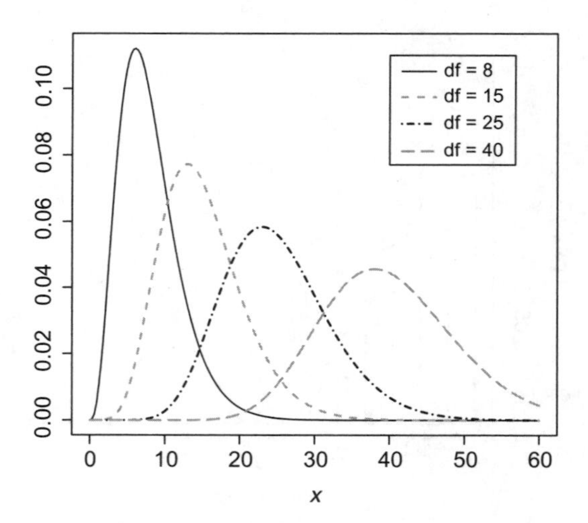

FIGURE 3.9 Densities for the chi-square distribution.

Like exponential distributions, chi-square distributions are a family of distributions parameterized by a number m, which is called the *degree of freedom*. We will write $X \sim \chi^2_m$ to denote random variables that follow this distribution. Some are shown in Figure 3.9.

R Note:

The command `pchisq` computes cumulative probabilities for the chi-square distribution: the syntax `pchisq(x, m)` gives $P(\chi^2_m \leq x)$.

For the death penalty example above,

```
> 1 - pchisq(23.45,3)
[1] 0.0001029331
```

3.5 CHI-SQUARE TEST OF INDEPENDENCE

Here, we describe the chi-square test of independence for a two-way table more formally.

Consider two categorical variables A and B with I and J levels, respectively. We use $A_1, A_2, \ldots, A_I, B_1, B_2, \ldots, B_J$ to denote the levels of A and B, respectively. For example, in the GSS Case Study, the education variable has $I = 5$ levels and the death penalty variable has $J = 2$ levels. A_1 would be those who have a graduate degree, while B_1 would be those who favor the death penalty.

From a sample of n randomly selected individuals, let the random variable N_{ij} denote the number classified by the ith level of A and the jth level of B, respectively,

TABLE 3.6 Observed Counts

	B				
A	B_1	B_2	$\cdots$	B_J	Row Sum
A_1	N_{11}	N_{12}	$\cdots$	N_{1J}	R_1
A_2	N_{21}	N_{22}	$\cdots$	N_{2J}	R_2
$\vdots$					
A_I	N_{I1}	N_{I2}	$\cdots$	N_{IJ}	R_I
Column sum	C_1	C_2	$\cdots$	C_J	n

with observed value n_{ij}. Also, let p_{ij} denote the probability that a randomly selected individual from some population is classified by the ith level of A, and the jth level of B. The N_{ij} are multinomial random variables with $\mathrm{E}\left[N_{ij}\right] = np_{ij}$. (See Section B.2 for more information on the multinomial distribution).

The row and column sums of the contingency table (Table 3.6) are

$$R_i = \sum_{j=1}^{J} N_{ij} \quad \text{and} \quad C_j = \sum_{i=1}^{I} N_{ij}.$$

In addition, let

$$p_{i.} = \sum_{j=1}^{J} p_{ij} \quad \text{and} \quad p_{.j} = \sum_{i=1}^{I} p_{ij}$$

denote the marginal probabilities; for example, $p_{1.}$ is the population proportion in the first row. Clearly,

$$\sum_{i=1}^{I} p_{i.} = \sum_{j=1}^{J} p_{.j} = 1.$$

The hypotheses to test whether or not the categorical variables are independent is given by

$$\begin{aligned} &H_0\colon p_{ij} = p_{i.}p_{.j}, \quad i = 1, 2, \ldots, I, \; j = 1, 2, \ldots, J, \\ &H_A\colon p_{ij} \neq p_{i.}p_{.j}, \quad \text{for some } i, j. \end{aligned} \tag{3.2}$$

Now, $p_{i.}$ and $p_{.j}$, $i = 1, 2, \ldots, I$; $j = 1, 2, \ldots, J$, are unknown, but it seems reasonable to estimate them by the sample proportions

$$\hat{p}_{i.} = \frac{R_i}{n} \quad \text{and} \quad \hat{p}_{.j} = \frac{C_j}{n}.$$

Thus, an estimate of the expected value of N_{ij} is

$$\hat{\mathrm{E}}[N_{ij}] = n\hat{p}_{i.}\hat{p}_{.j} = n\frac{R_i}{n}\frac{C_j}{n} = \frac{R_iC_j}{n}.$$

The random test statistic is

$$C = \sum_{i=1}^{I} \sum_{j=1}^{J} \frac{(N_{ij} - \hat{\mathrm{E}}[N_{ij}])^2}{\hat{\mathrm{E}}[N_{ij}]}, \tag{3.3}$$

with observed value

$$c = \sum_{i=1}^{I} \sum_{j=1}^{J} \frac{(n_{ij} - R_i C_j / n)^2}{R_i C_j / n}. \tag{3.4}$$

In 1924, R. A. Fisher proved that under the assumption of independence, $p_{ij} = p_{i.}p_{.j}$ for all i and j, the distribution of T approaches a chi-square distribution with $(I-1)(J-1)$ degrees of freedom as the sample size goes to infinity. In practice, this means that if the expected counts are large, then the chi-square approximation is reasonably accurate, except in the tails of the distribution.

Example 3.7 Two researchers studied whether or not being bullied in school was associated with being short. The table below summarizes their findings for 209 pupils, from "Bullying in school: are short pupils at risk? Questionnaire study in a cohort" (Voss and Mulligan (2000)).

	Bullied?	
Height	Yes	No
Short	42	50
Not short	30	87

The hypotheses to be tested are as follows:
H_0: Being bullied is independent of height (there is no association between being bullied and height).
H_A: Being bullied is not independent of height (there is an association between being bullied and height).
Here are the expected counts.

	Bullied?	
Height	Yes	No
Short	31.7	60.3
Not short	40.3	76.7

Thus, the test statistic is

$$c = \frac{(42-31.7)^2}{31.7} + \frac{(50-60.3)^2}{60.3} + \frac{(30-40.3)^2}{40.3} + \frac{(87-76.7)^2}{76.7} = 9.14.$$

We compare this value to a chi-square distribution with $(2-1)(2-1) = 1$ degree of freedom. The P-value is 0.0025, so the data support the hypothesis that being bullied and height are not independent! □

Remark For a test of independence, the hypotheses are usually stated in the form of the above example as opposed to Equation 3.2. ||

R Note:

The command `chisq.test` will compute the chi-square test statistic and compare to a χ^2 distribution to find the P-value when given two factor variables.

Using the vectors `Education` and `DeathPenalty` created earlier,

```
> chisq.test(Education, DeathPenalty)
        Pearson's Chi-squared test

data:  Education and DeathPenalty
X-squared = 23.4509, df = 4, p-value = 0.0001029
```

Remark

- The chi-square test statistic has an *approximate* chi-square distribution if the null hypothesis is true. There are various rules of thumb that give conditions under which usage of this test is appropriate. One rule suggests that all the expected cell counts should be larger than, say, 5. Another recommends that no more than 20% of the cells have expected count less than 5 (Cochran (1954)).
- In R, the `chisq.test` can also compute P-values by simulation. For instance, `chisq.test(X,Y, simulate.p.value=TRUE)` ||

3.6 TEST OF HOMOGENEITY

In the death penalty example in the previous section, we drew a sample from a single population, created a contingency table based on values of two-factor variables, and tested for independence between the factors. In other situations, we may instead draw samples from two or more different populations, classify the observations by the levels of a single-factor variable, and test whether that factor has the same distribution in each population.

For example, suppose a candy company wants to know whether boys or girls differ in their taste preferences for three new candy flavors that will be sold next year (Table 3.7). The company obtains a random sample of 100 boys and a random sample of 110 girls, gets each child to taste the three flavors, and asks them to name their favorite. We want to know if boys and girls have the same distribution of favorites.

TABLE 3.7 Counts of Candy Preferences

	Candy			
Gender	Flavor 1	Flavor 2	Flavor 3	Row Sum
Boys	42	20	38	100
Girls	33	27	50	110
Column sum	75	47	88	210

This seems very similar to the death penalty example, except that here the number of boys and girls is fixed in advance; in the death penalty example, the number of people in each education level depended on the original random survey (in other words, a different random sample would have resulted in a different number of people in each education level). This distinction turns out not to matter—we will use exactly the same test statistic and procedures for computing P-values and null distributions.

There is a second difference, that we may express the parameters and hypotheses differently. In the earlier example, we worked with a matrix of parameters p_{ij} that added to 1 for the whole table. Here we work with parameters for each population.

Let π_{ij} denote the proportion of gender i that prefer candy j, where $i = B, G$; $j = 1, 2, 3$; within each row, these probabilities add to 1. We test

$$H_0\text{: } \pi_{B1} = \pi_{G1}, \quad \pi_{B2} = \pi_{G2}, \quad \pi_{B3} = \pi_{G3};$$
$$H_{\text{A}}\text{: at least one of the equalities does not hold.}$$

If the null hypothesis is true, there is a single set of preferences $\pi_1 = \pi_{B1} = \pi_{G1}$, $\pi_2 = \pi_{B2} = \pi_{G2}$, and $\pi_3 = \pi_{B3} = \pi_{G3}$. We estimate these probabilities using the sample proportions

$$\hat{p}_1 = \frac{75}{210} = 0.3571, \quad \hat{p}_2 = \frac{47}{210} = 0.2238, \quad \hat{p}_3 = \frac{88}{210} = 0.4190.$$

The estimated expected counts of the six cells are obtained by multiplying these proportions by 100 (for the boys row) or 110 (for the girls) row.

In spite of the different parameterization, the expected counts are the same as using $R_i C_j / n$. For example, the expected number of boys who like flavor 1 is $\hat{p}_1 \times 100 = (75/210) \times 100 = R_1 C_1 / n = 35.71$. We use the same test statistic:

$$C = \sum_{\text{all cells}} \frac{(\text{observed} - \text{expected})^2}{\text{expected}} \tag{3.5}$$

$$= \frac{(42 - 35.71)^2}{35.71} + \frac{(20 - 22.38)^2}{22.38} + \frac{(38 - 41.90)^2}{41.90}$$
$$+ \frac{(33 - 39.281)^2}{39.281} + \frac{(27 - 24.618)^2}{24.618} + \frac{(50 - 46.09)^2}{46.09}$$
$$= 3.2902.$$

As before, we may compare this statistic to a permutation distribution, or a χ^2 distribution with $(2-1)(3-1) = 2$ degrees of freedom. The P-value using the χ^2 approximation is 0.193; so if the null hypothesis is true, then about 19.3% of samples from the same two populations would result in a test statistic as large or larger than this one. We conclude that the two populations could well be the same in their preferences for the three flavors. If there are differences, they are not large enough to distinguish from the random noise of sampling, with samples this small.

More generally, consider samples of size R_i, $i = 1, 2, \ldots, I$ from I independent populations. Suppose each individual can be classified as one of J different types and let N_{ij} denote the number of individuals from population i of type j. The data can be summarized as in Table 3.6.

Let π_{ij} be the probability that an observation from population i falls in column j (like the conditional probability of column j given row i). We test

$$H_0\colon \pi_{1j} = \pi_{2j} = \cdots = \pi_{Ij}, \quad j = 1, 2, \ldots, J,$$
$$H_A\colon \text{Equality does not hold for some } i, j$$

using the test statistic

$$c = \sum_{i=1}^{I} \sum_{j=1}^{J} \frac{(n_{ij} - R_i C_j / n)^2}{R_i C_j / n}. \tag{3.6}$$

If the null hypothesis is true and sample sizes are large, then the test statistic has an approximate χ^2 distribution with $(I-1)(J-1)$ degrees of freedom.

R Note:

The `chisq.test` command also accepts a contingency table as an argument. Use the `rbind` command to bind the values in Table 3.7 by row.

```
> candy.mat <- rbind(c(42,20,38), c(33,27,50)) # create matrix
> candy.mat                                     # check output
     [,1] [,2] [,3]
[1,]   42   20   38
[2,]   33   27   50
> chisq.test(candy.mat)
Pearson's chi-square test without Yates' continuity correction

data:  candy.mat
X-square = 3.2902, df = 2, p-value = 0.193
```

3.7 GOODNESS-OF-FIT: ALL PARAMETERS KNOWN

The chi-square statistic is useful in other situations to compare differences between observed and expected counts, including *goodness-of-fit tests*, to check whether the data fit a probability model.

Example 3.8 Barnsley et al. (1992) investigated the relationship between month of birth and achievement in sport. Birth dates were collected for players in teams competing in the 1990 World Cup soccer games.

	Birth Month			
	Aug–Oct	Nov–Jan	Feb–April	May–July
Observed	150	138	140	100

We wish to test whether these data are consistent with the hypothesis that birthdays of soccer players are uniformly distributed across the four quarters of the year.

Let p_i denote the probability of a birth occurring in the ith quarter; the hypotheses are as follows:

$$H_0\colon p_1 = 1/4, \quad p_2 = 1/4, \quad p_3 = 1/4, \quad p_4 = 1/4,$$
$$H_A\colon p_i \neq 1/4 \text{ for at least one } i.$$

There were a total of $n = 528$ players considered for this study, so the expected count for each quarter is $528/4 = 132$. Thus, the test statistic is

$$c = \frac{(150-132)^2}{132} + \frac{(138-132)^2}{132} + \frac{(140-132)^2}{132} + \frac{(100-132)^2}{132} = 10.97.$$

We cannot use permutation resampling to obtain a null distribution (there is nothing to permute, we are not testing the independence of two variables), so we will use a chi-square approximation, with $4 - 1 = 3$ degrees of freedom (we will discuss this below). The resulting P-value $P(C \geq 10.97) = 0.012$ supports the hypothesis that birthdays are not uniformly distributed across the four quarters. One explanation is that players born shortly after the yearly cutoff for school enrollment are relatively old for their grade and competing against younger classmates. They enjoy more success early, and ultimately do better in the sport. □

> The degree of freedom for a goodness-of-fit test with k cells and no parameters estimated from the data is $k - 1$.

Here is an explanation of the degrees of freedom. If no parameters are estimated from the data, the degrees of freedom would be $k - 1$: pick any $k - 1$ cells and numbers may be placed in these cells freely, but the number in the final cell must make the sum equal to the sample size, so there is no freedom in what number goes in that cell.

Example 3.9 Suppose you draw 100 numbers at random from an unknown distribution. Thirty values fall in the interval $(0, 0.25]$, 30 fall in $(0.25, 0.75]$, 22 fall in

(0.75, 1.25] and the rest fall in $(1.25, \infty)$. Your friend claims that the distribution is exponential with parameter $\lambda = 1$. Do you believe her?

Solution The hypotheses we wish to test are as follows:

H_0: The data are from an exponential distribution, $\lambda = 1$,
H_A: The data are not from an exponential distribution, $\lambda = 1$.

Let $X \sim \text{Exp}(1)$. The probabilities for each interval are as follows:

$$p_1 = P(0 \leq X \leq 0.25) = \int_0^{0.25} e^{-x}\,dx = 0.22,$$

$$p_2 = P(0.25 < X \leq 0.75) = \int_{0.25}^{0.75} e^{-x}\,dx = 0.306,$$

$$p_3 = P(0.75 < X \leq 1.25) = \int_{0.75}^{1.25} e^{-x}\,dx = 0.186,$$

$$p_4 = P(1.25 < X < \infty) = \int_{1.25}^{\infty} e^{-x}\,dx = 0.287.$$

Then, for a sample of $n = 100$ numbers, the expected counts are

Interval	(0, 0.25]	(0.25, 0.75]	(0.75, 1.25]	$(1.25, \infty)$
Observed count	30	30	28	18
Expected count np_i	22	30.6	18.6	28.7

Thus, the chi-square test statistic is

$$c = \frac{(30-22)^2}{22} + \frac{(30-30.6)^2}{30.6} + \frac{(22-28.6)^2}{28.6} + \frac{(18-28.7)^2}{28.7} = 8.43.$$

Under the null hypothesis, the test statistic comes from a chi-square distribution with $4 - 1 = 3$ degrees of freedom, so the P-value is $P(c \geq 8.43) = 0.038$. Thus, there is evidence that your data do not come from Exp(1). □

Example 3.10 Is it possible that the following 50 numbers are a random sample from a chi-square distribution with 10 degrees of freedom?

1.85	2.68	2.84	3.76	3.86	4.96	5.42	6.50	6.65	6.81
6.95	7.42	7.48	7.99	8.50	8.54	8.65	8.71	8.80	9.47
9.82	9.91	10.09	10.30	10.62	10.63	10.70	10.79	10.94	11.92
12.14	12.22	12.96	13.29	13.43	14.14	14.29	14.36	14.65	14.68
14.87	15.00	15.91	16.15	16.18	17.21	19.06	20.81	22.82	23.55

Solution We will use R to compute 0.2, 0.4, 0.6, and 0.8 quantiles of the chi-square distribution with 10 degrees of freedom—that is, those points that mark off probabilities (areas) equal to 0.2.

R Note:

The command qchisq computes quantiles of the chi-square distribution.

```
> qchisq(c(.2, .4, .6, .8), 10)
[1]  6.179079  8.295472 10.473236 13.441958
```

Thus, we expect to see 20% of the 50 values fall into each subinterval determined by the above quantiles. We will compare the expected count of 10 for each interval with the observed number of values that fall into each of the subintervals.

Interval	(0, 6.179]	(6.179, 8.295]	(8.295, 10.473]	(10.473, 13.442]	(13.442, ∞)
Observed	7	7	10	11	15
Expected	10	10	10	10	10

The chi-square test statistic is

$$c = \frac{(7-10)^2}{10} + \frac{(7-10)^2}{10} + \frac{(10-10)^2}{10} + \frac{(11-10)^2}{10} + \frac{(15-10)^2}{10} = 4.4.$$

If the data are from a chi-square distribution with 10 degrees of freedom, then the test statistic $c = 4.4$ comes from a chi-square distribution with $5 - 1 = 4$ degrees of freedom, so the P-value is $P(c \geq 4.4) = 0.355$. We cannot rule out the possibility that the data are a random sample from a chi-square distribution with 10 degrees of freedom. □

3.8 GOODNESS-OF-FIT: SOME PARAMETERS ESTIMATED

In other situations, some parameters must be estimated from the data. The testing procedure is similar, but with adjusted degrees of freedom.

Example 3.11 A home run in baseball is an exciting event, a majestic flight that can dramatically alter the course of a game, driving in as many as four runs with one swing of a bat. Table 3.8 displays a summary of the home run data for the Philadelphia Phillies in their 2009 season. For instance, the Phillies hit 2 home runs in 40 of the 162 games played, but 5 home runs in only 1 game.

Since we have counts data and home runs are relatively rare, the Poisson distribution is a natural candidate for modeling these data. If the random variable X denotes

TABLE 3.8 Counts of Home Runs (in 162 Games)

Number of Home Runs x	Number of Games	Proportion of Games
0	43	0.2653
1	52	0.3209
2	40	0.2469
3	17	0.1049
4	9	0.0555
5	1	0.0062

the number of home runs in a game, then the probability mass function is given by $f(x) = P(X = x) = (\lambda^x e^{-x})/x!$, $x = 0, 1, 2, \ldots$, and parameter $\lambda > 0$. Since λ is unknown, we may estimate it by the empirical average number of home runs per game, $224/162 = 1.3827$. In Chapter 6, we will see that this is a good choice. Thus, we will model the number of home runs per game with the probability density function $P(X = x) = (1.3827^x e^{-1.3827})/x!$, $x = 0, 1, 2, \cdots$.[3]

To assess this model, we consider the following hypotheses:

H_0: The home runs are modeled by a Poisson distribution,
H_A: The home runs are not modeled by a Poisson distribution.

We assume that the distribution of home runs is the same for each game and independent between games. Neither of these is exactly true in practice—some opposing pitchers are better than others, and a home run fest in one game might make the opponents pitch more conservatively in the next game—but for the current analysis, we ignore these effects.

Thus, under the null hypothesis, we compute the expected number of games in which there are x home runs, $x = 1, 2, \ldots$ by $P(X = x) \times 162 = (1.3827^x e^{-1.3827})/x! \times 162$ to obtain Table 3.9.

To test whether the actual counts are significantly different from the expected counts, we will use a chi-square statistic and calculate the P-value using a chi-square approximation. But this approximation is reasonable only if the expected counts are reasonably large, and an expected count of 1.7 does not qualify. The expected counts for 6, 7, and more home runs are even smaller, and we need to include them somewhere.

TABLE 3.9 Probabilities for Poisson Distribution with $\lambda = 1.3827$, and Expected Counts for 162 Games

$x =$	0	1	2	3	4	5
$P(X = x)$	0.251	0.347	0.240	0.111	0.038	0.011
Expected	40.6	56.2	38.9	17.9	6.2	1.7

[3]We will use the term probability density function (pdf) for both discrete and continuous random variables.

TABLE 3.10 Chi-Square Test for Poisson Goodness-of-Fit to Home Run Data

$x =$	0	1	2	3	4+
Observed count	43	52	40	17	10
Expected count	40.6	56.2	38.9	17.9	8.4
$(O - E)^2/E$	0.14	0.31	0.03	0.05	0.31

To work around this, we combine cells, with the final cell being four or more home runs per game. The observed, expected, and contributions to the chi-square statistic are as shown in Table 3.10.

The chi-square statistic is 0.84; the P-value is $P(\chi^2_3 > 0.084) = 0.84$. We conclude that the number of home runs per game is consistent with a Poisson distribution. □

> The degree of freedom for a goodness-of-fit test with k cells and ℓ parameters estimated from the data is $k - \ell - 1$.

Estimating parameters reduces the degrees of freedom, because estimated parameters tend to "overfit," making the model fit the data better than the true (unknown) parameters would.

R Note:

```
> Homeruns <- subset(Phillies, select = Homeruns, drop = T)
> lambda <- mean(Homeruns) # average number home runs/game
> dpois(0:4, lambda)       # theoretical model
[1] 0.25089618 0.34691818 0.23984466 0.11054569 0.03821332
> table(Homeruns)
Homeruns
 0  1  2  3  4  5
43 52 40 17  9  1
> table(Homeruns)/162      # empirical probabilities
HomeRuns
         0          1          2          3          4          5
0.26543210 0.32098765 0.24691358 0.10493827 0.05555556 0.00617284
```

3.9 EXERCISES

1. Suppose you conduct an experiment and inject a drug into three mice. Their times for running a maze are 8, 10, and 15 s; the times for two control mice are 5 and 9 s.
   (a) Compute the difference in mean times between the treatment group and the control group.

(b) Write out all possible permutations of these times to the two groups and calculate the difference in means.

(c) What proportion of the differences are as large or larger than the observed difference in mean times?

(d) For each permutation, calculate the mean of the treatment group only. What proportion of these means are as large or larger than the observed mean of the treatment group?

2. In the algorithms for conducting a permutation test, why do we add 1 to the number of replications N when calculating the P-value?

3. In the Flight Delays Case Study in Section 1.1,

   (a) The data contain flight delays for two airlines, American Airlines and United Airlines. Conduct a two-sided permutation test to see if the mean delay times between the two carriers are statistically significant.

   (b) The flight delays occured in May and June of 2009. Conduct a two-sided permutation test to see if the difference in mean delay times between the 2 months is statistically significant.

4. In the Flight Delays Case Study in Section 1.1, the data contain flight delays for two airlines, American Airlines and United Airlines.

   (a) Compute the proportion of times that each carrier's flights was delayed more than 20 min. Conduct a two-sided test to see if the difference in these proportions is statistically significant.

   (b) Compute the variance in the flight delay lengths for each carrier. Conduct a test to see if the variance for United Airlines is greater than that of American Airlines.

5. In the Flight Delays Case Study in Section 1.1, repeat Exercise 3 part (a) using three test statistics: (i) the mean of the United Airlines delay times, (ii) the sum of the United Airlines delay times, and (iii) the difference in means, and compare the P-values. Make sure all three test statistics are computed within the same `for` loop.

6. In the Flight Delays Case Study in Section 1.1,

   (a) Find the trimmed mean of the delay times for United Airlines and American Airlines.

   (b) Conduct a two-sided test to see if the difference in trimmed means is statistically significant.

7. In the Flight Delays Case Study in Section 1.1,

   (a) Compute the proportion of times the flights in May and in June were delayed more than 20 min, and conduct a two-sided test of whether the difference between months is statistically significant.

   (b) Compute the variance of the flight delay times in May and June and then conduct a two-sided test of whether the ratio of variances is statistically significantly different from 1.

8. In the Black Spruce Case Study in Section 1.9, seedlings were planted in plots that were either subject to competition (from other plants) or not. Use the data set `Spruce` to conduct a test to see if the mean difference in how much the seedlings grew over the course of the study under these two treatments is statistically significant.
9. The file `Phillies2009` contains data from the 2009 season for the baseball team of the Philadelphia Phillies. Import these data into `R`.
   (a) Compare the empirical distribution functions of the number of strikeouts per game (`StrikeOuts`) for games played at home and games played away (`Location`).
   (b) Find the mean number of strikeouts per game for the games played at home and games played away from home.
   (c) Perform a permutation test to see if the difference in means is statistically significant.
10. Researchers at the University of Nebraska conducted a study to investigate sex differences in dieting trends among a group of midwestern college students (Davy et al. (2006)). Students were recruited from an introductory nutrition course during one term. Below are data from one question asked to 286 participants.

	Tried a Low-Fat Diet?	
Gender	Yes	No
Women	35	146
Men	8	97

   (a) Write down the appropriate hypothesis to test to see if there is a relationship between gender and diet and then carry out the test.
   (b) Can the results be generalized to a population? Explain.
11. A national polling company conducted a survey in 2001 asking a randomly selected group of Americans of 18 years of age or older whether they supported limited use of marijuana for medicinal purposes. Here is a summary of the data:

	Response	
Age	For	Against
18–29 years old	172	52
30–49 years old	313	103
50 years or older	258	119

Write down the appropriate hypothesis to test whether there is a relationship between age and support for medicinal marijuana and carry out the test.

12. Two students went to a local supermarket and collected data on cereals; they classified cereals by their target consumer (children versus adults) and the placement of the cereal on the shelf (bottom, middle, and top). The data are given in `Cereals`.
    (a) Create a table to summarize the relationship between age of target consumer and shelf location.
    (b) Conduct a chi-square test using R's `chisq.test` command.
    (c) R returns a warning message. Compute the expected counts for each cell to see why.
    (d) Conduct a permutation test for independence, adapting the code on page 56.
13. The California Department of Game and Fish published a report on a study of jack mackerel fish from three different regions off the waters of California: near Guadalupe Island and Cedros Island off Baja California and near San Clemente Island in southern California (Gregory and Tasto (1976)). One characteristic of the fish that they were interested in was the number of rays on the second dorsal fin; in particular, is the number of rays different for fish from the different regions?

	Fin Ray Count					
Habitat	≥ 36	35	34	33	32	≤ 31
Guadalupe Island	14	30	42	78	33	14
Cedro Island	11	28	53	66	27	9
San Clemente Island	10	17	61	53	22	10

    (a) Does this setting call for a test of independence or a test of homogeneity?
    (b) Write down the appropriate hypothesis and carry out the test.
14. A researcher in Hong Kong conducted a study on children's perception of advertising and brands on television (Chan (2008)). For one part of the study, she analyzed surveys sent to 1481 children from rural areas of Mainland China and received responses from 726 boys and 755 girls. The responses to the question about their feeling of commercials on TV are summarized below:

	Feeling Towards TV Ads				
Sex	Like Very Much	Like	Neither	Dislike	Dislike Very Much
Boys	180	260	137	96	52
Girls	210	266	145	85	49

    (a) Will this be a test of independence or a test of homogeneity?

(b) Conduct the appropriate test to determine the relationship between sex and feelings toward TV commercials.

15. For the Flight Delays Case Study in Section 1.1, conduct a test of homogeneity to determine if there is a relationship between carrier and the number of flights delayed more than 30 min (`Delayed30`).

16. From the GSS 2002 Case Study in Section 1.6,
    (a) Create a table to summarize the relationship between gender and the person's choice for president in the 2000 election.
    (b) Test to see if a person's choice for president in the 2000 election is independent of gender (use `chisq.test` in R).
    (c) Repeat the test but use the permutation test for independence. Does your conclusion change? (Be sure to remove observations with missing values).

17. From the GSS 2002 Case Study in Section 1.6,
    (a) Create a table to summarize the relationship between gender and the person's general level of happiness (`Happy`).
    (b) Conduct a permutation test to see if gender and level of happiness are independent (Be sure to remove the observations with missing values).

18. From the GSS 2002 Case Study in Section 1.6,
    (a) Create a table to summarize the relationship between support for gun laws (`GunLaw`) and views on government spending on the military (`Spend-Military`).
    (b) Conduct a permutation test to see if support for gun laws and views on government spending on the military are independent (Be sure to remove observations with missing values).

19. For a given $r \times c$ contingency table, we have the test statistic C given by Equation 3.1.
    (a) What happens to the value of C if every entry in the contingency table is multiplied by the same integer $k > 1$? Do the marginal probabilities change? Does the degrees of freedom change?
    (b) What is the implication of this fact? That is, if the probabilities stay the same, but the actual counts in each cell increase (multiplicatively) by the same amount, what happens to our conclusion from the test?

20. Suppose you randomly draw 75 values from a distribution that your friend claims has pdf $f(y) = (1/9)y^2, 0 < y \leq 3$. If 2 of the values fall in the interval $(0, 1.25]$, 6 fall in the interval $(1.25, 1.75]$, 10 fall in the interval $(1.75, 2.25]$, 32 fall in the interval $(2.25, 2.75]$, and the rest fall in the interval $(2.75, 3]$, perform a goodness-of-fit test to see if your data support his claim.

21. Of a sample of 70 random numbers, 30 fall in the interval $[1, 1.5)$, 18 fall in the interval $[1.5, 2)$, 9 fall in the interval $[2, 3)$, 10 fall in the interval $[3, 5)$, and the rest are greater than 5. Is it plausible that these numbers were drawn from a distribution with pdf $f(x) = 2/x^3$ for $x \geq 1$?

22. Suppose you randomly draw 50 values from an unknown distribution.

1.28	4.53	5.50	7.91	8.23	9.67	9.82	10.28	10.45	11.91
12.57	13.75	13.80	14.00	14.05	16.02	16.18	16.25	16.58	16.68
16.87	17.61	17.63	17.71	18.13	18.42	18.43	18.44	19.62	20.401
20.73	20.74	21.29	21.51	21.66	21.87	22.67	23.11	24.40	24.55
24.66	25.30	25.46	25.91	26.12	26.61	26.72	29.28	31.93	36.94

Could these data have come from the normal distribution $N(22, 7^2)$?

(a) Use the `qnorm` command in `R` to find 0.2, 0.4, 0.6, and 0.8 quantiles of the normal distribution—that is, those points that mark off equal probabilities (equal areas) of 0.2.

(b) Use these quantiles to determine your intervals and count the number of values that fall in each interval.

(c) Finish the goodness-of-fit test.

23. Suppose you randomly draw 60 values from an unknown distribution.

16.21	16.96	17.07	17.81	19.66	21.16	21.95	22.76	23.81	23.94
24.12	24.26	25.10	25.15	25.22	25.47	25.62	25.91	27.34	27.51
28.05	28.67	28.76	28.89	28.93	29.45	29.54	29.64	30.38	30.60
31.49	31.52	32.25	32.26	32.40	32.52	32.54	32.66	33.01	33.02
33.91	34.32	34.83	34.88	34.93	35.05	35.33	35.84	36.18	36.33
37.27	37.84	38.24	38.33	38.42	38.74	38.83	40.87	41.77	43.91

(a) Conduct a test to see if these data are consistent with a normal distribution with $\mu = 25, \sigma = 10$.

(b) Suppose you suspect the data are from a normal distribution but do not know μ or σ. Using the sample mean of 30.328 and standard deviation 6.54 as estimates of μ and σ, conduct a goodness-of-fit test.

24. For the Philadelphia Phillies data (`Phillies2009`), consider the number of doubles hit per game. Model this using a Poisson distribution, and perform a goodness-of-fit test to compare the theoretical model with the empirical data.

25. California, like many states, sponsors lotteries to raise revenue. In one popular game, Fantasy 5, a player tries to match 5 numbers chosen from 1 through 39. For instance, on August 15, 2010, the five winning numbers were 29, 19, 37, 34 and 07. California uses a random mechanism to draw the numbers each day; how good is it? The file `Lottery` contains the winning numbers for the daily games from May 5, 2010 through August 15, 2010 (from `http://www.calottery.com/Games/FantasyFive/`). Determine whether or not the winning numbers are randomly drawn.

26. Researchers conduct a pilot study to test the effectiveness of a drug in preventing a certain disease. Of 20 patients in the study, 10 are randomly assigned to receive the drug and 10 to receive a placebo. After 1 year, suppose five patients in the control group contract the disease, while two patients who took the drug contract the disease.

	Response	
Outcome	Disease	No Disease
Drug	2	8
Placebo	5	5

(a) For a test of homogeneity, what are the expected cell counts?

(b) If the drug is not effective, then every patient is equally likely to contract the disease. In that case, if 7 patients out of 20 contract the disease, what is the probability that 2 of them are in the treatment group?

(c) In that case, what is the probability that two or fewer of them are in the treatment group?

27. At a university, 15 juniors and 20 seniors volunteer to serve as a special committee that requires 8 members. A lottery is used to select the committee from among the volunteers. Suppose the chosen students consists of six juniors and two seniors.
    (a) For a test of homogeneity, what are the expected counts?
    (b) If the selection had been random, what is the probability of the committee having exactly two seniors?
    (c) If the selection had been random, what is the probability that the committee would have two or fewer seniors?
    (d) Is there evidence that the selection was not random?

28. In the sampling version of permutation testing, the one-sided P-value is $\hat{P} = (X+1)/(N+1)$, where X is the number of permutation test statistics that are as large or larger than the observed test statistic. Suppose the true P-value (for the exhaustive test, conditional on the observed data) is p.
    (a) What is the variance of $\hat{P}$?
    (b) What is the variance of $\hat{P}_2$ for the two-sided test (assuming that p is not close to 0.5, where p is the true one-sided P-value?)

29. Consider a 2×2 contingency table. Using the notation of Table 3.6,
    (a) Show that $(N_{ij} - \hat{E}[N_{ij}])^2$ has the same value for all i, j.
    (b) Using (a), show that $C = (n(N_{11}N_{22} - N_{12}N_{21})^2)/R_1R_2C_1C_2$.
    (c) Verify that (b) yields the same value of C as Equation 3.3 for the 2×2 table:

	B_1	B_2
A_1	6	8
A_2	10	12

30. Consider a test for independence of two variables that have the following 2×2 table:

	B_1	B_2
A_1	m	10
A_2	10	m

What value(s) of m would lead to a conclusion that the two variables are not independent at the $\alpha = 0.05$ significance level?

4

SAMPLING DISTRIBUTIONS

During an election year, pollsters may want to gauge how the voters feel about a particular issue. For instance, what proportion p of registered voters in a state intend to vote "Yes" on a referendum? From one random sample of size $n = 1000$, the pollsters might calculate a sample proportion of $\hat{p} = 0.47$. However, a different sample of the same size might have yielded $\hat{p} = 0.49$, or maybe even 0.37. To gauge the accuracy of the original estimate, we need to understand how these proportions $\hat{p}$ vary from sample to sample.

In Chapter 3, we saw examples of a permutation distribution: all possible values of a test statistic obtained by permuting the data among two groups. By comparing the observed test statistic to the null distribution, we could quantify how unusual the observed test statistic was. There are many other situations where we want to know something about how a statistic varies due to random sampling.

4.1 SAMPLING DISTRIBUTIONS

Toss a fair coin $n = 10$ times and note the proportion of heads $\hat{p}$. If you repeat the experiment, you probably would not get exactly the same proportion of heads. If you toss 50 sets of 10 coin flips, you might see outcomes (i.e., proportions of heads, $\hat{p}$) such as those shown in Figure 4.1.

Mathematical Statistics with Resampling and R, First Edition. By Laura Chihara and Tim Hesterberg.

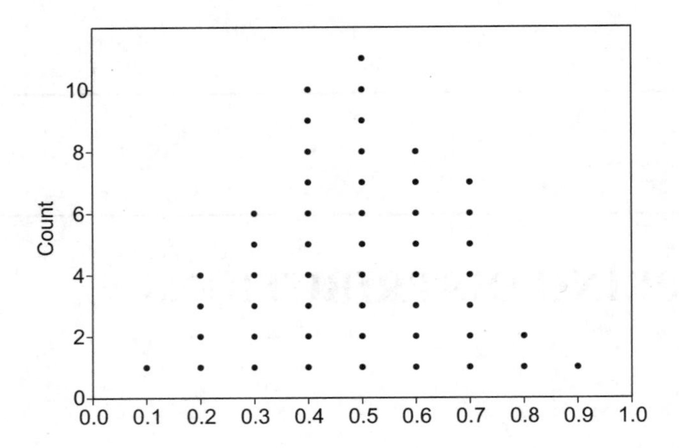

FIGURE 4.1 Distribution of $\hat{p}$ after 50 sets of 10 tosses.

Although proportions between 0.3 and 0.7 occur most often, we see there is a proportion as low as 0.1 heads or as high as 0.9. If you do this yourself, you might be "lucky" and obtain a sample proportion as low as 0, or as high as 1!

Figure 4.2 shows the result from 5000 sets of 10 tosses done using a computer.

This distribution is (an approximation to) the *sampling distribution of* $\hat{p}$. The most likely outcome is 0.5, followed by 0.4 and 0.6, and so on—values farther from 0.5 are less likely. The distribution is bell shaped and centered approximately at 0.50; the sample mean is 0.504. The standard deviation is 0.157; we call the standard deviation of a statistic a *standard error*.

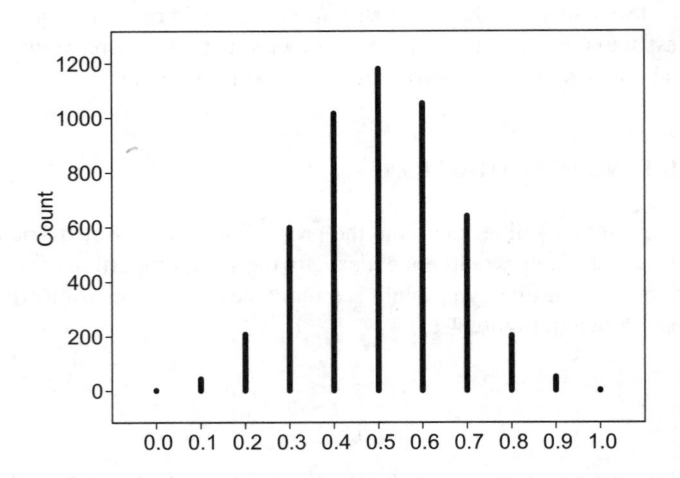

FIGURE 4.2 Distribution of $\hat{p}$ after 5000 sets of 10 tosses.

Definition 4.1 Let $\mathbf{X}$ be a random sample (e.g., $X_1, X_2, \ldots, X_n$ in the case of a sample from a single population) and let $T = h(\mathbf{X})$ denote some statistic. The *sampling distribution of* T is its probability distribution. ||

The permutation distributions in Chapter 3 are sampling distributions, as is the above example. The key point is that a sampling distribution is the distribution of a *statistic* that summarizes a data set, and represents how the statistic varies across many random data sets. A histogram of one set of observations drawn from a population does not represent a sampling distribution. A histogram of permutation means, each from one sample, does represent a sampling distribution.

Definition 4.2 The standard deviation of a sampling distribution is called the *standard error*. We use the notation $\mathrm{SE}[T]$ to denote the standard error for the sampling distribution of a statistic T and $\widehat{\mathrm{SE}}[T]$ to denote any estimate of the standard error for the sampling distribution of T. ||

Let us consider another example.

Example 4.1 Suppose a population consists of four numbers, $w_1 = 3$, $w_2 = 4$, $w_3 = 6$, and $w_4 = 6$; the population mean and standard deviation are $\mu = 4.75$ and $\sigma = 1.299$, respectively. If we draw samples of size $n = 2$ (with replacement), there are 16 unique samples, with the following sample means:

Sample	w_1, w_1	w_1, w_2	w_1, w_3	w_1, w_4	w_2, w_1	w_2, w_2	w_2, w_3	w_2, w_4
Mean	3	3.5	4.5	4.5	3.5	4	5	5
Sample	w_3, w_1	w_3, w_2	w_3, w_3	w_3, w_4	w_4, w_1	w_4, w_2	w_4, w_3	w_4, w_4
Mean	4.5	5	6	6	4.5	5	6	6

The sampling distribution for the mean of samples of size 2 (with replacement) from the given population is shown in Figure 4.3. The range is from 3 to 6, with mean

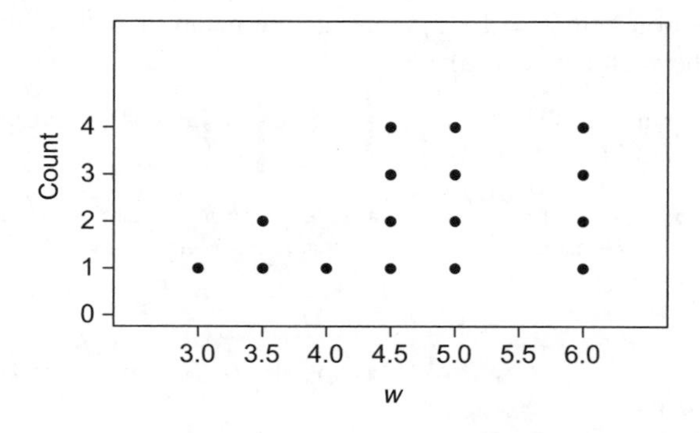

FIGURE 4.3 Dot plot for sampling distribution of sample means.

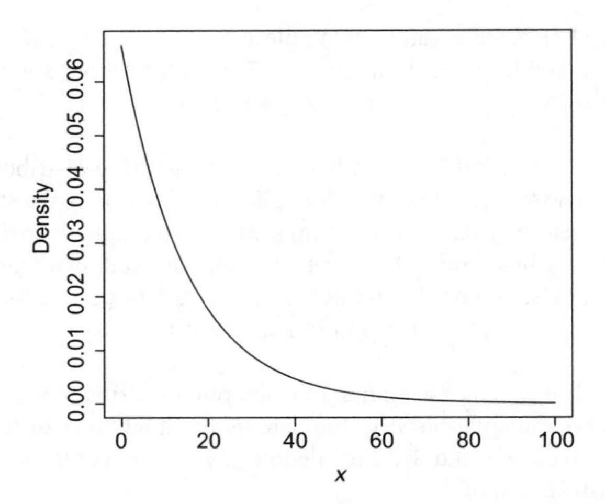

FIGURE 4.4 Density for the exponential distribution with $\lambda = 1/15$.

4.75 and standard deviation (standard error) $1.299/\sqrt{2} = 0.9186$ (both agreeing with Theorem A.5). □

Example 4.2 Let us simulate the sampling distribution for the mean of an exponential distribution with pdf $f(x) = 1/15e^{-x/15}$ ($\lambda = 1/15$) (see Figure 4.4). Recall that the mean and the standard deviation are both $1/\lambda = 15$.

We draw samples of size $n = 100$ from this distribution. From Theorem A.5, the mean of the sampling distribution of $\bar{x}$ is 15 and the standard error is $15/\sqrt{100}$.

R Note:

The following commands draw 1000 random samples of size 100 from the exponential distribution with $\lambda = 1/15$, compute the mean of each sample, and store this mean in the vector `my.means`.

```
my.means <- numeric(1000)      # space for results (vector of 0's)
for (i in 1:1000)
{
   x <- rexp(100, rate = 1/15) # draw random sample of size 100
   my.means[i] <- mean(x)      # compute mean, save in position i
}
hist(my.means)

dev.new()                      # open new graphics device
qqnorm(my.means)
qqline(my.means)
```

We can see how close the simulation-based mean and standard deviation are to Theorem A.5:

```
> mean(my.means)
[1] 15.0489
> sd(my.means)
[1] 1.567628
```

In contrast to the original distribution, the sampling distribution of $\bar{X}$ seen in Figure 4.5 is nearly bell shaped, with the normal quantile plot indicating a hint of skewness. From Theorem A.5, the mean of the sampling distribution is 15. Does the mean obtained by our simulation approximate this reasonably well?

Also, compare the estimated standard error (the standard deviation of the sampling distribution) to the theoretical standard error, $\sigma/\sqrt{n} = 15/10 = 1.5$, and the standard deviation of the population, 15. The standard error measures how much the sample means deviate from the population mean, when drawing samples of size 100 from the exponential distribution with $\lambda = 1/15$. □

Let us look at the sampling distribution of a different statistic.

Example 4.3 We draw random samples of size 12 from the uniform distribution on the interval [0, 1] and take the maximum value of each sample. We simulate the sampling distribution of the maximum by taking 1000 samples of size 12 from the uniform distribution Unif[0, 1].

R Note:

In R, the command `runif` gives random samples from Unif[0, 1].

```
my.max <- numeric(1000)
for (i in 1:1000)
{
   y <- runif(12)         # draw random sample of size 12
   my.max[i] <- max(y)    # find max, save in position i
}

hist(my.max)
```

The sampling distribution is shown in Figure 4.6. For samples of size 12 from the Unif[0, 1] distribution, the maximum is usually larger than 0.8. Rarely is the maximum less than 0.6. □

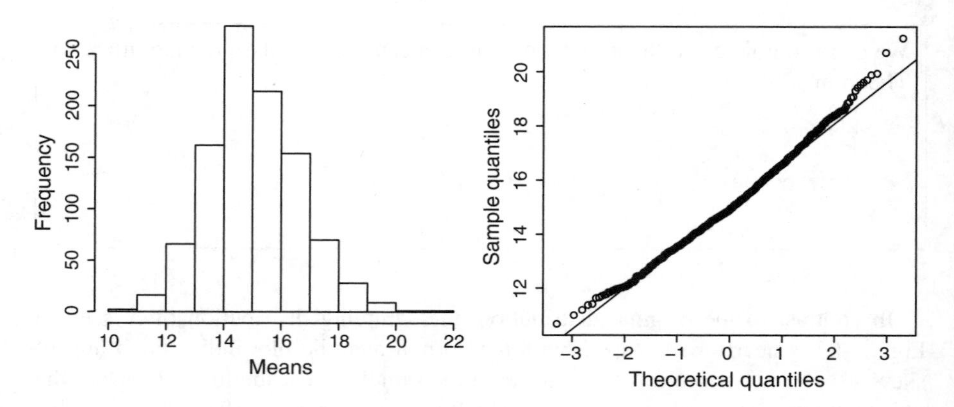

FIGURE 4.5 Simulated sampling distribution of $\bar{x}$ for $n = 100$ from Exp(1/15).

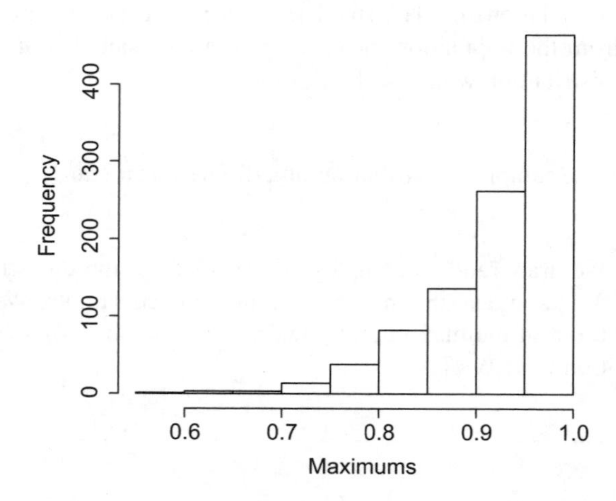

FIGURE 4.6 Simulated sampling distribution of the maximum of a sample.

4.2 CALCULATING SAMPLING DISTRIBUTIONS

There are three basic approaches for calculating sampling distributions and standard errors—exact calculations (by exhaustive calculation or formulas), simulation, and formula approximations. The mice example in Section 3.1 was small enough to calculate the exact permutation distribution by exhaustive calculation, but we approximated the Verizon (Example 3.3) and hot wings (Section 3.3) permutation distributions using simulation. Earlier in this chapter, we used simulation and exact calculation for coin tosses and sampling from a small population, respectively. In some cases, we can obtain exact answers by formulas rather than exhaustive calculation.

Example 4.4 In Example 4.3, we simulated the sampling distribution of the maximum for a sample of size 12 from a uniform distribution. We can also obtain the distribution as follows:

First note that for $X_i \sim \text{Unif}[0, 1]$, if $0 \leq a \leq 1$, then $P(X_i \leq a) = a$. Therefore, for $X_1, X_2, \ldots, X_{12} \overset{i.i.d.}{\sim} \text{Unif}[0, 1]$, the cdf of the maximum of $X_1, X_2, \ldots, X_{12}$ is

$$\begin{aligned} F(a) &= P(\max\{X_1, X_2, \ldots, X_{12}\} \leq a) \\ &= P(X_1 \leq a, X_2 \leq a, \ldots, X_{12} \leq a) \\ &= P(X_1 \leq a)P(X_2 \leq a) \ldots P(X_{12} \leq a) \qquad \text{by independence} \\ &= a^{12} \end{aligned}$$

Thus, the pdf of the sampling distribution of the maximum of $X_1, X_2, \ldots, X_{12}$ is $f(a) = F'(a) = 12a^{11}$. □

The maximum or minimum of a set of random variables comes up frequently, so we state this result more formally.

Theorem 4.1 *Suppose we have continuous random variables $X_1, X_2, \ldots, X_n$ that are i.i.d. with pdf f and cdf F. Define their minimum and maximum to be the random variables*

$$X_{\min} = \min\{X_1, X_2, \ldots, X_n\},$$
$$X_{\max} = \max\{X_1, X_2, \ldots, X_n\}.$$

Then, the pdf's for X_{min} and X_{max} are

$$f_{\min}(x) = n\Big(1 - F(x)\Big)^{n-1} f(x), \tag{4.1}$$

$$f_{\max}(x) = nF^{n-1}(x)f(x). \tag{4.2}$$

Proof Exercise. □

In particular,

Corollary 4.1 *Let $X_1, X_2, \ldots, X_n \overset{i.i.d.}{\sim}$ Unif$[0, \beta]$. The pdf's for the minimum and maximum of $X_1, X_2, \ldots, X_n$ are*

$$f_{\min}(x) = n\left(1 - \frac{x}{\beta}\right)^{n-1} \frac{1}{\beta}, \tag{4.3}$$

$$f_{\max}(x) = \frac{n}{\beta^n} x^{n-1}. \tag{4.4}$$

Proof The cdf for random variables from Unif$[0, \beta]$ is $F(x) = \dfrac{x}{\beta}$. □

Example 4.5 Let $X_1, X_2, \ldots, X_{10}$ be a random sample from a distribution with pdf $f(x) = 2/x^3, x \geq 1$. Let $X_{\min}$ denote the minimum of this sample. Find the probability that $X_{\min}$ is less than or equal to 1.2.

Solution First, we compute the cdf F corresponding to f:

$$F(x) = \int_1^x \frac{2}{t^3}\, dt = 1 - \frac{1}{x^2}.$$

By Theorem 4.1, we have

$$f_{\min}(x) = 10\left(1 - \left(1 - \frac{1}{x^2}\right)\right)^9 \frac{2}{x^3} = \frac{20}{x^{21}},$$

for $x \geq 1$. Thus,

$$P(X_{\min} \leq 1.2) = \int_1^{1.2} \frac{20}{x^{21}}\, dx = 1 - \frac{1}{1.2^{20}} = 0.974.$$

□

Example 4.6 Let $X_1, X_2, \ldots, X_n \overset{i.i.d.}{\sim} \text{Pois}(\lambda)$. Then by Theorem B.6, the sampling distribution of $X = X_1 + X_2 + \cdots + X_n$ is $\text{Pois}(n\lambda)$. □

For another example of a sampling distribution, let $X_1, X_2, \ldots, X_n$ be a random sample from a normal distribution $N(\mu, \sigma^2)$ and let $\bar{X}$ denote the sample mean. The sampling distribution of $\bar{X}$ is $N(\mu, \sigma^2/n)$ by Corollary A.2.

In other cases, we estimate sampling distributions by approximations, such as the chi-square approximation for chi-square statistics for contingency tables in Section 3.4 and goodness-of-fit in Sections 3.7 and 3.8. But the most common approximations are based on normal distributions, for which the Central Limit Theorem (CLT) plays a central role.

4.3 THE CENTRAL LIMIT THEOREM

The sampling distributions were approximately normally distributed in a number of previous examples. That is not a fluke. A wide variety of statistics have approximately normal sampling distributions, if sample sizes are large enough and some other conditions are met. Here, we look at the most common statistic, the mean.

We already know that if populations are normal, then the sampling distribution of $\bar{X}$ is normal, by Corollary A.2—if $X_1, X_2, \ldots, X_n$ are a sample of independent observations from a normal distribution with mean μ and standard deviation σ, the sampling distribution of $\bar{X}$ is also normal with mean μ and standard deviation $\sigma/\sqrt{n}$. We also observe that for exponential populations, the sampling distribution is approximately normal in Example 4.2. This is true for many other distributions by the CLT:

Theorem 4.2 (The Central Limit Theorem) *Let $X_1, X_2, \ldots, X_n$ be independent, identically distributed random variables with mean μ and variance σ^2, both finite. Then for any constant z,*

$$\lim_{n\to\infty} P\left(\frac{\bar{X}-\mu}{\sigma/\sqrt{n}} \leq z\right) = \Phi(z),$$

where Φ is the cdf of the standard normal distribution (Equation A.5).

See Casella and Berger (2001); Ghahramani (2004); and Ross (2009) for a proof.

The Central Limit Theorem means that for n "sufficiently large," the sampling distribution of $\bar{X}$ is approximately normal with mean μ and standard error $\sigma/\sqrt{n}$, regardless of the distribution from which the sample was drawn. Thus, the standardized random variable

$$Z = \frac{\bar{X}-\mu}{\sigma/\sqrt{n}} = \frac{\bar{X}-\mathrm{E}\left[\bar{X}\right]}{\mathrm{SE}[\bar{X}]}$$

is approximately standard normal. How large is enough? It primarily depends on how skewed the population is, whether the population is continuous or discrete, and how accurate the answers should be. We will come back to this in Sections 4.3.2 and 4.3.3.

Example 4.7 Suppose $X_1, X_2, \ldots, X_{30}$ are a random sample from the gamma distribution with parameters $r = 5$, and $\lambda = 2$. Use the CLT to estimate the probability $P(\bar{X} > 3)$.

Solution From Theorem B.10, we have $\mathrm{E}[X_i] = 5/2$ and $\mathrm{SD}[X_i] = \sqrt{5/2^2}$, $i = 1, 2, \ldots, 30$. The sampling distribution of $\bar{X}$ is approximately normal with mean $\mathrm{E}\left[\bar{X}\right] = 5/2$ and standard error $\mathrm{SE}[\bar{X}] = \sqrt{5/2^2}/\sqrt{30}$ (Figure 4.7).

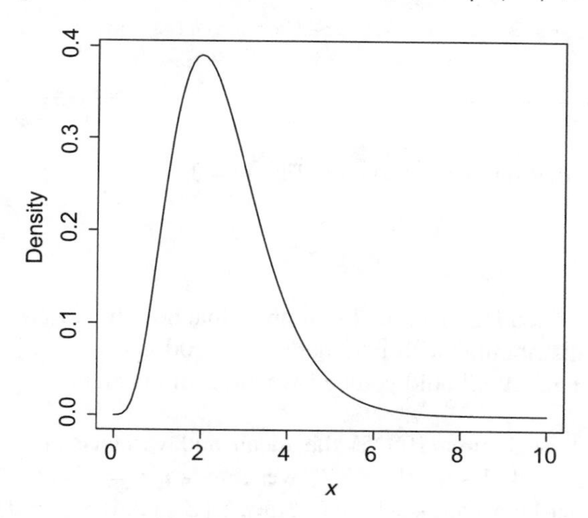

FIGURE 4.7 Density for the gamma distribution $r = 5$, and $\lambda = 2$.

Hence,

$$\begin{aligned} P(\bar{X} > 3) &= P\left(\frac{\bar{X} - 5/2}{\sqrt{(5/2^2)}/\sqrt{30}} > \frac{3 - 5/2}{\sqrt{(5/2^2)}/\sqrt{30}}\right) \\ &\approx P(Z > 2.4495) \\ &= 0.0072 \end{aligned}$$

We can also simulate the sampling distribution in R.

R Note:

```
my.means <- numeric(1000)
for (i in 1:1000)
{
  x <- rgamma(30, shape = 5, rate = 2)
  my.means[i] <- mean(x)
}

hist(my.means)
dev.new()                    # open new graphics device
qqnorm(my.means)
qqline(my.means)
```

The approximate mean, standard error, and an empirical check of the probability that the sample mean is greater than 3 are given by:

```
> mean(my.means)          # output will vary
[1] 2.484278
> sd(my.means)
[1] 0.2030441
> mean(my.means > 3)      # empirical check of P(mean > 3)
[1] 0.007
```

The sampling distribution is shown in Figure 4.8. □

Example 4.8 A friend has a set of 17 numbers that he claims were randomly drawn from a gamma distribution with parameters $r = 100$ and $\lambda = 5$. You compute the sample mean to be 21.9. Should you be suspicious of his claim?

Solution For $X \sim \text{Gamma}(100, 5)$ the mean and variance are $\mu = 100/5 = 20$ and $\sigma^2 = 100/5^2 = 4$. Using the CLT, we compare $z = (21.9 - 20)/(2/\sqrt{17}) = 2.4533$ to a standard normal distribution. Since $P(Z \geq 2.4533) = 0.0071$, a random sample of size 17 with mean 21.9 or larger from Gamma(100, 5) is relatively rare. Yes, you are justified in being suspicious of your friend's claim. □

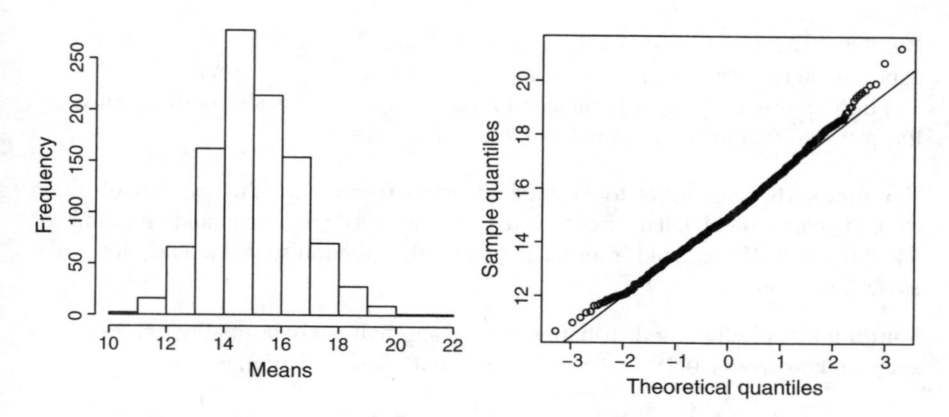

FIGURE 4.8 Sampling distribution of $\bar{X}$ for Gamma(5, 2).

4.3.1 CLT for Binomial Data

An important special case is the use of the CLT for binomial data.

Example 4.9 Toss a fair coin 300 times. Find the approximate probability of getting at most 160 heads.

Solution Let $X_1, X_2, \ldots, X_{300}$ denote Bernoulli random variables (1 for heads, 0 for tails). Then $E[X_i] = 1/2$ and $\text{Var}[X_i] = 1/4$, $i = 1, 2, \ldots, 300$. Now $\bar{X}$, which gives the proportion of heads, is approximately normal with mean 1/2 and standard error $(1/2)/\sqrt{300} = 0.029$.

Thus,

$$P(\bar{X} \leq 160/300) = P\left(\frac{\bar{X} - 0.5}{0.0289} \leq \frac{0.5333 - 0.5}{0.0289}\right) = P(Z \leq 1.1522) = 0.875.$$

□

We generally take a shortcut in problems such as these. Note that $X = \sum_{i=1}^{n} X_i$ is a binomial random variable, and $\hat{p} = \bar{X} = X/n$. So rather than expressing the problem as a sum of the Bernoulli random variable, we work directly with the binomial variable, using the following corollary to the CLT:

Corollary 4.2 *Let $X \sim$ Binom(n, p) be a binomial random variable and $\hat{p} = X/n$, the proportion of successes. Then for sufficiently large n, the sampling distribution of $\hat{p}$ is approximately normal with mean p and standard deviation $\sqrt{p(1-p)/n}$.*

Similarly, the sampling distribution of X is approximately normal with mean np and standard deviation $\sqrt{np(1-p)}$.

Proof A binomial variable X can be written as $X = \sum_{i=1}^{n} X_i$, where $X_1, X_2, \ldots, X_n \overset{i.i.d.}{\sim}$ Bern(p) with $E[X_i] = p$ and $\text{Var}[X_i] = p(1-p)$. Then $\bar{X} = (X_1 + X_2 + \cdots + X_n)/n$ is the proportion of 1's, more commonly denoted by $\hat{p}$.

By the CLT, for n sufficiently large, the sampling distribution of $\bar{X} = \hat{p}$ is approximately normal with mean p and standard error $\sqrt{p(1-p)/n}$.

Furthermore, if $\hat{p}$ is approximately normal, then $X = n\hat{p}$ is also normal, since any linear transformation of a normal variable is also normal. □

Example 4.10 According to the 2004 American Community Survey, 28% of adults over 25 years old in Utah have completed a bachelor's degree. In a random sample of 64 adults over 25 years old from Utah, what is the probability that at least 30% have a bachelor's degree?

Solution The sampling distribution of $\hat{p}$ is approximately normal with mean 0.28 and standard error $\sqrt{0.28(1-0.28)/64} = 0.056$. Standardizing, we find

$$P(\hat{p} \geq 0.30)) = P(Z \geq \frac{0.30 - 0.28}{0.056}) = P(Z \geq 0.356) = 0.361.$$

That is, the probability that at least 30% of those in the sample have a bachelor's degree is 0.361. □

Example 4.11 Let X be a binomial random variable, $X \sim \text{Binom}(120, 0.3)$. Compute $P(X \leq 25)$.

Solution An exact solution would require calculating $P(X \leq 25) = \sum_{k=0}^{25} \binom{120}{k} 0.3^k 0.7^{120-k}$. We use Corollary 4.2 instead.

$$\begin{aligned} P(X \leq 25) &= P(\frac{X}{120} \leq \frac{25}{120}) \\ &= P\left(\frac{\hat{p} - 0.3}{\sqrt{0.3(1-0.3)/120}} \leq \frac{0.2083 - 0.3}{\sqrt{0.3(1-0.3)/120}}\right) \\ &\approx P(Z \leq -2.1913) = 0.014. \end{aligned}$$

The exact probability is 0.0159, so the CLT approximation underestimates the exact probability by about 0.0017. This is a large error, over 10% of the exact answer. We need to do better, so we turn to a continuity correction.

R Note:

To compute $\binom{n}{k} p^k (1-p)^{n-k}$, use the `dbinom` command.
For instance, $\binom{120}{25} 0.3^{25} 0.7^{95}$,

```
> dbinom(25, 120, .3)
[1] [1] 0.006807598
```

The command `pbinom` computes cumulative probabilities for the binomial distribution; that is, it is the cumulative mass function. For instance, to calculate $\sum_{k=0}^{25} \binom{120}{k} 0.3^k 0.7^{120-k}$,

```
> pbinom(25, 120, .3)
[1] 0.01593170
```

□

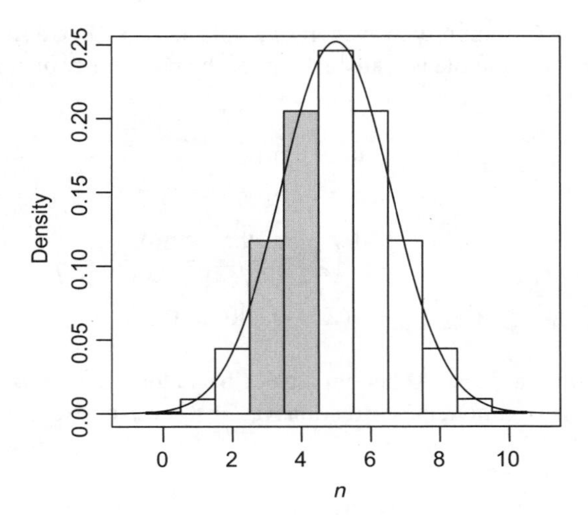

FIGURE 4.9 Binomial distribution with $n = 10$ and $p = 0.5$, with CLT approximation, and $P(X = 3 \text{ or } 4)$ highlighted.

4.3.2 Continuity Correction for Discrete Random Variables

A binomial variable X is a discrete random variable, but the CLT approximation uses a continuous density. We can improve the approximation of the CLT for binomial and other discrete data using a *continuity correction*. This is best explained visually.

Consider the case of tossing 10 coins and estimating the probability that either 3 or 4 coins come up. $X \sim \text{Binom}(10, 0.5)$, and our goal is to estimate $P(3 \leq X \leq 4)$. Consider Figure 4.9 that shows the exact probabilities for the binomial distribution, overlaid with the CLT normal approximation density. The desired probability corresponds to the area of the two shaded bars, each with width 1.0 and heights $f(3)$ and $f(4)$, respectively. To approximate the shaded region using an area under the normal curve, it would be better to use the integral $\int_{2.5}^{4.5} g(x)dx$ rather than $\int_{3}^{4} g(x)dx$, where g is the corresponding normal density.

To do the corresponding calculations, we express each inequality in two different ways and then split the difference, ultimately adding or subtracting 0.5 to each end. In this case, there are two inequalities inside the probability, $3 \leq X$ and $X \leq 4$; we handle these separately. Remember that X is discrete, so $P(3 \leq X) = P(2 < X) = P(2.5 < X)$; similarly, $P(X \leq 4) = P(X < 5) = P(X < 4.5)$. In each case, we split the difference between the two ways to express the boundaries using whole numbers. This yields

$$P(3 \leq X \leq 4) = P(2.5 < X < 4.5)$$
$$\approx P\left(\frac{2.5 - 5}{1.58} \leq Z \leq \frac{4.5 - 5}{1.58}\right) = 0.319.$$

This is not far from the exact answer of 0.322. In contrast, without the continuity correction, the estimate would be 0.161, only about half as large as desired.

In the American Community Survey in Example 4.10, 30% of a sample of size 64 is 19.2 people, so the example is really asking for the probability of at least 20 people having a bachelor's degree.

$$\begin{aligned}P(X \geq 19.2) &= P(X \geq 20)\\ &= P(X > 19.5)\\ &\approx P\left(Z > \frac{19.5 - (64)(0.28)}{\sqrt{64(0.28)(0.72)}}\right)\\ &= P(Z > 0.440) = 0.330.\end{aligned}$$

The probability that at least 20 people have a bachelor's degree is about 33%; in contrast, the estimate without the correction was 36%, and the exact answer is 32.4%.

In the Example 4.11,

$$\begin{aligned}P(X \leq 25) &= P(X < 25.5)\\ &\approx P\left(Z < \frac{25.5 - 36}{\sqrt{120(0.3)(0.7)}}\right)\\ &= P(Z \leq -2.092) = 0.018.\end{aligned}$$

This is even further off than without the continuity correction! It turns out that the CLT is not very accurate for skewed distributions unless sample sizes are much larger, especially in the tails of distributions, and with a z-score of -2.092, we are fairly far in the tail. We will discuss the accuracy of the CLT in more detail later. In this case, we fixed a small error in a direction that exacerbated a larger error. Applying the same method elsewhere in the distribution would generally not be so unlucky.

Example 4.12 According to the Centers for Disease Control, about 22.9% of adults in Kentucky are everyday smokers. In a random sample of 700 adults in Kentucky, what is the probability that between 150 and 170 of them are everyday smokers?

Solution Let X denote the number of everyday smokers in the sample. Then $X \sim 700 \times 0.229$, with expected value $np = 700 \times 0.229 = 160.3$ and standard error $\sqrt{700(0.229)(1 - 0.229)} = 11.117$. We find

$$\begin{aligned}P(150 \leq X \leq 170) &= P(149.5 < X < 170.5)\\ &= P\left(\frac{149.5 - 160.3}{11.117} \leq Z \leq \frac{170.5 - 160.3}{11.117}\right)\\ &\approx \Phi(0.9175) - \Phi(-0.9715) = 0.655.\end{aligned}$$

□

4.3.3 Accuracy of the Central Limit Theorem

The usual rule of thumb, found in most textbooks, is that the CLT is reasonably accurate if $n \geq 30$, unless the data are quite skewed. For binomial data, the common rule of thumb is to use the CLT, with continuity correction, if both $np \geq 10$ and $n(1-p) \geq 10$.

These rules are wishful thinking, dating to a pre-computer age when one had few realistic alternatives to using the CLT because most other methods were computationally infeasible. We can obtain better approximations using simulation-based methods, including permutation tests (Chapter 3) and bootstrapping (Chapter 5); in addition, these methods provide a way to check the accuracy of the CLT, based on a set of data. The CLT is exact if the population is normal (Corollary A.2). For nonnormal populations, the biggest problem is skewness, followed by discreteness. If the population is symmetric, the sampling distribution of the mean may be very close to normal for quite small n. If not, it may be nonnormal even for very large n. There is an expanded version of the CLT approximation for continuous data (obtained using a particular Taylor series, known as an Edgeworth approximation):

$$P\left(\frac{\bar{X}-\mu}{\sigma/\sqrt{n}} \leq z\right) \approx \Phi(z) + \frac{\kappa_3}{6\sqrt{n}}(z^2-1)\Phi'(z), \tag{4.5}$$

where $\Phi(z)$ is the standard normal density, $\Phi'(z)$ is its derivative, and $\kappa_3 = E(X-\mu)^3)/\sigma^3$. This can be used in its own right, or the term

$$\frac{\kappa_3}{6\sqrt{n}}(z^2-1)\Phi'(z) \tag{4.6}$$

can be used to estimate the error of the CLT approximation. We can use this approximation to estimate necessary sample sizes for prescribed accuracy for two examples considered above, the exponential and binomial.

We will focus on the error of the CLT approximation at $z = 2.33$ since the probability $P(X > x) \approx 0.01$ is important in statistical practice ($x = \mu + 2.33\sigma/\sqrt{n}$). What size n ensures an error of at most 10%, that is, that the true probability be between 0.009 and 0.011? For exponential distributions, $\kappa_3 = 2$. Setting $0.001 = \frac{2}{6\sqrt{n}}(1-z^2)\Phi'(z)$ and solving for n in Equation 4.6 results in $n \geq 1536$, a far cry from the rule $n \geq 30$! The approximation Equation 4.5 is not far off; exact calculations indicate that $n \geq 1541$ is required.

In the binomial case, again assuming an error of 0.001, Equation 4.5 can be used to obtain the rule that $np \geq 384$ and $n(1-p) \geq 384$ for $z = 2.33$. In either case, the point is that for skewed samples, the old rules of thumbs for when the CLT is appropriate are not necessarily valid. The CLT is often fine for quick-and-dirty approximations, but use caution when accuracy is important.

4.3.4 CLT for Sampling Without Replacement

When working with finite populations, one typically samples without replacement. Thus, a random sample $X_1, X_2, \ldots, X_n$ is not independent and we cannot invoke the usual CLT to say that $\bar{X}$ is asymptotically normally distributed. In fact, $n \to \infty$ is impossible because the sample size n is bounded above by the size N of the population! However, in 1960, Jaroslav Hajek proved a version of the CLT for sampling without replacement. The accuracy depends on both n and the number of nonsampled observations $(N - n)$. The standard error for the sample mean requires a *finite population correction factor*, $\sqrt{(N-n)/(N-1)}$, though we may skip this if it is close to 1 (see Exercise 27). Courses on survey sampling discuss issues related to finite populations in more detail.

4.4 EXERCISES

1. Consider the population $\{1, 2, 5, 6, 10, 12\}$. Find (and plot) the sampling distribution of medians for samples of size 3 without replacement. Compare the median of the population to the mean of the medians.
2. Consider the population $\{3, 6, 7, 9, 11, 14\}$. For samples of size 3 without replacement, find (and plot) the sampling distribution of the minimum. What is the mean of the sampling distribution? The statistic is an estimate of some parameter—what is the value of that parameter?
3. Let A denote the population $\{1, 3, 4, 5\}$ and B the population $\{5, 7, 9\}$. Let X be a random value from A, and Y a random value from B.
   (a) Find the sampling distribution of $X + Y$.
   (b) In this example, does the sampling distribution depend on whether you sample with or without replacement? Why or why not?
   (c) Compute the mean of the values for each of A and B. Compute the mean of the values in the sampling distribution of $X + Y$. How are the means related?
   (d) Suppose you draw a random value from A and a random value from B. What is the probability that the sum is 13 or larger?
4. Let $X_1, X_2, \ldots, X_{25}$ be a random sample from some distribution and $W = T(X_1, X_2, \ldots, X_{25})$ be a statistic. Suppose the *sampling distribution* of W has a pdf given by $f(x) = 2/x^2$, $1 < x < 2$. Find the probability that $W < 1.5$.
5. Let $X_1, X_2, \ldots, X_n$ be a random sample from some distribution and suppose $Y = T(X_1, X_2, \ldots, X_n)$ is a statistic. Suppose the sampling distribution of Y has pdf $f(y) = (3/8)y^2$ for $0 \le y \le 2$. Find $P(0 \le Y \le 1/5)$.
6. Suppose the heights of boys in a certain large city follows a distribution with mean 48 in. and variance 9^2. Use the CLT approximation to estimate the probability that in a random sample of 30 boys, the mean height is more than 51 in.
7. Let $X_1, X_2, \ldots, X_{36} \sim \text{Bern}(0.55)$ be independent and let $\hat{p}$ denote the sample proportion. Use the CLT approximation with continuity correction to find the probability that $\hat{p} \le 0.50$.
8. A random sample of size $n = 20$ is drawn from a distribution with mean 6 and variance 10. Use the CLT approximation to estimate $P(\bar{X} \le 4.6)$.
9. A random sample of size $n = 244$ is drawn from a distribution with pdf $f(x) = (3/16)(x-4)^2$, $2 \le x \le 6$. Use the CLT approximation to estimate $P(\bar{X} \ge 4.2)$.

10. According to the 2000 census, 28.6% of the U.S. adult population received a high school diploma. In a random sample of 800 U.S. adults, what is the probability that between 220 and 230 (inclusive) people have a high school diploma? Use the CLT approximation with continuity correction and compare to the exact probability (use `pbinom` in `R`).
11. If $X_1, \ldots, X_n$ are i.i.d. from Unif[0, 1], how large should n be so that $P(|\bar{X} - 1/2| < 0.05) \geq 0.90$, that is, there is at least a 90% chance that the sample mean is within 0.05 of 1/2? Use the CLT approximation.
12. A friend claims that she has drawn a random sample of size 30 from the exponential distribution with $\lambda = 1/10$. The mean of her sample is 12.
    (a) What is the expected value of a sample mean?
    (b) Run a simulation by drawing 1000 random samples, each of size 30, from Exp(1/10) and then compute the mean. What proportion of the sample means are as large as or larger than 12?
    (c) Is a mean of 12 unusual for a sample of size 30 from Exp(1/10)?
13. Let $X_1, X_2, \ldots, X_{10} \overset{i.i.d.}{\sim} N(20, 8^2)$ and $Y_1, Y_2, \ldots, Y_{15} \overset{i.i.d.}{\sim} N(16, 7^2)$. Let $W = \bar{X} + \bar{Y}$.
    (a) Give the exact sampling distribution of W.
    (b) Simulate the sampling distribution in `R` and plot your results. Check that the simulated mean and standard error are close to the theoretical mean and the standard error.
    (c) Use your simulation to find $P(W < 40)$. Calculate an exact answer and compare.

```
W <- numeric(1000)
for (i in 1:1000)
{
  x <- rnorm(10, 20, 8)       # draw 10 from N(20, 8^2)
  y <- rnorm(15, 16, 7)       # draw 15 from N(16, 7^2)
  W[i] <- mean(x) + mean(y) # save sum of means
}
hist(W)
mean(W < 40)
```

14. Let $X_1, X_2, \ldots, X_9 \overset{i.i.d.}{\sim} N(7, 3^2)$ and $Y_1, Y_2, \ldots, Y_{12} \overset{i.i.d.}{\sim} N(10, 5^2)$. Let $W = \bar{X} - \bar{Y}$.
    (a) Give the exact sampling distribution of W.
    (b) Simulate the sampling distribution of W in `R` and plot your results (adapt code from the previous exercise). Check that the simulated mean and the standard error are close to the theoretical mean and the standard error.
    (c) Use your simulation to find $P(W < -1.5)$. Calculate an exact answer and compare.

15. Let $X_1, X_2, \ldots, X_n$ be a random sample from $N(0, 1)$. Let $W = X_1^2 + X_2^2 + \cdots + X_n^2$. Describe the sampling distribution of W by running a simulation, using $n = 2$. What is the mean and variance of the sampling distribution of W? Repeat using $n = 4$, $n = 5$. What observations or conjectures do you have for general n?
16. Let X be a uniform random variable on the interval $[40, 60]$ and Y a uniform random variable on $[45, 80]$. Assume that X and Y are independent.
    (a) Compute the expected value and variance of $X + Y$ (see Theorem B.7).
    (b) The following code simulates the sampling distribution of $X + Y$:

```
X <- runif(1000, 40, 60) # Draw 1000 values from Unif[40,60]
Y <- runif(1000, 45, 80) # Draw 1000 values from Unif[45,80]
total <- X + Y           # Add them coordinate-wise
hist(total)              # Distribution of the sums
```

    Describe the graph of the distribution of $X + Y$. Compute the mean and variance of the sampling distribution (i.e., of `total`) and compare this to the theoretical mean and variance.
    (c) Suppose the time (in minutes) Jack takes to complete his statistics homework is Unif[40, 60] and the time Jill takes is Unif[45, 80]. Assume they work independently. One day they announce that their total time to finish an assignment was less than 90 min. How likely is this? (Use the simulated sampling distribution in part (b)).
17. Let $X_1, X_2, \ldots, X_{20} \overset{i.i.d.}{\sim} \text{Exp}(2)$. Let $X = \sum_{i=1}^{20} X_i$.
    (a) Simulate the sampling distribution of X in R.
    (b) From your simulation, find $E[X]$ and $\text{Var}[X]$.
    (c) From your simulation, find $P(X \leq 10)$.
18. Let $X_1, X_2, \ldots, X_{30} \overset{i.i.d.}{\sim} \text{Exp}(1/3)$ and let $\bar{X}$ denote the sample mean.
    (a) Simulate the sampling distribution of $\bar{X}$ in R.
    (b) Find the mean and standard error of the sampling distribution and compare to the theoretical results.
    (c) From your simulation, find $P(\bar{X} \leq 3.5)$.
    (d) Estimate $P(\bar{X} \leq 3.5)$ by assuming that the CLT approximation holds. Compare this result with the one in part (c).
19. Consider the exponential distribution with density $f(x) = (1/20)e^{-x/20}$, with mean and standard deviation 20.
    (a) Calculate the median of this distribution.
    (b) Using R, draw a random sample of size 50 and graph the histogram. Describe the distribution of your *sample*. What are the mean and the standard deviation of your sample?

(c) Run a simulation to find the (approximate) sampling distribution for the median of samples of size 50 from the exponential distribution and describe it. What is the mean and the standard error of this sampling distribution?

(d) Repeat the above but use sample sizes $n = 100$, 500, and 1000. How does sample size affect the sampling distribution?

20. Prove Theorem 4.1.

21. Let $X_1, X_2 \overset{i.i.d.}{\sim} F$ with corresponding pdf $f(x) = 2/x^2$, $1 \leq x \leq 2$.
    (a) Find the pdf of $X_{\max}$.
    (b) Find the expected value of $X_{\max}$.

22. Let $X_1, X_2, \ldots, X_n \overset{i.i.d.}{\sim} F$ with corresponding pdf $f(x) = 3x^2$, $0 \leq x \leq 1$.
    (a) Find the pdf for $X_{\min}$.
    (b) Find the pdf for $X_{\max}$.
    (c) If $n = 10$, find the probability that the largest value, $X_{\max}$, is greater than 0.92.

23. Compute the pdf of the sampling distribution of the maximum of samples of size 10 from a population with an exponential distribution with $\lambda = 12$.

24. Let $X_1, X_2, \ldots, X_n \overset{i.i.d.}{\sim} \text{Exp}(\lambda)$ with pdf $f(x) = \lambda e^{-\lambda x}$, $\lambda > 0$, $x > 0$.
    (a) Find the pdf $f_{\min}(x)$ for the sample minimum $X_{\min}$. Recognize this as the pdf of a known distribution.
    (b) Simulate in R the sampling distribution of $X_{\min}$ of samples of size $n = 25$ from the exponential distribution with $\lambda = 7$. Compare the theoretical expected value of $X_{\min}$ to the simulated expected value.

25. Let $X_1, X_2, \ldots, X_{10} \overset{i.i.d.}{\sim} \text{Pois}(3)$. Let $X = \sum_{i=1}^{10} X_i$. Find the pdf for the sampling distribution of X.

26. Let X_1 and X_2 be independent exponential random variables, both with parameter $\lambda > 0$. Find the cumulative distribution function for the sampling distribution of $X = X_1 + X_2$.

27. This simulation illustrates the CLT for a finite population.

```
N <- 400    # population size
n <- 5      # sample size
finpop <- rexp(N, 1/10) # Create a finite pop. of size N=400 from
                        # Exp(1/10)
hist(finpop)        # distribution of your finite pop.
mean(finpop)        # mean (mu) of your pop.
sd(finpop)          # stdev (sigma) of your pop.

sd(finpop)/sqrt(n)       # theoretical standard error of sampling
                         # dist. of mean(x), with replacement
sd(finpop)/sqrt(n)  * sqrt((N-n)/(N-1)) # without replacement
```

```
my.means <- numeric(1000)
for (i in 1:1000)
{
  x <- sample(finpop, n) # Random sample of size n (w/o replacement)
  my.means[i] <- mean(x) # Find mean of sample, store in my.means
}

hist(my.means)
dev.new()                # new graphics device
qqnorm(my.means)
qqline(my.means)

mean(my.means)
sd(my.means)             # estimated standard error of sampling
                         # distribution
```

(a) Does the sampling distribution of sample means appear approximately normal?

(b) Compare the mean and standard error of your simulated sampling distribution to the theoretical ones.

(c) Calculate $(\sigma/\sqrt{n})\sqrt{(N-n)/(N-1)}$, where σ is the standard deviation of the finite population and compare with the (estimated) standard error of the sampling distribution.

(d) Repeat for larger n, say $n = 20$, and $n = 100$.

28. Let $X_1, X_2, \ldots, X_n$ be independent random variables from $N(\mu, \sigma^2)$. We are interested in the sampling distribution of the variance. Run the following script in R, which draws random samples of size 20 from $N(25, 7^2)$ and calculates the variance for each sample.

```
my.vars <- numeric(1000)
for (i in 1:1000)
{
  x <- rnorm(20, 25, 7)
  my.vars[i] <- var(x)
}

mean(my.vars)
var(my.vars)

hist(my.vars)
dev.new()          # new graphics device
qqnorm(my.vars)
qqline(my.vars)
```

Does the sampling distribution appear to be normally distributed? Repeat with $n = 50$ and $n = 200$.

29. A random sample of size $n = 100$ is drawn from a distribution with pdf $f(x) = 3(1-x)^2$, $0 \leq x \leq 1$.
    (a) Use the CLT approximation to estimate $P(\bar{X} \leq 0.27)$.
    (b) Use the expanded CLT to estimate the same probability. The function `dnorm` computes the normal density in R. *Hint*: $E((X-\mu)^3) = 1/160$.
    (c) If $X_1, X_2, X_3 \overset{i.i.d.}{\sim} \text{Unif}[0, 1]$, then the minimum has density f given above. Use simulation to estimate the probability. *Hints*: `pmin(runif(100), runif(100), runif(100))` gives 100 values from the density, $E((X-\mu)^3) = 1/160$.

5

THE BOOTSTRAP

In the previous chapters we learned about sampling distributions and some ways to compute or estimate them. A common feature of previous examples is that the relevant populations were *known*—for example, a binomial distribution with specified p or exponential distribution with specified λ.

You may protest—What about permutation distributions or goodness-of-fit tests? The populations were not known there. But even in such situations, we were concerned only with sampling distributions when the null hypothesis is true. For example, we assumed that the true means (and spreads and shapes) of two populations are same, or the distribution of birthdays are uniform across quarters, or the home run counts come from a Poisson distribution. These assumptions provided enough additional information that our sampling was from known populations. Thus, in permutation testing, under the null hypothesis, we could then pool the data and proceed to draw samples without replacement, using the pooled data as the known population.

We now move from the realm of probability to statistics, from situations where the population is known to where it is unknown. If all we have are data and a statistic estimated from the data, we need to estimate the sampling distribution of the statistic. In this chapter, we introduce one way to do so, the bootstrap.

5.1 INTRODUCTION TO THE BOOTSTRAP

For the North Carolina data (Case Study in Section 1.2), the mean weight of the 1009 babies in the sample is 3448.26 g. We are interested in μ, the true mean birth weight

---

Mathematical Statistics with Resampling and R, First Edition. By Laura Chihara and Tim Hesterberg.

for all North Carolina babies born in 2004; this is probably not the same as the sample mean. We have already mentioned that different samples of the same size will yield different sample means, so how can we gauge the accuracy of 3448.26 as an estimate to μ?

If we knew the sampling distribution of sample means for samples of size 1009 from the population of all 2004 North Carolina births, then this would give us an idea of how means vary from sample to sample, for the standard error of the sampling distribution tells us how far means deviate from the population mean μ. But of course, since we do not have *all* the birth weights, we cannot generate the sampling distribution (and if we did have all the weights, we would know the true μ!).

The bootstrap is a procedure that uses the given sample to create a new distribution, called the bootstrap distribution, that approximates the sampling distribution for the sample mean (or for other statistics).

We will begin by considering only a small subset of the birth weights—three observations, 3969, 3204, and 2892.

To find the bootstrap distribution of the mean, we draw samples (called *resamples* or *bootstrap samples*) of size n, with replacement, from the original sample and then compute the mean of each resample. In other words, we now treat the original sample as the population. For instance, Table 5.1 shows all $3^3 = 27$ samples of size 3, taking order into account (the notation x^* indicates a resampled observation, and $\bar{x}^*$ or $\hat{\theta}^*$ the statistic for a bootstrap sample).

The idea behind the bootstrap is that if the original sample is representative of the population, then the bootstrap distribution of the mean will look approximately like the sampling distribution of the mean; that is, have roughly the same spread and shape. However, the mean of the bootstrap distribution will be same as the mean of the *original sample* (Theorem A.5), not necessarily that of the original population.

THE BOOTSTRAP IDEA

The original sample approximates the population from which it was drawn. So resamples from this sample approximate what we would get if we took many samples from the population. The bootstrap distribution of a statistic, based on many resamples, approximates the sampling distribution of the statistic, based on many samples.

Thus, the standard deviation of all the resample means listed in Table 5.1 is 266 and we use this value as an estimate of the actual standard error (standard deviation of the true sampling distribution).

Of course, it is hard for three observations to accurately approximate the population. Let us work with the full data set; we draw resamples of size $n = 1009$ from the 1009 birth weights and calculate the mean for each.

There are now 1009^{1009} samples, too many for exhaustive calculation. Instead, we draw samples randomly, of size 1009 with replacement from the data, and calculate the mean for each. We repeat this many times, say 10,000, to create the bootstrap distribution. You can imagine that there is a table like Table 5.1 with 1009^{1009} rows, and we are randomly picking 10, 000 rows from that table.

TABLE 5.1 All Possible Samples of Size 3 from 3969, 3204, and 2892.

x_1^*	x_2^*	x_3^*	$\bar{x}^*$
3969	3969	3969	3969
3969	3969	3204	3714
3969	3969	2892	3610
3969	3204	3969	3714
3969	3204	3204	3459
3969	3204	2892	3355
3969	2892	3969	3610
3969	2892	3204	3355
3969	2892	2892	3251
3204	3969	3969	3714
3204	3969	3204	3459
3204	3969	2892	3355
3204	3204	3969	3459
3204	3204	3204	3204
3204	3204	2892	3100
3204	2892	3969	3355
3204	2892	3204	3100
3204	2892	2892	2996
2892	3969	3969	3610
2892	3969	3204	3355
2892	3969	2892	3251
2892	3204	3969	3355
2892	3204	3204	3100
2892	3204	2892	2996
2892	2892	3969	3251
2892	2892	3204	2996
2892	2892	2892	2892

BOOTSTRAP FOR A SINGLE POPULATION

Given a sample of size n from a population,

1. Draw a resample of size n with replacement from the sample. Compute a statistic that describes the sample, such as the sample mean.
2. Repeat this resampling process many times, say 10,000.
3. Construct the bootstrap distribution of the statistic. Inspect its spread, bias, and shape.

Remark One technical point—there are n^n samples where order matters, but only $\binom{2n-1}{n}$ unordered samples (see Exercise 5), a much smaller number. For exhaustive calculations we could use unordered samples and be careful to keep track of the

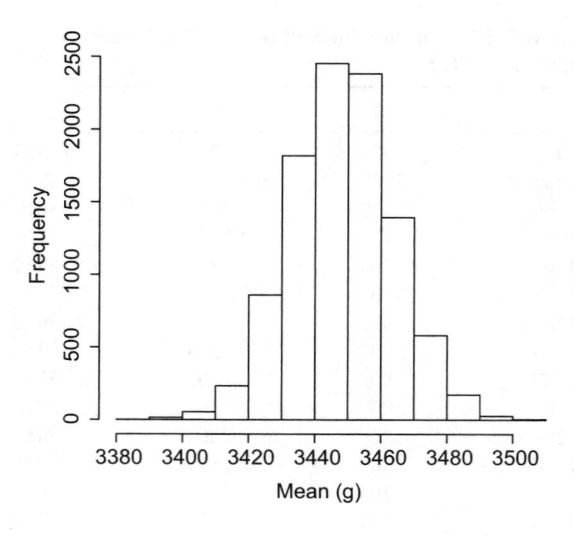

FIGURE 5.1 Bootstrap distribution of means for the North Carolina birth weights.

different probabilities for each sample. But when programming the random sampling procedure, it is easier to sample as if order matters. ||

In Figure 5.1, we note that the bootstrap distribution is approximately normal. Second, with mean 3448.206, it is centered at approximately the same location as the original mean, 3448.26. Third, we get a rough idea of the amount of variability. We can quantify the variability by computing the standard deviation of the bootstrap distribution, in this case 15.379. This is the bootstrap standard error.

For comparison, the standard deviation of the data is 487.736. The bootstrap standard error is smaller—this reflects the fact that an average of 1009 observations is more accurate (less variable) than is a single observation.

> **BOOTSTRAP STANDARD ERROR**
>
> The *bootstrap standard error* of a statistic is the standard deviation of the bootstrap distribution of that statistic.

To highlight some key features of the bootstrap distribution, we begin with two examples in which the theoretical sampling distributions of the mean are known.

Example 5.1 Consider a random sample of size 50 drawn from $N(23, 7^2)$. From Corollary A.2, we know the sampling distribution of the sample means is normal with mean 23 and standard error $\sigma/\sqrt{n} = 7/\sqrt{50} = 0.99$. Figure 5.2 shows the distribution of one such random sample with sample mean and standard deviation $\bar{x} = 24.13$, $s = 6.69$, respectively.

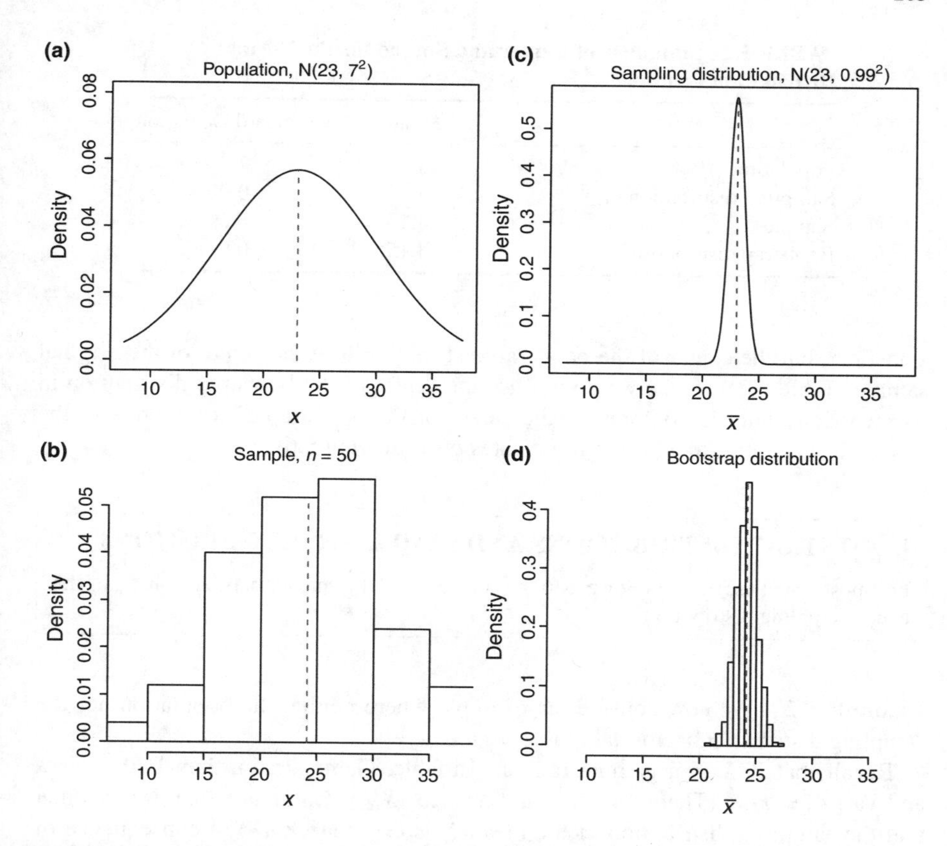

FIGURE 5.2 Sampling and bootstrap distributions of the mean for $N(23, 7^2)$. (a) The population distribution, $N(23, 7^2)$. (b) The distribution of one sample of size 50 from $N(23, 7^2)$. (c) The theoretical sampling distribution of $\bar{X}$, $N(23, 7^2/50)$. (d) The bootstrap distribution. Vertical lines mark the means.

In this example, we use software to run the algorithm on page 101 drawing 1000 resamples of size 50 from the original sample, and computing the mean of each resample. The bootstrap distribution is then the distribution of these 1000 resample means. The bootstrap standard error is the standard deviation of these 1000 resample means. We will also look at the center and shape of the bootstrap distribution.

From Figure 5.2, we can see that the bootstrap distribution has roughly the same spread and shape as the theoretical sampling distribution, but the centers are different. □

This example illustrates some important features of the bootstrap that hold for other statistics besides the mean: the bootstrap distribution of a particular statistic $\hat{\theta}$ has approximately the same spread and shape as the sampling distribution of the

TABLE 5.2 Summary of Center and Spread for the Normal Distribution Example

	Mean	Standard Deviation
Population	23	7
Sampling distribution of $\bar{X}$	23	0.99
Sample	24.13	6.69
Bootstrap distribution	24.15	0.92

statistic $\hat{\theta}$, but the center of the bootstrap distribution is at the center of the original sample (Table 5.2). Hence we do not use the center of the bootstrap distribution in its own right, but we do compare the center of the bootstrap distribution with the observed statistic; if they differ, it indicates *bias* (Section 5.6).

BOOTSTRAP DISTRIBUTIONS AND SAMPLING DISTRIBUTIONS

For most statistics, bootstrap distributions approximate the spread, bias, and shape of the actual sampling distribution.

Example 5.2 We now consider an example where neither the population nor the sampling distribution is normal (Table 5.3).

Recall that if X is a gamma random variable, Gamma(r, λ), then $\mathrm{E}[X] = r/\lambda$ and $\mathrm{Var}[X] = r/\lambda^2$ (Theorem B.10). Let $X_1, \ldots, X_n \sim$ Gamma(r, λ). It is a fact that that the sampling distribution of the mean $\bar{X}$ is Gamma(nr, $n\lambda$) (a consequence of Theorem B.11 and Proposition B.3).

We draw a random sample of size $n = 16$ from the gamma distribution Gamma(1,1/2) (population mean 2, standard deviation 2.) Figure 5.3a and c shows a graph of the population and the distribution of one random sample ($\bar{x} = 2.73$ and $s = 2.61$), respectively. Figure 5.3b displays the theoretical sampling distribution of the means, Gamma(16, 8). Figure 5.3d displays the bootstrap distribution, based on 10,000 resamples.

Even though the distribution of the sample does not exactly match the population distribution, the bootstrap distribution is similar to the sampling distribution: it has

TABLE 5.3 Summary of Center and Spread for the Gamma Distribution Example

	Mean	Standard Deviation
Population	2	2
Sampling distribution of $\bar{X}$	2	0.5
Sample	2.74	2.61
Bootstrap distribution	2.73	0.642

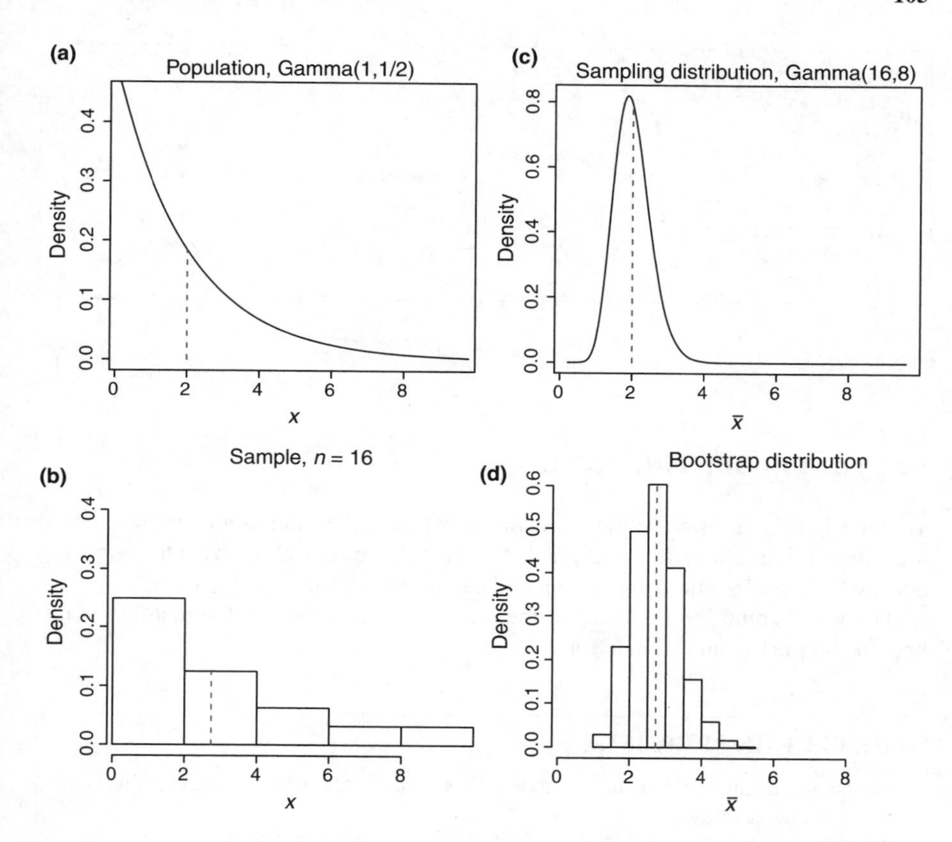

FIGURE 5.3 Sampling and bootstrap distribution from a gamma distribution. (a) The population distribution Gamma(1, 1/2). (b) A single sample of size 16 from Gamma(1, 1/2). (c) The theoretical sampling distribution of $\bar{X}$. (d) The bootstrap distribution for means of size 16, drawn from the sample.

roughly the same shape, a slightly larger spread (because the data have a slightly larger standard deviation than does the population), and the mean of the bootstrap distribution matches the empirical distribution rather than the population.

R Note:

Draw a random sample of size 16 from Gamma(1, 1/2):

```
my.sample <- rgamma(16, 1, 1/2)
```

The following simulates a bootstrap distribution based on 10^5 resamples.

```
N <- 10^5
```

```
my.boot <- numeric(N)
for (i in 1:N)
{
  x <- sample(my.sample, 16, replace = TRUE) # draw resample
  my.boot[i] <- mean(x)           # compute mean, store in my.boot
}
hist(my.boot)
mean(my.boot)
sd(my.boot)
```

□

5.2 THE PLUG-IN PRINCIPLE

We have hinted that we use the bootstrap to estimate the sampling distribution or at least some things about the sampling distribution. Let us talk about what the bootstrap does, why it works, and what we can and cannot do with it.

The idea behind the bootstrap is the *plug-in principle*—that if something is unknown, we plug in an estimate for it.

THE PLUG-IN PRINCIPLE

To estimate a parameter, a quantity that describes the population, use the statistic that is the corresponding quantity for the sample.

This principle is used all the time in statistics. For example, the standard error for $\bar{X}$ calculated from i.i.d. observations from a population with standard deviation σ is $\sigma/\sqrt{n}$; when σ is unknown, we plug in an estimate s to obtain the usual standard error $s/\sqrt{n}$.

What is different in the bootstrap is that we plug in an estimate for the whole population, not just for a numerical summary of the population. We use the observed data as an estimate of the whole population; we will come to this in Section 5.2.1, and alternatives in Chapter 11, but for now we will continue with the main idea, that we plug in an estimate and what follows from that.

Our goal is to estimate a sampling distribution of some statistic. The sampling distribution depends on

1. the underlying population(s),
2. the sampling procedure (e.g., sampling with or without replacement), and
3. the statistic, such as $\bar{X}$.

Figure 5.4 contains a diagram of this process.

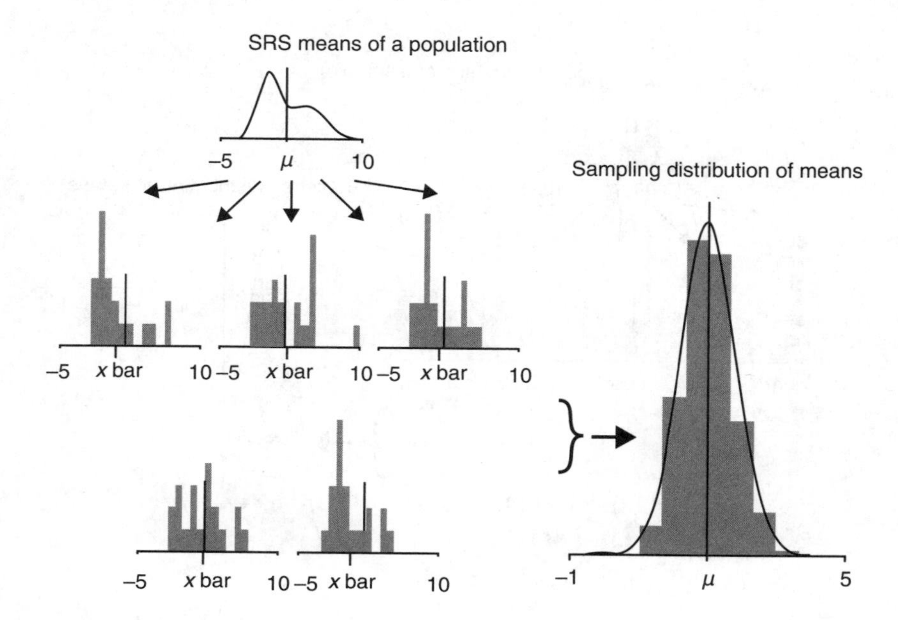

FIGURE 5.4 Diagram of the process of creating a sampling distribution. Many (infinitely many) samples are drawn from the population, a statistic like $\bar{X}$ is calculated for each. The distribution of the statistics is the sampling distribution.

The sampling distribution of a statistic is the result of drawing many samples from the population and calculating the statistic for each. The problem in most statistical applications is that the population is unknown.

The bootstrap principle is to plug in an estimate for the population and then mimic the real-life sampling procedure and statistic calculation (Figure 5.5). The bootstrap distribution depends on

1. an estimate for the population(s),
2. the sampling procedure, and
3. the statistic, such as $\bar{X}$.

In this chapter, we use the empirical distribution as an estimate for the population; that is, we use the empirical cumulative distribution function (Section 2.5) as an estimate for the unknown cumulative distribution function. This corresponds to sampling from the data. Let us look at this more closely.

5.2.1 Estimating the Population Distribution

We are using the data as an estimate for the population. Let us look more carefully at this; we will also see other alternatives in Chapter 11.

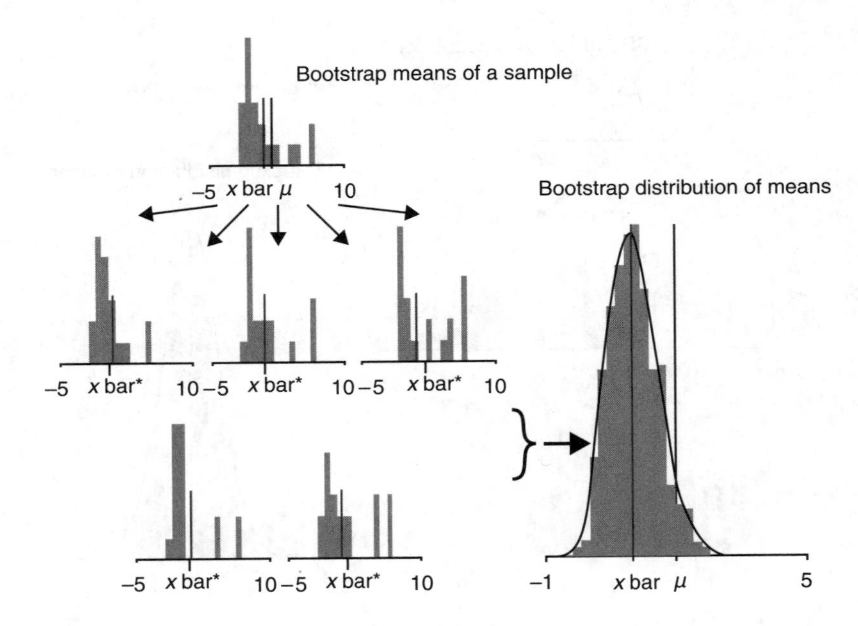

FIGURE 5.5 Diagram of the process of creating a bootstrap distribution. This is like Figure 5.4, except that the original data take the place of the population. We draw many samples from the original data, calculate $\bar{X}^*$ or another statistic for each, and collect the statistics to form the bootstrap distribution.

Let F and f denote the cdf and pdf for some unknown distribution and let $x_1, x_2, \ldots, x_n$ denote a random sample.

If we were willing to make assumptions about the population, say that it followed an exponential distribution, we could estimate the parameter λ from the data and then draw bootstrap samples from an exponential distribution with the estimated λ. This would be a parametric bootstrap, discussed in Section 11.2.

But most often when bootstrapping, we want to make as few assumptions as possible about the population. We want the data to tell us what it can, not introduce bias by making assumptions that may be wrong. So we resort to the empirical distribution, introduced in Section 2.5:

$$\hat{F}(s) = \frac{1}{n}\{\text{number of points } \leq s\}.$$

This is a discrete distribution, with probability $1/n$ at each of the observed data points, and empirical mass function

$$\hat{f}(s) = \frac{1}{n}\{\text{number of points } = s\}.$$

For instance, if the sample is $5, 5, 6, 10, 11, 11, 11, 12$, then $\hat{f}(5) = 2/8$, $\hat{f}(6) = 1/8$, $\hat{f}(10) = 1/8$, $\hat{f}(11) = 3/8$, $\hat{f} = 1/8$, and $\hat{f}(s) = 0$ for all other values s.

For most bootstrap applications, we never bother to define or write down $\hat{F}$ and $\hat{f}$—instead we just need to know how to draw samples, which we do by sampling from the original observations, with equal probabilities $1/n$ for each.

In some cases we need the mean and variance of $\hat{F}$. For comparison, the mean for F is

$$E_F[X] = \mu_F = \int_{-\infty}^{\infty} x\, f(x)\, dx$$

or

$$E_F[X] = \mu_F = \sum_x x\, f(x),$$

depending whether the population is continuous or discrete, where the subscript F indicates expected value based on F. Since $\hat{F}$ is discrete, we calculate the expected value using summation,

$$\begin{aligned} E_{\hat{F}}[X] &= \mu_{\hat{F}} \\ &= \sum_x x\, \hat{f}(x) \\ &= \sum_{i=1}^{n} x_i(1/n) = \bar{x}. \end{aligned}$$

Similarly, the population variance under $\hat{F}$ is

$$\begin{aligned} \text{Var}_{\hat{F}}[X] &= \sigma^2_{\hat{F}} \\ &= E_{\hat{F}}[(X - \mu_{\hat{F}})^2] \\ &= \sum_{i=1}^{n} (x_i - \bar{x})^2(1/n). \end{aligned}$$

This is like the sample variance s^2 (Equation 2.2), but with a denominator of n instead of $n - 1$.

5.2.2 How Useful Is the Bootstrap Distribution?

A fundamental question is how well the bootstrap distribution approximates the sampling distribution. We discuss this question in greater detail in Section 5.8, but note a few key points here.

First, the statistics that we bootstrap are generally *estimators*, statistics that estimate a parameter. For example, $\bar{X}$ is an estimator for μ, whereas a chi-squared test statistic (Equation 3.3) is not an estimator.

Definition 5.1 If $X, X_2, \ldots, X_n$ are random variables from a distribution with parameter θ and $g(X_1, X_2, \ldots, X_n)$ an expression used to estimate θ, then we call this function an *estimator*. ||

For most common estimators and under fairly general distribution assumptions, the following need to be noted:

Center The center of the bootstrap distribution is *not* an accurate approximation for the center of the sampling distribution. For example, the center of the bootstrap distribution for $\bar{X}$ is centered at approximately $\bar{x} = \mu_{\hat{F}}$, the mean of the sample, whereas the sampling distribution is centered at μ.

Spread The spread of the bootstrap distribution does reflect the spread of the sampling distribution.

Bias The bootstrap bias estimate (see Section 5.6) does reflect the bias of the sampling distribution. Bias occurs if a sampling distribution is not centered at the parameter.

Skewness The skewness of the bootstrap distribution does reflect the skewness of the sampling distribution.

The first point bears emphasis. It means that *the bootstrap is not used to get better parameter estimates* because the bootstrap distributions are centered around statistics $\hat{\theta}$ calculated from the data (e.g., $\bar{x}$) rather than the unknown population values (e.g., μ). Drawing thousands of bootstrap observations from the original data is not like drawing observations from the underlying population, it does not create new data.

Instead, the bootstrap sampling is useful for *quantifying the behavior of a parameter estimate*, such as its standard error, skewness, bias, or for calculating confidence intervals.

Example 5.3 Arsenic is a naturally occurring element in the groundwater of Bangladesh. However, much of this groundwater is used for drinking water by rural populations, so arsenic poisoning is a serious health issue. Figure 5.6 displays the distribution of arsenic concentrations from 271 wells in Bangladesh.[1]

The sample mean and standard deviation are $\bar{x} = 125.31$ and $s = 297.98$, respectively (measured in micrograms per liter).

We draw resamples of size 271 with replacement from the data and compute the mean for each resample. Figure 5.7 shows a histogram and normal quantile plot of the bootstrap distribution. The bootstrap distribution looks quite normal, with some skewness. This is the Central Limit Theorem (CLT) at work—when the sample size is large enough, the sampling distribution for the mean is approximately normal, even if the population is not normal.

[1]Data provided solely for illustrative purposes and to enable statistical analysis. Full data are available from the British Geological Survey web site `http://www.bgs.ac.uk/arsenic/bphase2/datadownload.htm`.

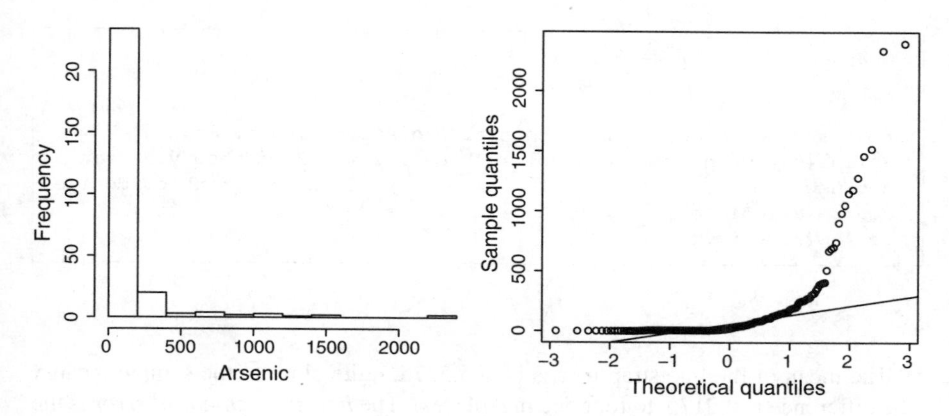

FIGURE 5.6 Arsenic levels in 271 wells in Bangladesh.

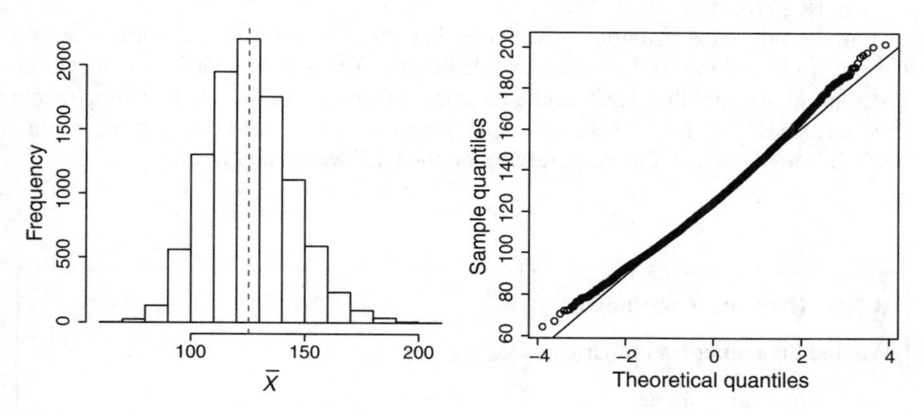

FIGURE 5.7 Histogram and QQ plot of the bootstrap distribution for the arsenic concentrations.

R Note:

Import the data set `Bangladesh` into `R`, then:

```
Arsenic <- Bangladesh$Arsenic
hist(Arsenic)
dev.new()              # New graphics device
qqnorm(Arsenic)
qqline(Arsenic)

n <- length(Arsenic)
N <- 10^4
arsenic.mean <- numeric(N)
for (i in 1:N)
{
  x <- sample(Arsenic, n, replace = TRUE)
```

```
  arsenic.mean[i] <- mean(x)
}

hist(arsenic.mean, main = "Bootstrap distribution of means")
abline(v = mean(Arsenic), col = "blue", lty = 2) # observed mean
dev.new()                                  # open new graphics device
qqnorm(arsenic.mean)
qqline(arsenic.mean)
```

The mean of the bootstrap means is 125.5375, quite close to the sample mean $\bar{x}$ (the difference is 0.2176, to four decimal places). The *bootstrap standard error* is the standard deviation of the bootstrap distribution; in this case, the bootstrap standard error is 18.2576.

For the normal distribution, we know that the 2.5 and 97.5 percentiles are at the mean plus or minus 1.96 standard deviations. But for this particular bootstrap distribution, we find that 1.5% of the resample means are below the bootstrap mean minus 1.96SE, and 3.4% of the resample means are above the bootstrap mean plus 1.96SE (see below). In this case, relying on the CLT would be inaccurate.

R Note (Arsenic, Continued):

We find the numeric summaries:

```
> mean(arsenic.mean)                  # bootstrap mean
[1] 125.5375
> mean(arsenic.mean)-mean(Arsenic)   # bias
[1] 0.2175773
> sd(arsenic.mean)                    # bootstrap SE
[1] 18.25759
```

Compute the points that are 1.96 standard errors from the mean of the bootstrap distribution:

```
> 125.5375-1.96*18.25759              # mark of 1.96SE from mean
[1] 89.75262
> 125.5375+1.96*18.25759
[1] 161.3224
> sum(arsenic.mean > 161.3224)/N
[1] 0.0337
> sum(arsenic.mean < 89.75262)/N
[1] 0.0153
```

□

5.3 BOOTSTRAP PERCENTILE INTERVALS

The sample mean $\bar{x}$ gives an estimate of the true mean μ, but it probably does not hit it exactly. It would be nice to have a *range* of values for the true μ that we are 95% sure includes the true μ.

In the North Carolina birth weights case study, the bootstrap distribution (Figure 5.1) shows roughly how sample means vary for samples of size 1009. If most of the sample means are concentrated within a certain interval of the bootstrap distribution, it seems reasonable to assume that the true mean is most likely somewhere in that same interval. Thus, we can construct what is called a 95% confidence interval by using the 2.5 and 97.5 percentiles of the bootstrap distribution as endpoints. We would then say that we are 95% confident that the true mean lies within this interval. These are *bootstrap percentile confidence intervals*.[2]

BOOTSTRAP PERCENTILE CONFIDENCE INTERVALS

The interval between 2.5 and 97.5 percentiles of the bootstrap distribution of a statistic is a 95% *bootstrap percentile confidence interval* for the corresponding parameter.

For the NC birth weights, the interval marked by the 2.5 and 97.5 percentiles is (3419, 3478). Thus, we would state that we are 95% confident that the true mean weight of NC babies born in 2004 is between 3419 and 3478 g.

In the arsenic example, the 2.5% and 97.5% points of the bootstrap distribution give us the interval (92.9515, 164.4418), so we are 95% confident that the true mean arsenic level is between 92.95 and 164.44 μg/L. Note that with $\bar{x} = 125.5375$, this interval can be written $(\bar{x} - 32.586, \bar{x} + 38.9043)$; in particular, this interval is *not* symmetric about the mean, reflecting the asymmetry of the bootstrap distribution.

R Note:

```
> quantile(arsenic.mean, c(0.025, 0.975))
    2.5%    97.5%
 92.9515 164.4418
```

The arsenic data illustrate an interesting point. A good confidence interval for the mean need not necessarily be symmetric: an endpoint will be farther from the sample mean in the direction of any outliers. A confidence interval is an insurance policy: rather than relying on a single statistic, the sample mean, as an estimate of μ, we give a range of plausible values for μ. We can see that there are some extremely

[2]We will discuss the logic of confidence intervals more formally in Chapter 7.

large arsenic measurements: of the 271 observations, 8 are above 1000 μg/L and 2 are above 2200 μg/L (remember, the sample mean is only 125.31!). What we do not know is just how huge arsenic levels in the population can be, or how many huge ones there are. It could be that huge observations are *underrepresented* in our data set. In order to protect against this—that is, to have only a 2.5% chance of missing a true big mean, the interval of plausible values for μ must stretch far to the right. Conversely, there is less risk of missing the true mean on the low side, so the left endpoint need not be as far away from the mean.

5.4 TWO SAMPLE BOOTSTRAP

We now turn to the problem of comparing two samples. In general, bootstrapping should mimic how the data were obtained. So if the data correspond to independent samples from two populations, we should draw two samples that way. Then we proceed to compute the same statistic comparing the samples as for the original data, for example, difference in means or ratio of proportions.

BOOTSTRAP FOR COMPARING TWO POPULATIONS

Given independent samples of sizes m and n from two populations,

1. Draw a resample of size m with replacement from the first sample and a separate resample of size n from the second sample. Compute a statistic that compares the two groups, such as the difference between the two sample means.
2. Repeat this resampling process many times, say 10,000.
3. Construct the bootstrap distribution of the statistic. Inspect its spread, bias, and shape.

Example 5.4 A high school student was curious about the total number of minutes devoted to commercials during any given half-hour time period on basic and extended cable TV channels (Rodgers and Robinson (2004)). Table 5.4 contains some data that he collected to study this.

The means of the basic and extended channel commercial times are 9.21 and 6.87 min, respectively, so on average, commercials on basic channels are 2.34 min longer than on extended channels. Is this difference of 2.34 min statistically significant? The poor student could only stand to watch 10 h of random TV, so his observations may not accurately represent the TV universe.

TABLE 5.4 Length of Commercials (Minutes) During Random Half-Hour Periods from 7a.m. to 11p.m.

Basic	7.0	10.0	10.6	10.2	8.6	7.6	8.2	10.4	11.0	8.5
Extended	3.4	7.8	9.4	4.7	5.4	7.6	5.0	8.0	7.8	9.6

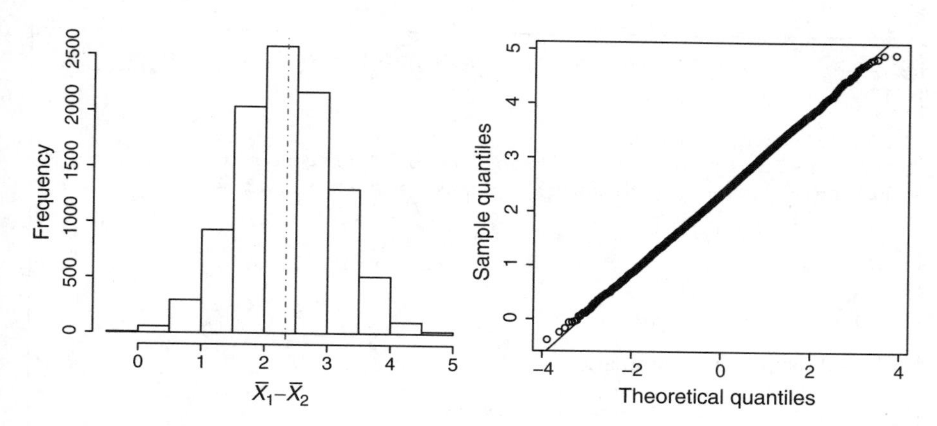

FIGURE 5.8 Histogram and normal quantile plot of the bootstrap distribution for the difference in mean advertisement time in basic versus extended TV channels. The vertical line in the histogram marks the observed mean difference.

The original data are simple random samples of size 10 from two populations. We draw a bootstrap sample from the basic channel data and independently draw a bootstrap sample from the extended channel data, compute the means for each sample, and take the difference.

Figure 5.8 shows the bootstrap distribution of the difference of sample means. As in the single-sample case, we see that the bootstrap distribution is approximately normal and centered at the original statistic (the difference in sample means). We also get a quick idea of how much the difference in sample means varies due to random sampling. We may quantify this variation by computing the bootstrap standard error, which is 0.76. Again, the bootstrap standard error is the standard error of the sampling distribution.

The right panel of Figure 5.8 shows a normal quantile plot for the bootstrap distribution: the distribution is very close to normal.

R Note:

Import the TV data into R, then

```
times.Basic <- subset(TV, select = Times,
                      subset = Cable == "Basic", drop = T)
times.Ext <- subset(TV, select = Times,
                    subset = Cable == "Extended", drop = T)
N <- 10^4
times.diff.mean <- numeric(N)
for (i in 1:N)
{
  Basic.sample <- sample(times.Basic, 10, replace = TRUE)
  Ext.sample <- sample(times.Ext, 10, replace = TRUE)
```

```
  times.diff.mean[i] <- mean(Basic.sample) - mean(Ext.sample)
}

hist(times.diff.mean,
      main = "Bootstrap distribution of difference in means")
abline(v = mean(times.Basic) - mean(times.Ext), col = "blue",
       lty = 2)

dev.new()                     # Open new graphics device
qqnorm(times.diff.mean)
qqline(times.diff.mean)
```

We find the numeric summaries:

```
> mean(times.Basic) - mean(times.Ext)
[1] 2.34
> mean(times.diff.mean)
[1] 2.344409
> sd(times.diff.mean)
[1] 0.747343
> quantile(times.diff.mean, c(0.025, 0.975))
 2.5% 97.5%
 0.89  3.80
> mean(times.diff.mean) - (mean(times.Basic) -
                           mean(times.Ext)) # bias
[1] 0.004409
```

We will discuss bias in Section 5.6.

The 95% bootstrap percentile confidence interval for the difference in means (basic−extended) is (0.89, 3.80). Thus, we are 95% confident that commercial times on basic channels are, on average, between 0.89 and 3.80 min longer than on extended channels (per half-hour time periods).

We can also conduct a permutation test of the hypothesis that the mean commercial times for the two cable options are the same versus the hypothesis that mean times are not. Figure 5.9 shows the permutation distribution for the difference in mean advertisement time between basic and extended TV channels.

Recall that in permutation testing, we sample *without* replacement from the pooled data. The permutation distribution corresponds to sampling in a way that is consistent with the null hypothesis that the population means are the same. Thus, the permutation distribution is centered at 0. But in bootstrapping, we sample *with* replacement from the individual sample. However, the bootstrap has no restriction in regard to any null hypothesis, so its distribution is centered at the original difference in means.

The permutation distribution is used for a single purpose: calculate a P-value to see how extreme an observed statistic is if the null hypothesis is true. The bootstrap is used for estimating standard errors and for answering some other questions we will raise below.

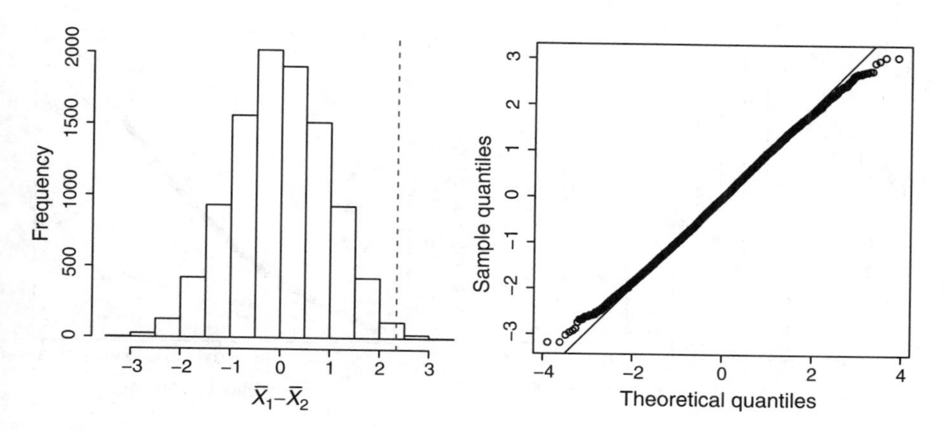

FIGURE 5.9 Histogram and normal quantile plot of the permutation distribution for the difference in mean advertisement time in basic versus extended TV channels. The vertical line in the histogram marks the observed mean difference.

The permutation test for this example results in a P-value of 0.0055; thus, we conclude that the mean commercial times are not the same between the two types of cable TV channels. □

Example 5.5 We return again to the Verizon example in Section 3.3. The distribution of the original data is shown in Figure 3.4, the permutation distribution for the difference in means is shown in Figure 3.5, and a permutation test of the difference in medians and trimmed means shown in Figure 3.6.

The bootstrap distribution for the larger ILEC data set ($n = 1664$) is shown in Figure 5.10. The distribution is centered around the sample mean of 8.4, has a relatively narrow spread primarily due to the large sample size, with a bootstrap SE of

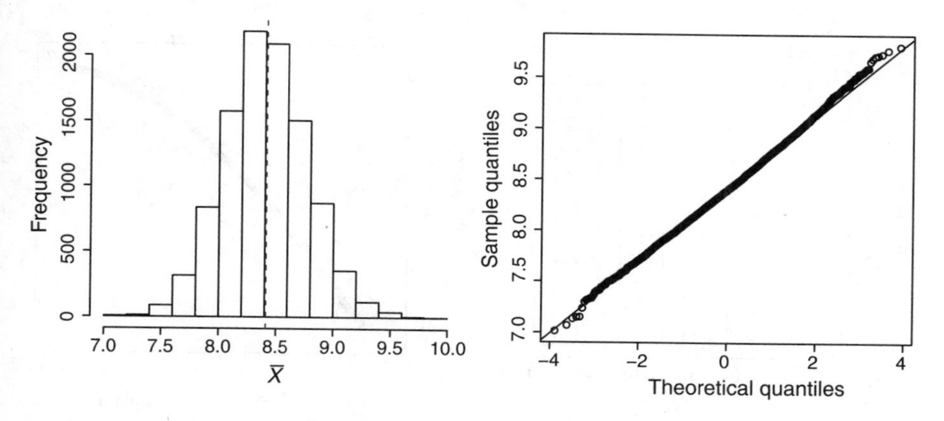

FIGURE 5.10 Bootstrap distribution for the sample mean of the Verizon ILEC data set, $n = 1664$. The vertical line is at the observed mean.

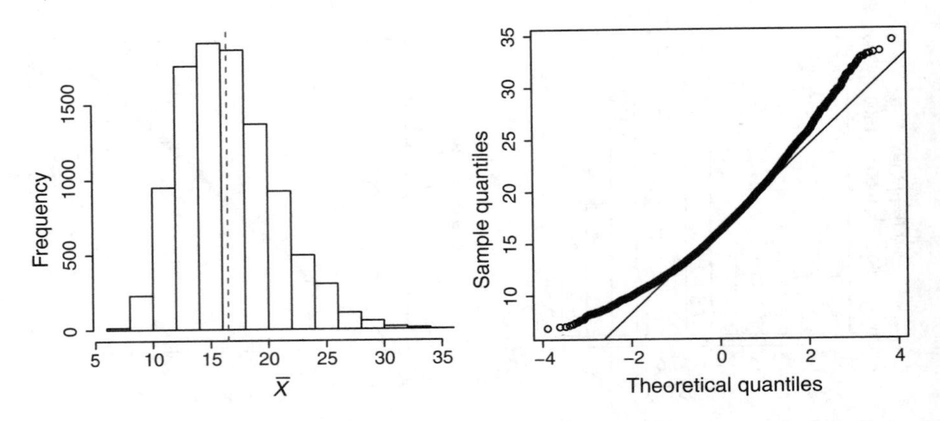

FIGURE 5.11 Bootstrap distribution for the sample mean of the Verizon CLEC data set, $n = 23$. The vertical line is at the observed mean.

0.36 and a 95% bootstrap percentile interval of (7.7, 9.1). The distribution is roughly symmetric, with little skewness.

The bootstrap distribution for the smaller CLEC data set ($n = 23$) is shown in Figure 5.11. The distribution is centered around the sample mean of 16.5, has a much larger spread due to the small sample size, with a bootstrap SE of 3.98 and a 95% bootstrap percentile interval of (10.1, 25.6). The distribution is very skewed.

The bootstrap distribution for the difference in means is shown in Figure 5.12. Note the strong skewness in the distribution. The mean of the bootstrap distribution is -8.1457 with a standard error of 4.0648. A 95% bootstrap percentile confidence interval for the difference in means (ILEC–CLEC) is given by $(-17.1838, -1.6114)$ and so we would say that with 95% confidence, the repair times for ILEC customers are, on average, 1.61–17.18 h shorter than the repair times for CLEC customers. □

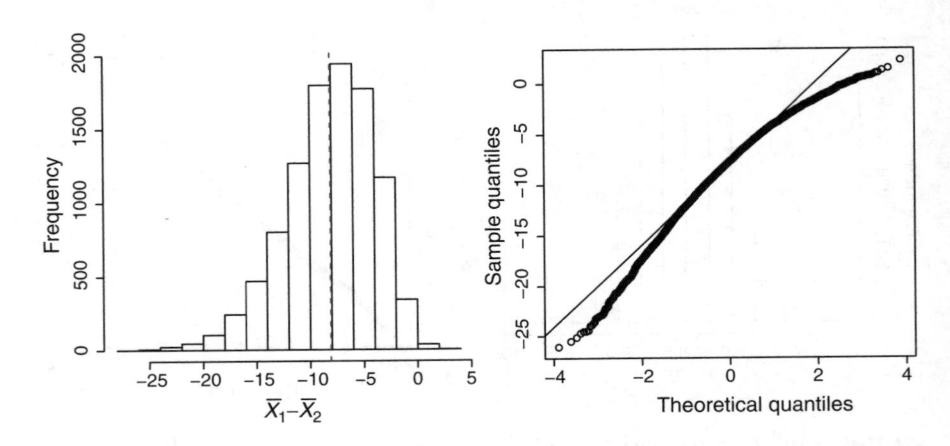

FIGURE 5.12 Bootstrap distribution for the difference in means. The vertical line in the histogram is at the observed mean difference.

TABLE 5.5 Partial View of Price Data in File `Cameras`

Item	J & R	B & H
Canon PowerShot A3000	129.99	149.99
Canon PowerShot A495	99.88	96.95
Casio EX-FC150	241.88	241.19
Kodak EasyShare C142	74.94	79.59
⋮		

5.4.1 The Two Independent Populations Assumption

Savvy consumers will often compare prices at different stores before making a purchase. Are some stores really better bargains consistently than others? We compiled the prices of a sample of point-and-shoot digital cameras from two electronic stores with an online presence, J & R and B & H (Table 5.5). The mean price of the cameras was \$155.42 at J & R and \$152.62 at B& H. Does this imply cameras are more expensive at J & R, or could the difference in mean price (\$2.81) be chance variation?

Now, it may be tempting to proceed as in the TV commercials or Verizon repair times examples, looking at the prices as coming from two populations, B & H and J & R. But note that the data are *not independent*! For each camera priced at B & H, we matched it with a price for the *same* camera at JR. Thus, the data are called *matched pairs* or *paired data*.

In this case, for each camera, we compute the difference in price between the two stores (J & R price − B & H price). We then have one variable—the price differences—and we are back to the one-sample setting given on page 101. The price differences are shown in Figure 5.13.

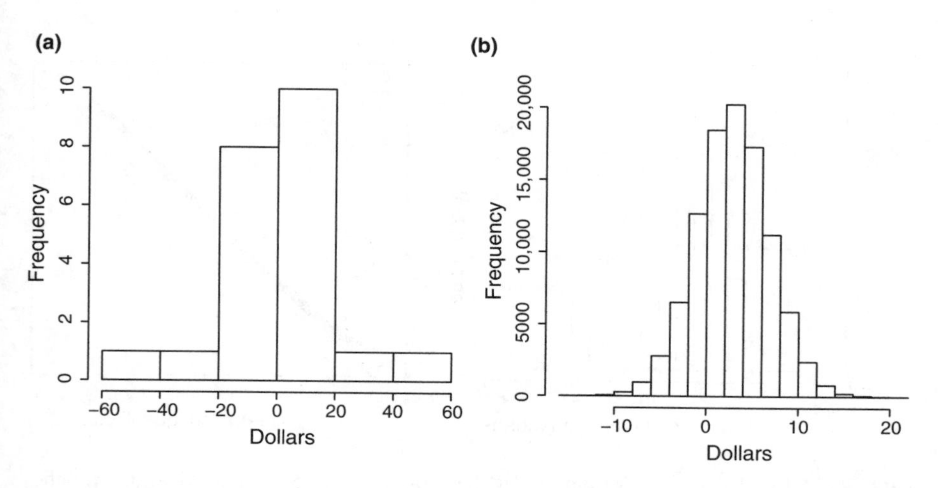

FIGURE 5.13 (a) Distribution of price differences. (b) Bootstrap distribution of price differences. The vertical line is at the observed mean price difference of \$2.81.

Performing a one sample bootstrap with 10^5 resamples, we find a 95% bootstrap percentile interval for the mean price difference to be $(-4.91, 10.62)$. Since 0 is contained in the interval, we cannot conclude that the mean prices for digital point-and-shoot cameras differ between the two stores.

5.5 OTHER STATISTICS

As with permutation testing, when bootstrapping, we are not limited to simple statistics like the simple mean. Once we have drawn a bootstrap sample, we can calculate any statistic for that sample.

For example, instead of a sample mean, we can use more robust statistics that are less sensitive to extreme observations. Figure 5.14 shows the bootstrap distribution for the difference in trimmed means, in this case 25% trimmed means, also known as the *midmean*, the mean of the middle 50% of observations. Compared to the bootstrap difference in ordinary means (Figure 5.12), this distribution has a much smaller spread.

The bootstrap procedure may be used with a wide variety of statistics—means, medians, trimmed means, correlation coefficients, and so on—using the same procedure. This is a major advantage of the bootstrap. It allows statistical inferences such as confidence intervals to be calculated even for statistics for which there are no easy formulas. It offers hope of reforming statistical practice—away from simple but nonrobust estimators like a sample mean or least-squares regression (Chapter 9), in favor of robust alternatives.

Example 5.6 In the Verizon data, rather than looking at the difference in means, suppose we look at the ratio of means. The sample ratio is 0.51, so for ILEC customers, repair times are about half of that for CLEC customers.

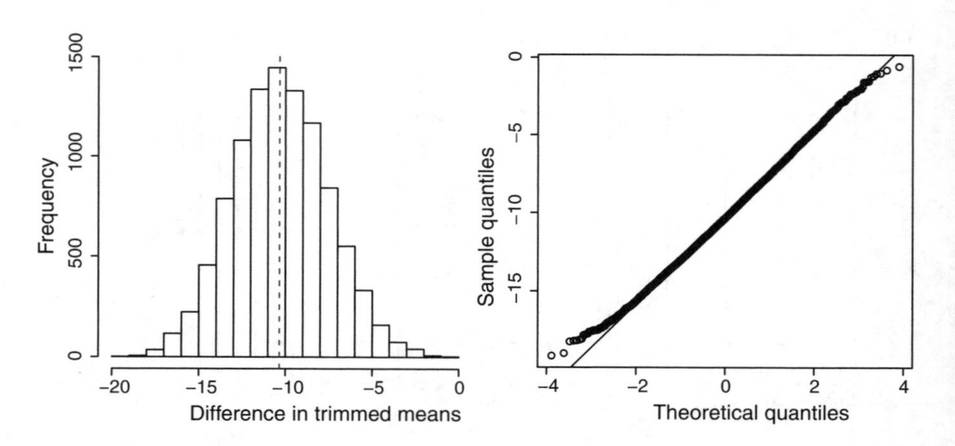

FIGURE 5.14 Bootstrap distribution for the difference in 25% trimmed means for the Verizon data.

R Note:

First create two vectors, one with the time information for the ILEC customers, one for the CLEC customers.

```
Time.ILEC <- subset(Verizon, select = Time, Group == "ILEC",
                    drop = T)
Time.CLEC <- subset(Verizon, select = Time, Group == "CLEC",
                    drop = T)
N <- 10^4
time.ratio.mean <- numeric(N)
for (i in 1:N)
{
  ILEC.sample <- sample(Time.ILEC, 1664, replace = TRUE)
  CLEC.sample <- sample(Time.CLEC, 23, replace = TRUE)
  time.ratio.mean[i] <- mean(ILEC.sample)/mean(CLEC.sample)
}

hist(time.ratio.mean,
     main="Bootstrap distribution of ratio of means")
abline(v=mean(time.ratio.mean, col = "red", lty = 2)
abline(v=mean(Time.ILEC)/mean(Time.CLEC), col = "blue", lty = 4)

dev.new()                    # open new graphics device
qqnorm(time.ratio.mean)
qqline(time.ratio.mean)
```

As in the difference of means example, the bootstrap distribution of the ratio of means exhibits skewness (Figure 5.15).

The 95% bootstrap percentile confidence interval for the ratio of means (ILEC/CLEC) is (0.3258, 0.8415), so with 95% confidence, the true mean repair

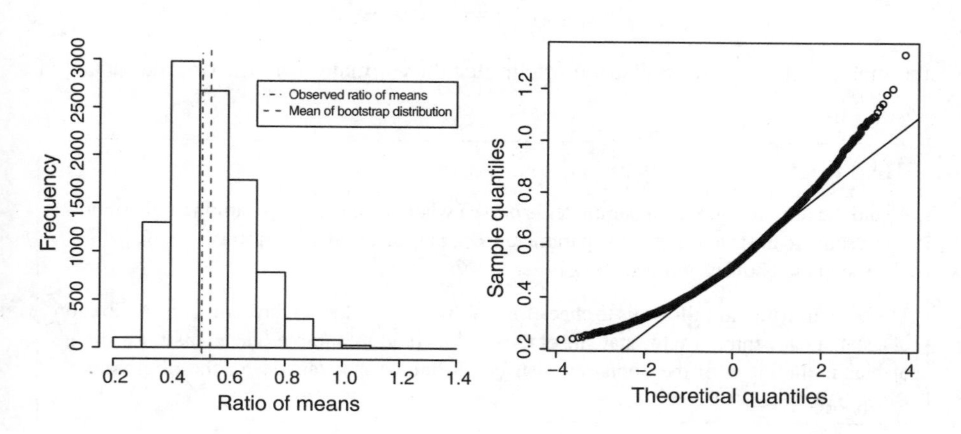

FIGURE 5.15 Bootstrap distribution for the ratio of means.

times for ILEC customers is between 0.33 and 0.84 times less than that for CLEC customers.

R Note (Verizon, Continued):

For the numeric summaries:

```
> mean(time.ratio.mean)
[1] 0.53878
> sd(time.ratio.mean)
[1] 0.1349371
> quantile(time.ratio.mean, c(0.025, 0.975))
  2.5%      97.5%
  0.3258330 0.8415138
> mean(time.ratio.mean) - mean(Time.ILEC)/mean(Time.CLEC) # bias
[1] 0.0292674
```

We will discuss bias in the next section.

□

5.6 BIAS

An estimator $\hat{\theta}$ is biased if, on average, it tends to be too high or too low, relative to the true value of θ. Formally, this is defined using expected values:

Definition 5.2 The *bias* of an estimator $\hat{\theta}$ is

$$\text{Bias}[\hat{\theta}] = \text{E}\left[\hat{\theta}\right] - \theta.$$

The bootstrap estimate of bias is

$$\text{Bias}_{\text{boot}}[\hat{\theta}^*] = \text{E}\left[\hat{\theta}^*\right] - \hat{\theta},$$

the mean of the bootstrap distribution, minus the estimate from the original data. ||

BIAS

A statistic used to estimate a parameter is *biased* when the mean of its sampling distribution is not equal to the true value of the parameter. The bias of a statistic $\hat{\theta}$ is $\text{Bias}[\hat{\theta}] = \text{E}\left[\hat{\theta}\right] - \theta$.
A statistic is **unbiased** if its bias is zero.

The bootstrap method allows us to check for bias by seeing whether the bootstrap distribution of a statistic is centered at the statistic of the original random sample. The bootstrap estimate of bias is the mean of the bootstrap distribution minus the statistic for the original data, $\text{Bias}_{\text{boot}}[\hat{\theta}] = \hat{\text{E}}[\hat{\theta}^*] - \hat{\theta}$.

We have already proven (Theorem A.5) that the sample mean is an unbiased estimator of the population mean μ. In addition, the difference of sample means is also an unbiased estimator of the difference of population means. However, the ratio of sample means is not generally an unbiased estimator of the ratio of population means. The bootstrap distribution for the ratio of means has a long right tail, large observations that occur when the denominator is small. Consequently, the mean of the resample ratio of means is large, causing positive mean bias.

Let us compare the ratio of bias/SE for different examples. For the arsenic example (page 111), the ratio is only $0.2176/18.2576 = 0.0119$. For the TV example (page 115), the ratio is only $0.0044/0.7473 = 0.0059$, so the bias is less than 0.6% of the standard error. On the other hand, for the Verizon ratio of means (page 121), the ratio is $0.0293/0.1349 = 0.2172$, so the bias is about 22% of the standard error.

If the ratio of bias/SE exceeds ± 0.10, then it is large enough to potentially have a substantial effect on the accuracy of confidence intervals. In applications where accuracy matters, there are other more accurate bootstrap confidence intervals rather than the relatively quick-and-dirty bootstrap percentile intervals. (As it turns out, bootstrap percentile intervals are actually reasonably accurate for the ratio of means.)

The next example also shows noticeable bias.

Example 5.7 A major study of the association between blood pressure and cardiovascular disease found that 55 out of 3338 ($\hat{p}_1 = 0.0165$) men with high blood pressure died of cardiovascular disease during the study period, compared to 21 out of 2676 ($\hat{p}_2 = 0.0078$) with low blood pressure. The estimated *relative risk* is $\hat{\theta} = \hat{p}_1/\hat{p}_2 = 0.0165/0.0078 = 2.12$. Thus, we would say that the risk of cardiovascular disease for men with high blood pressure is 2.12 times greater than the risk for men with low blood pressure.

To bootstrap the relative risk, we draw samples of size $n_1 = 3338$ with replacement from the first group, independently draw samples of size $n_2 = 2676$ from the second group, and calculate the relative risk $\hat{\theta}^*$. In addition, we record the individual proportions $\hat{p}_1^*$ and $\hat{p}_2^*$. The bootstrap distribution for relative risk is shown in Figure 5.16. It is highly skewed, with a long right tail caused by denominator values relatively close to zero. The standard error, from a sample of 10^4 observations, is 0.6188. The theoretical (exhaustive) bootstrap standard error is undefined because some of the $n_1^{n_1} n_2^{n_2}$ bootstrap samples have $\hat{\theta}^*$ undefined: this occurs when the denominator $\hat{p}_2^*$ is zero; this is rare enough to ignore.

The average of the resample relative risks is larger than the sample relative risk, indicating bias. The estimated bias is $2.2107 - 2.10 = 0.1107$, so the ratio of bias to the standard error is 0.1784. While the bias does not appear large in the figure, this amount of bias can have a huge impact on formula-based confidence intervals. While the bootstrap percentile interval is fine, some common symmetric confidence intervals would miss by falling under the true value about twice as often as they should.

Figure 5.17 shows the joint bootstrap distribution of $\hat{p}_1^*$ and $\hat{p}_2^*$. Each point corresponds to one bootstrap resample, and the relative risk is the slope of the line

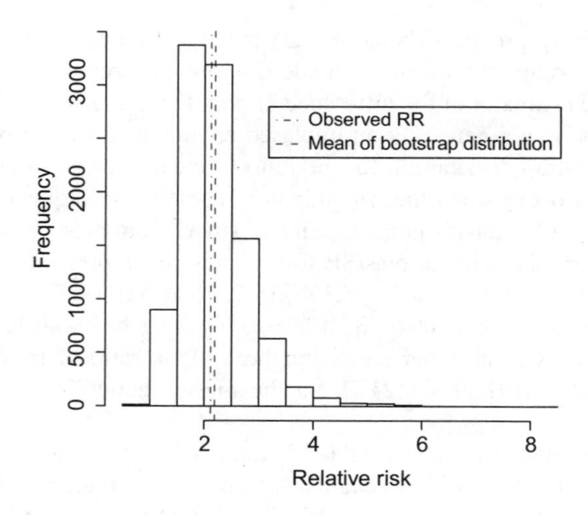

FIGURE 5.16 Bootstrap distribution of relative risk.

between the origin and the point. The original data are at the intersection of horizontal and vertical lines. The 95% bootstrap confidence interval for the true relative risk is (1.3118, 3.6877). Thus, for one bootstrap sample, if the relative risk satisfies $\hat{p}_1^*/\hat{p}_2^* < 1.3118$, then $\hat{p}_1^* < 1.3118\hat{p}_2^*$. Similarly, if $\hat{p}_1^*/\hat{p}_2^* > 3.6877$, then $\hat{p}_1^* > 3.6877\hat{p}_2^*$. This is shown by the points outside the region bounded by the dashed lines of slopes 1.3118 and 3.6877. □

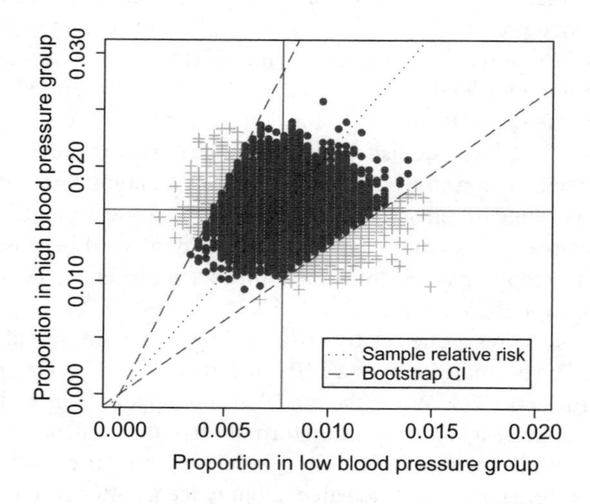

FIGURE 5.17 Bootstrapped proportions of the high blood pressure group against the proportions of the low blood pressure group.

5.7 MONTE CARLO SAMPLING: THE "SECOND BOOTSTRAP PRINCIPLE"

The second bootstrap "principle" is that the bootstrap is implemented by random sampling. This is not actually a principle but an implementation detail.

The name *Monte Carlo* dates from the 1940s, when physicists and applied mathematicians working on the Manhattan Project at Los Alamos Laboratory in New Mexico encountered difficult integrals with no closed form solutions. Stanislaw Ulam and John von Neumann proposed using computer simulations to estimate these integrals. Their conceptual leap was in using a random method to solve a deterministic problem. Because of the use of randomness, they named the method after the casino in Monaco.

Given that we are drawing i.i.d. samples of size n from the observed data, there are at most n^n possible samples ($\binom{2n-1}{n}$, if we disregard the order of observations), and ties in the data can further reduce the number of unique samples. In small samples, we could create all possible bootstrap samples, deterministically. In practice, n is usually too large for that to be feasible, so we use random sampling.

Let N be the number of bootstrap samples used, for example, $N = 10^4$. The resulting N resample statistic values represent a random sample of size N with replacement from the *theoretical bootstrap distribution* consisting of n^n values.

In some cases, we can calculate some aspects of the sampling distribution without simulation. When the statistic is the sample mean, for example, and for the ordinary one-sample bootstrap, the mean and standard deviation of the theoretical bootstrap distribution are $\bar{x}$ and $\hat{\sigma}/\sqrt{n}$, respectively, where $\hat{\sigma}^2 = 1/n \sum(x_i - \bar{x})^2$.

The use of Monte Carlo sampling adds additional unwanted variability that may be reduced by increasing the value of N. We discuss how large N should be in Section 5.9.

5.8 ACCURACY OF BOOTSTRAP DISTRIBUTIONS

How accurate is the bootstrap? This entails two questions:

- How accurate is the theoretical bootstrap?
- How accurately does the Monte Carlo implementation approximate the theoretical bootstrap?

SOURCES OF VARIATION IN A BOOTSTRAP DISTRIBUTION

Bootstrap distributions and conclusions based on them include two sources of random variation:

1. The original sample is chosen at random from the population.
2. Bootstrap resamples are chosen at random from the original sample.

We begin this section with a series of pictures intended to illustrate both questions. We conclude this section with a discussion of cases where the theoretical bootstrap is

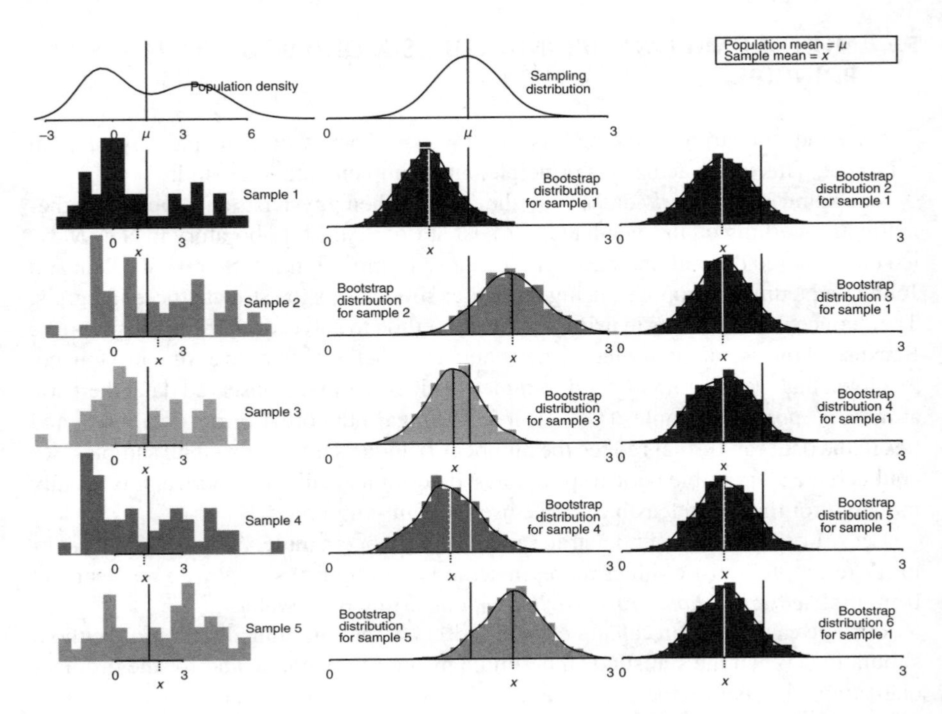

FIGURE 5.18 Bootstrap distribution for the mean, $n = 50$. The left column shows the population and five samples. The middle column shows the sampling distribution and bootstrap distributions from each sample. The right column shows five more bootstrap distributions from the first sample, with $N = 1000$ or $N = 10^4$.

not accurate and remedies. In Section 5.9, we return to the question of Monte Carlo accuracy.

5.8.1 Sample Mean: Large Sample Size

Figure 5.18 shows a population and five samples of size 50 from the population in the left column. The middle column shows the sampling distribution for the mean and bootstrap distributions from each sample, based on $N = 1000$ bootstrap samples. Each bootstrap distribution is centered at the statistic ($\bar{x}$) from the corresponding sample rather than being centered at the population mean μ. The spreads and shapes of the bootstrap distributions vary a bit but not a lot.

This informs what the bootstrap distributions may be used for. The bootstrap does not provide a better estimate of the population parameter μ, because no matter how many bootstrap samples are used, they are centered at $\bar{x}$ (plus random variation), not μ. On the other hand, the bootstrap distributions are useful for estimating the spread and shape of the sampling distribution.

The right column shows five more bootstrap distributions from the first sample, using $N = 1000$ resamples. These illustrate the Monte Carlo variation in the bootstrap. This variation is much smaller than the variation due to different original samples. For many uses, such as quick-and-dirty estimation of standard errors or approximate

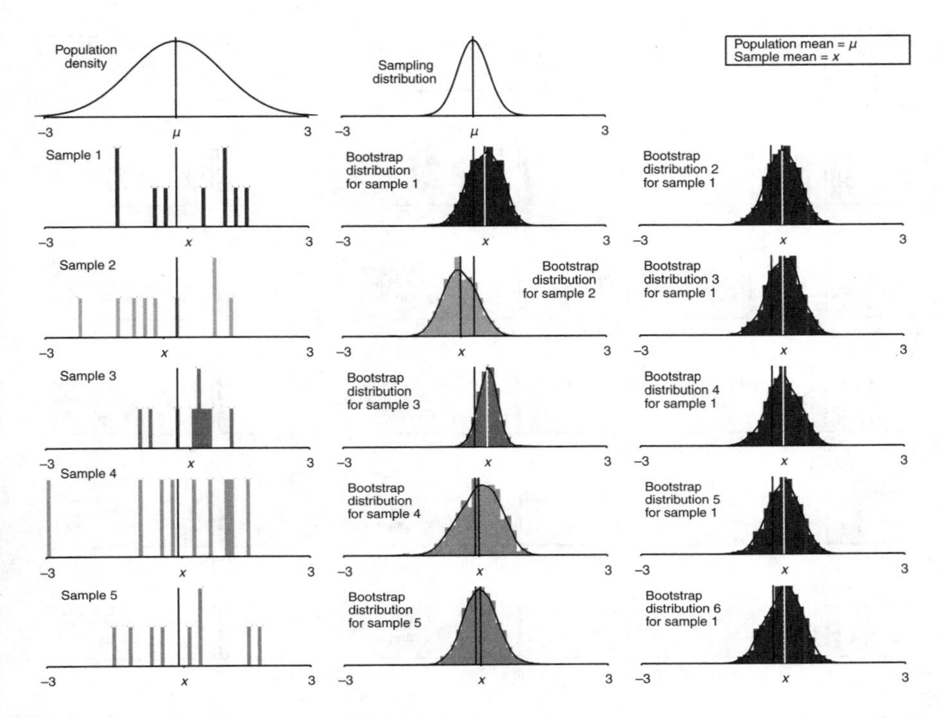

FIGURE 5.19 Bootstrap distributions for the mean, $n = 9$. The left column shows the population and five samples. The middle column shows the sampling distribution and bootstrap distributions from each sample. The right column shows five more bootstrap distributions from the first sample, with $N = 1000$ or $N = 10^4$.

confidence intervals, $N = 1000$ resamples is adequate. However, there is noticeable variability, particularly in the tails of the bootstrap distributions; so when accuracy matters, $N = 10^4$ or more samples should be used.

5.8.2 Sample Mean: Small Sample Size

Figure 5.19 is similar to Figure 5.18, but for a smaller sample size, $n = 9$ (and a different population). As before, the bootstrap distributions are centered at the corresponding sample means, but now the spreads and shapes of the bootstrap distributions vary substantially, because the spreads and shapes of the samples vary substantially. As a result, bootstrap confidence interval widths vary substantially (this is also true of nonbootstrap confidence intervals). As before, the Monte Carlo variation is small and may be reduced with more samples.

5.8.3 Sample Median

Now turn to Figure 5.20 where the statistic is the sample median. Here, the bootstrap distributions are poor approximations of the sampling distribution. In contrast, the sampling distribution is continuous, but the bootstrap distributions are discrete with the only possible values being values in the original sample (here n is odd). The

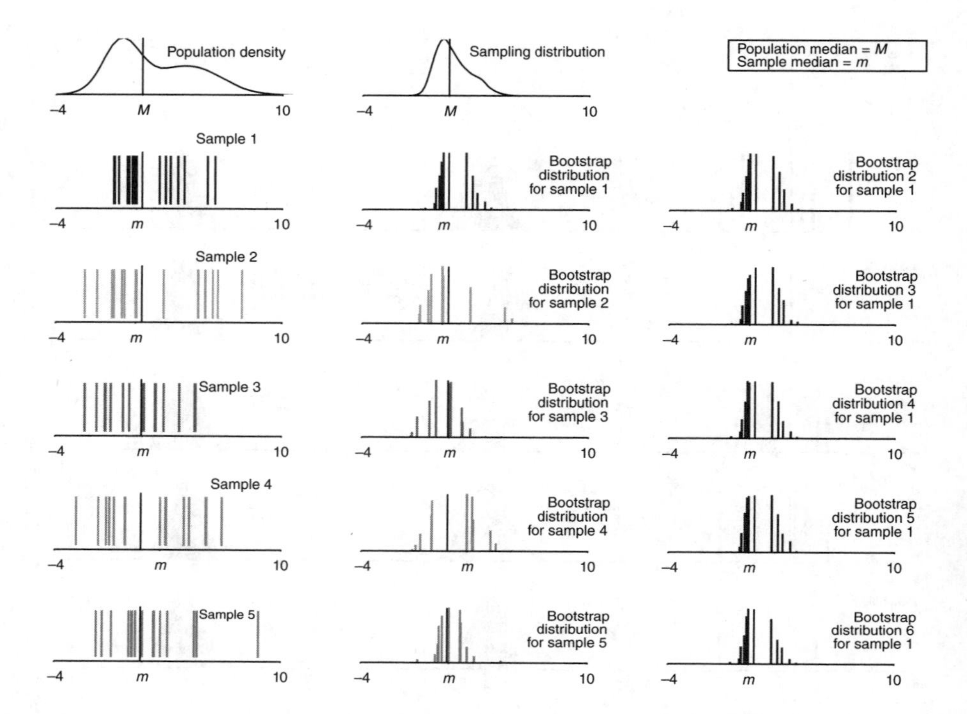

FIGURE 5.20 Bootstrap distributions for the median, $n = 15$. The left column shows the population and five samples. The middle column shows the sampling distribution and bootstrap distributions from each sample. The right column shows five more bootstrap distributions from the first sample.

bootstrap distributions are very sensitive to the sizes of gaps among the observations near the center of the sample (see Exercise 9).

The ordinary bootstrap tends not to work well for statistics such as the median or other quantiles that depend heavily on a small number of observations out of a larger sample.

VARIATION IN BOOTSTRAP DISTRIBUTIONS

For most statistics, almost all the variation in bootstrap distributions comes from the selection of the original sample from the population. Reducing this variation requires collecting a larger original sample.

Bootstrapping does not overcome the weakness of small samples as a basis for inference. Some bootstrap procedures are more accurate than others (we will discuss this later) and more accurate than common nonbootstrap procedures, but still they may not be accurate for very small samples. Use caution in any inference—including bootstrap inference—from a small sample.

The bootstrap resampling process using 1000 or more resamples introduces little additional variation, but for good accuracy use 10,000 or more.

5.9 HOW MANY BOOTSTRAP SAMPLES ARE NEEDED?

We suggested in Section 5.8 that 1000 bootstrap samples are enough for rough approximations, but more are needed for greater accuracy. We elaborate on this here. The focus here is on Monte Carlo accuracy—how well the usual random sampling implementation of the bootstrap approximates the theoretical bootstrap distribution.

A bootstrap distribution based on N random samples corresponds to drawing N observations with replacement from the theoretical bootstrap distribution.

Brad Efron, inventor of the bootstrap, suggested in 1993 that $N = 200$, or even as few as $N = 25$, suffices for estimating standard errors and that $N = 1000$ is enough for confidence intervals (Efron and Tibshirani (1993)).

We argue that more resamples are appropriate, on two grounds. First, those criteria were developed when computers were much slower; with faster computers it is much easier to take more resamples.

Second, those criteria were developed using arguments that combine the random variation due to the original sample with the random variation due to bootstrap sampling. We prefer to treat the data as given and look just at the variability due to bootstrap sampling. For typical 95% bootstrap percentile confidence intervals, to reduce Monte Carlo variability to the point that a supposed 95% confidence interval has a high probability of missing between 2.5% and $\pm$ 0.25% on each side requires about 15,000 bootstrap samples.

So for routine practice we recommend at least 10,000 bootstrap resamples and more when accuracy matters.

5.10 EXERCISES

For all exercises that ask you to perform exploratory data analysis (EDA), you should plot the data (histogram, normal quantile plots), describe the shape of the distribution (bell-shaped, symmetric, skewed, etc.), and provide summary statistics (mean, standard deviation). For bootstrapping questions, always provide plots and describe the shape, spread, and bias of the distribution.

1. Consider the sample 1–6. Use a six-sided die to obtain three different bootstrap samples and their corresponding means.
2. Consider the sample 1, 3, 4, 6 from some distribution.
   (a) For one random bootstrap sample, find the probability the mean is 1.
   (b) For one random bootstrap sample, find the probability the maximum is 6.
   (c) For one random bootstrap sample, find the probability that exactly two elements in the sample are less than 2.

   Assume order matters.
3. Consider the sample 1–3.
   (a) List all the bootstrap samples from this sample. How many are there?

(b) How many *distinct* bootstrap samples are there? For example, $\{1, 2, 2\}$ and $\{2, 1, 2\}$ are considered to be the same.

(c) How many bootstrap samples have one occurrence of 1 and two occurrences of 3? Is this the same number of bootstrap samples that have each of 1, 2, and 3 occurring exactly once?

(d) Is the probability of obtaining a bootstrap sample with one 1 and two 3's the same as the probability of obtaining a bootstrap sample with each of 1, 2, and 3 occurring exactly once?

4. Consider the sample 1, 3, 3, 5 from some distribution.

   (a) How many bootstrap samples are there?

   (b) List the distinct bootstrap samples assuming order does not matter.

5. (a) A bakery sells five types of cookies: sugar, chocolate chip, oatmeal, peanut butter, and ginger snap. Show that the number of ways to order five cookies is $\binom{9}{5}$. *Hint:* Represent an order of, say, 1 sugar, 2 chocolate chips, 0 oatmeal, 0 peanut butter, and 2 ginger snaps as $x/xx///xx$. How many symbols are there? How many ways to place the x's?

   (b) Show that the number of sets of size n (order does not matter) drawn with replacement from (the distinct) $a_1, a_2, \ldots, a_n$ is $\binom{2n-1}{n}$.

   (c) Conclude that the number of distinct bootstrap samples from the set $\{a_1, a_2, \ldots, a_n\}$ is $\binom{2n-1}{n}$ (the a_i's are distinct).

6. Let $k_1, k_2, \ldots, k_n$ denote nonnegative integers satisfying $k_1 + k_2 + \cdots + k_n = n$ and suppose the elements in the set $\{a_1, a_2, \ldots, a_n\}$ are distinct.

   (a) Show that the number of bootstrap samples with k_1 occurrences of a_1, k_2 occurrences of $a_k, \ldots, k_n$ occurrences of a_n is $\binom{n}{k_1, k_2, \ldots, k_s}$, the multinomial coefficient (Section B.2).

   (b) Compute the probability that a randomly drawn bootstrap sample will have k_i occurrences of a_i, $k = 1, 2, \ldots, n$.

7. Consider a population that has a normal distribution with mean $\mu = 36$, standard deviation $\sigma = 8$.

   (a) The sampling distribution of $\bar{X}$ for samples of size 200 will have what distribution, mean, and standard error?

   (b) Use R to draw a random sample of size 200 from this population. Conduct EDA on your sample.

   (c) Compute the bootstrap distribution for your sample and note the bootstrap mean and standard error.

   (d) Compare the bootstrap distribution to the theoretical sampling distribution by creating a table like Table 5.2.

   (e) Repeat for sample sizes of $n = 50$ and $n = 10$. Carefully describe your observations about the effects of sample size on the bootstrap distribution.

8. Consider a population that has a gamma distribution with parameters $r = 5$, $\lambda = 1/4$.

(a) Use simulation to generate an approximate sampling distribution of the mean; plot and describe the distribution.

(b) Now, draw one random sample of size 200 from this population. Create a histogram of your sample and find the mean and standard deviation.

(c) Compute the bootstrap distribution of the mean for your sample, plot it, and note the bootstrap mean and standard error.

(d) Compare the bootstrap distribution to the approximate theoretical sampling distribution by creating a table like Table 5.2.

(e) Repeat for sample sizes of $n = 50$ and $n = 10$. Describe carefully your observations about the effects of sample size on the bootstrap distribution.

9. We investigate the bootstrap distribution of the median. Create random samples of size n for various n and bootstrap the median. Describe the bootstrap distribution.

```
ne <- 14 # n even
no <- 15 # n odd

wwe <- rnorm(ne) # draw random sample of size ne
wwo <- rnorm(no) # draw random sample of size no

N <- 10^4
even.boot <- numeric(N) # save space
odd.boot <- numeric(N)
for (i in 1:N)
{
  x.even <- sample(wwe, ne, replace = TRUE)
  x.odd <- sample(wwo, no, replace = TRUE)
  even.boot[i] <- median(x.even)
  odd.boot[i]  <- median(x.odd)
}

par(mfrow = c(2,1))                # set figure layout
hist(even.boot, xlim = c(-2,2))    # same x range in
                                   # both plots
hist(odd.boot, xlim = c(-2,2))
par(mfrow = c(1,1))                # reset layout
```

Change the sample sizes to 36 and 37; 200 and 201; 10,000 and 10,001. Note the similarities/dissimilarities, trends, and so on. Why does the parity of the sample size matter? (*Note:* Adjust the x limits in the plots as needed.)

10. Import the data from data set `Bangladesh`. In addition to arsenic concentrations for 271 wells, the data set contains cobalt and chlorine concentrations.

(a) Conduct EDA on the chlorine concentrations and describe the salient features.

(b) Bootstrap the mean.

(c) Find and interpret the 95% bootstrap percentile confidence interval.

(d) What is the bootstrap estimate of the bias? What fraction of the bootstrap standard error does it represent?

The `Chlorine` variable has some missing values. The following code will remove these entries:

```
chlorine <- subset(Bangladesh, select = Chlorine, subset =
                   !is.na(Chlorine), drop = T)
```

11. Consider Bangladesh chlorine (concentration). Bootstrap the trimmed mean (say, trim the upper and lower 25%) and compare your results with the usual mean (previous exercise).

12. The data set `FishMercury` contains mercury levels (parts per million) for 30 fish caught in lakes in Minnesota.
    (a) Create a histogram or boxplot of the data. What do you observe?
    (b) Bootstrap the mean and record the bootstrap standard error and the 95% bootstrap percentile interval.
    (c) Remove the outlier and bootstrap the mean of the remaining data. Record the bootstrap standard error and the 95% bootstrap percentile interval.
    (d) What effect did removing the outlier have on the bootstrap distribution, in particular, the standard error?

13. In Section 3.3, we performed a permutation test to determine if men and women consumed, on average, different amounts of hot wings.
    (a) Bootstrap the difference in means and describe the bootstrap distribution.
    (b) Find a 95% bootstrap percentile confidence interval for the difference of means and give a sentence interpreting this interval.
    (c) How do the bootstrap and permutation distributions differ?

14. Import the data from `Girls2004` (see Section 1.2).
    (a) Perform some exploratory data analysis and obtain summary statistics on the weights of baby girls born in Wyoming and Arkansas (do separate analyses for each state).
    (b) Bootstrap the difference in means, plot the distribution, and give the summary statistics. Obtain a 95% bootstrap percentile confidence interval and interpret this interval.
    (c) What is the bootstrap estimate of the bias? What fraction of the bootstrap standard error does it represent?
    (d) Conduct a permutation test to calculate the difference in mean weights and state your conclusion.
    (e) For what population(s), if any, does this conclusion hold? Explain.

15. Do chocolate and vanilla ice creams have the same number of calories? The data set `IceCream` contains calorie information for a sample of brands of

chocolate and vanilla ice cream. Use the bootstrap to determine whether or not there is a difference in the mean number of calories.

16. Import the data from Flight Delays Case Study in Section 1.1 data into `R`. Although the data are on all UA and AA flights flown in May and June of 2009, we will assume these represent a sample from a larger population of UA and AA flights flown under similar circumstances. We will consider the ratio of means of the flight delay lengths, $\mu_{\text{UA}}/\mu_{\text{AA}}$.
    (a) Perform some exploratory data analysis on flight delay lengths for each of UA and AA flights.
    (b) Bootstrap the mean of flight delay lengths for each airline separately and describe the distribution.
    (c) Bootstrap the ratio of means. Provide plots of the bootstrap distribution and describe the distribution.
    (d) Find the 95% bootstrap percentile interval for the ratio of means. Interpret this interval.
    (e) What is the bootstrap estimate of the bias? What fraction of the bootstrap standard error does it represent?
    (f) For inference in this text, we assume that the observations are independent. Is that condition met here? Explain.
17. Two college students collected data on the price of hardcover textbooks from two disciplinary areas: Mathematics and the Natural Sciences, and the Social Sciences (Hien and Baker (2010)). The data are in the file `BookPrices`.
    (a) Perform some exploratory data analysis on book prices for each of the two disciplinary areas.
    (b) Bootstrap the mean of book price for each area separately and describe the distributions.
    (c) Bootstrap the ratio of means. Provide plots of the bootstrap distribution and comment.
    (d) Find the 95% bootstrap percentile interval for the ratio of means. Interpret this interval.
    (e) What is the bootstrap estimate of the bias? What fraction of the bootstrap standard error does it represent?

6

ESTIMATION

In Section 3.8, we modeled the number of home runs with a Poisson distribution. There, we used the sample mean as an estimate of the parameter λ, the expected value of the Poisson random variable. Elsewhere, we have used the sample mean as an estimate for the true population mean and the sample proportion as an estimate of the true population proportion. Intuitively, these all seem to be very reasonable choices. In this chapter, we will consider more rigorously the nature of these choices and consider some general procedures for estimating parameters. In addition, we will consider properties that we would want an estimator to have.

6.1 MAXIMUM LIKELIHOOD ESTIMATION

We begin with a very general procedure, maximum likelihood estimation (MLE). This has a number of nice properties, of which we will mention one now: the answers it gives are reasonable, typically giving the answers that just make sense, and it never gives impossible answers.

Suppose a friend tells you she has a total of 25 chocolate chip or oatmeal raisin cookies in a bag. She also tells you that the number of chocolate chip cookies is either 2 or 20. If you draw out a cookie at random from the bag and see it is chocolate chip, what do you think is more likely the truth, that there are 2 or 20 chocolate cookies? Based solely on your one data point (the chocolate chip cookie), it seems that 20 chocolate cookies is more likely than 2 chocolate cookies. This is the idea behind

Mathematical Statistics with Resampling and R, First Edition. By Laura Chihara and Tim Hesterberg.

maximum likelihood estimation—what value of a parameter is most consistent with the data, in that it makes the data most likely?

6.1.1 Maximum Likelihood for Discrete Distributions

Suppose you buy a weighted coin in a magic shop. It looks like a typical coin with heads and tails, but the coin is not necessarily fair. You flip the coin eight times and observe the sequence HHHTHTHT. What would be a good estimate for p, the probability of heads, based on this sequence? Let $X_i \sim \text{Bern}(p)$, where $X_i = 1$ indicates heads. We will assume that the flips are independent. Then,

$$\begin{aligned} P(X_1 = 1, X_2 = 1, X_3 = 1, X_4 = 0, X_5 = 1, X_6 = 0, X_7 = 1, X_8 = 0) \\ &= P(X_1 = 1)P(X_2 = 1)P(X_3 = 1)P(X_4 = 0)P(X_5 = 1)P(X_6 = 0) \\ &\quad \times P(X_7 = 1)P(X_8 = 0) \\ &= p^5(1-p)^3. \end{aligned}$$

Since you want to find the p that makes the observed sequence most likely, it seems reasonable to find the p that maximizes the function $L(p) = p^5(1-p)^3$. Using calculus, compute the derivative $L'(p) = 5p^4(1-p)^3 + p^5 3(1-p)2(-1)$ and set this equal to 0. Solving for p yields $p = 5/8$ as the most likely candidate for p.

The function $L(p) = p^5(1-p)^3$ is called the *likelihood function* for the parameter p (Figure 6.1) and the estimate $\hat{p} = 5/8$ is the *maximum likelihood estimate* for p.

More generally,

Definition 6.1 Let $f(x; \theta)$ denote the probability mass function for a discrete distribution with associated parameter θ. Suppose $X_1, X_2, \ldots, X_n$ are a random sample

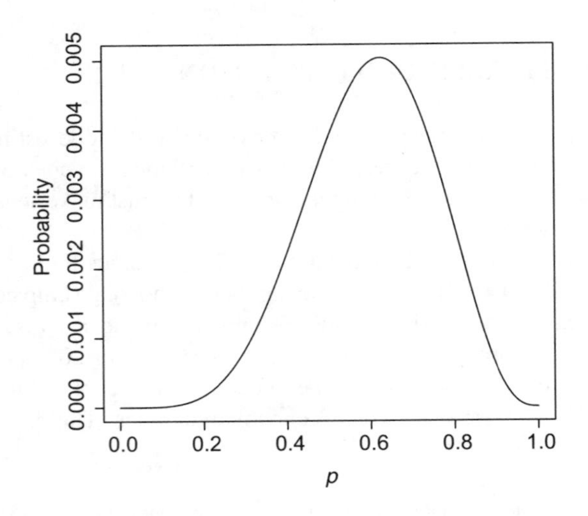

FIGURE 6.1 Likelihood for p.

from this distribution and $x_1, x_2, \ldots, x_n$ are the actual observed values. Then, the *likelihood function* of θ is

$$\begin{aligned} L(\theta \mid x_1, x_2, \ldots, x_n) &= P(X_1 = x_1, X_2 = x_2, \ldots, X_n = x_n) \\ &= P(X_1 = x_1)P(X_2 = x_2)\cdots P(X_n = x_n) \\ &= f(x_1;\theta)f(x_2;\theta)\cdots f(x_n;\theta) \\ &= \prod_{i=1}^{n} f(x_i;\theta). \end{aligned}$$

The likelihood function is a function of θ and is sometimes written as $L(\theta) = L(\theta \mid x_1, x_2, \ldots, x_n)$.

A maximum likelihood estimate is a value $\hat{\theta}$ such that $L(\hat{\theta}) \geq L(\theta)$ for all θ. ||

LIKELIHOOD FUNCTION, MAXIMUM LIKELIHOOD ESTIMATE

The likelihood function $L(\theta) = L(\theta \mid x_1, x_2, \ldots, x_n)$ gives the likelihood of θ, given the data. A maximum likelihood estimate $\hat{\theta}_{\text{MLE}}$ is a value of θ that maximizes the likelihood.

One point we wish to clarify is that even though we have $L(\theta \mid x_1, x_2, \ldots, x_n)$ equal to an expression involving $f(x_i;\theta)$, the two functions are thought of differently. When we consider the density $f(x;\theta)$, we consider x to be variable and θ to be fixed, whereas when we consider the likelihood $L(\theta|x_1, x_2, \ldots, x_n)$, we consider θ to be variable and the x_i's to be fixed.

Proposition 6.1 *Let $X_1, X_2, \ldots, X_n$ denote n independent Bernoulli random variables,* Bern(p), $0 < p < 1$. *Let $X = \sum_{i=1}^{n} X_i$, the number of* 1*'s. The maximum likelihood estimator of p is $\hat{p} = X/n$.*

Proof Assume $x_i = 1$, with probability p, $i = 1, 2, \ldots, x_n$, and let $x = \sum_{i=1}^{n} x_i$.

$$\begin{aligned} L(p) &= P(X_1 = x_1, X_2 = x_2, \ldots, X_n = x_n) \\ &= \prod_{i=1}^{n} P(X_i = x_i) \\ &= \prod_{x_i=1} p^{x_i} \prod_{x_i=0} (1-p)^{x_i} \\ &= p^{\sum_{i=1}^{n} x_i}(1-p)^{n-\sum_{i=1}^{n} x_i} \\ &= p^x(1-p)^{n-x}. \end{aligned}$$

Solving $L'(p) = 0$ for p yields $\hat{p} = x/n$ as a critical value of L. A check of the graph of $L(p)$ or the first derivative test confirms that $\hat{p}$ is the value that maximizes $L(p)$. Thus, $\hat{p} = X/n$ is the maximum likelihood estimator of p. □

Remark We use the term *estimator* for a function of random variables (see Definition 5.1), but we call the result for a set of observations, $X_1 = x_1, X_2 = x_2, \ldots, X_n = x_n$, an *estimate*.

An estimator like $\bar{X}$ is a rule that can be applied to new random data; an estimate like $\bar{x}$ is a specific value for a given set of data. ||

In many cases, the likelihood function is difficult to differentiate, so we employ logarithmic differentiation to find the maximum. Note that since the logarithm is an increasing function, if $\hat{\theta}$ is a maximum of $\ln(L(\theta))$, then $\hat{\theta}$ will be a maximum of $L(\theta)$.

Example 6.1 Suppose $x_1 = 3, x_2 = 4, x_3 = 3, x_4 = 7$ come from a Poisson distribution with λ unknown. The probability mass function is $f(x; \lambda) = \lambda^x e^{-\lambda}/x!$, $x = 0, 1, 2, \ldots$, so the likelihood function is

$$\begin{aligned} L(\lambda) &= f(x_1;\lambda) f(x_2;\lambda) f(x_3;\lambda) f(x_4;\lambda) \\ &= \frac{\lambda^3 e^{-\lambda}}{3!} \frac{\lambda^4 e^{-\lambda}}{4!} \frac{\lambda^3 e^{-\lambda}}{3!} \frac{\lambda^7 e^{-\lambda}}{7!} \\ &= \frac{\lambda^{17} e^{-4\lambda}}{3!\,4!\,3!\,7!}. \end{aligned}$$

Take the logarithm of each side, then differentiate with respect to λ:

$$\begin{aligned} \ln(L(\lambda)) &= 17\ln(\lambda) - 4\lambda - \ln(3!\,4!\,3!\,7!) \\ \frac{L'(\lambda)}{L(\lambda)} &= 17\frac{1}{\lambda} - 4. \end{aligned}$$

Setting $L'(\lambda) = 0$, we find the maximum of $L(\lambda)$ occurs at $\hat{\lambda} = 17/4$, the average of the four observed values. The first derivative test confirms that this is a global maximum. □

More generally,

Proposition 6.2 *Let $x_1, x_2, \ldots, x_n$ be a random sample from a Poisson distribution with unknown parameter λ. Then the maximum likelihood estimate of λ is $\hat{\lambda} = \bar{x}$, the sample mean.*

Proof Exercise. □

Remark Referring to the Philadelphia Phillies home runs example in Section 3.8, we let x_i denote the number of home runs hit in game number i, $i = 1, 2, \ldots, 162$, and assume these values come from a Poisson distribution with λ unknown. We noted earlier that $\sum_{i=1}^{162} x_i = 224$, so $\hat{\lambda} = \bar{x} = 224/162 = 1.3827$. Thus, we model the number of home runs per game with $P(X = x) = 1.383^x e^{-1.383}/x!$, $x = 0, 1, 2, \ldots$. ||

6.1.2 Maximum Likelihood for Continuous Distributions

Now consider the case of data $X_1 = x_1, X_2 = x_2, \ldots, X_n = x_n$ from a continuous distribution. If we were to mimic the discrete case, we would compute $P(X_1 = x_1)P(X_2 = x_2) \cdots P(X_n = x_n)$. But $P(X = a) = 0$ for continuous random variables!

Let $f(x)$ denote the pdf of the continuous random variable X and recall the interpretation of the integral as area under the graph of $y = f(x)$. Then, for small $h > 0$,

$$P(a - h < X < a + h) = \int_{a-h}^{a+h} f(x)\,dx \approx 2h\,f(a).$$

Thus, $P(a - h < X < a + h)$ is approximately proportional to $f(a)$ (Figure 6.2).

Definition 6.2 Let $f(x;\theta)$ denote the pdf of a continuous random variable with associated parameter θ. Suppose $X_1, X_2, \cdots, X_n$ are a random sample from this distribution and $X_1 = x_1, X_2 = x_2, \ldots, X_n = x_n$ are actual observed values. Then, the *likelihood function* of θ is

$$L(\theta \mid x_1, x_2, \ldots, x_n) = f(x_1;\theta)f(x_2;\theta) \cdots f(x_n;\theta). \tag{6.1}$$

||

A maximum likelihood estimate is a statistic $\hat{\theta}$ that maximizes L: $L(\hat{\theta}) \geq L(\theta)$ for all θ.

Example 6.2 Let $X_1 = x_1, X_2 = x_2, X_3 = x_3, X_4 = x_4$ be independent and drawn from the exponential distribution with pdf $f(x;\lambda) = \lambda e^{\lambda x}$, $x \geq 0$. Find a maximum likelihood estimate for λ.

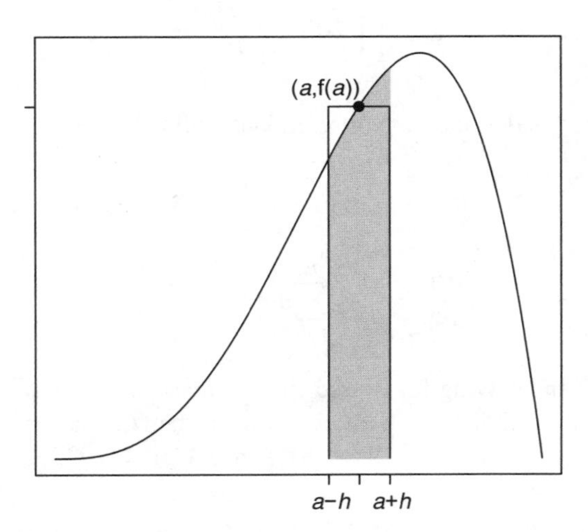

FIGURE 6.2 Estimate of area under $y = f(x)$ by rectangle.

Solution

$$\begin{aligned} L(\lambda \mid x_1, x_2, x_3, x_4) &= f(x_1;\lambda)f(x_2;\lambda)f(x_3;\lambda)f(x_4;\lambda) \\ &= \lambda e^{-\lambda x_1}\,\lambda e^{-\lambda x_2}\,\lambda e^{-\lambda x_3}\,\lambda e^{-\lambda x_4} \\ &= \lambda^4 e^{-\lambda(x_1+x_2+x_3+x_4)}. \end{aligned}$$

Thus, differentiating $L(\lambda)$ with respect to λ, we have

$$\begin{aligned} L'(\lambda) &= 4\lambda^3 e^{-\lambda \sum_{i=1}^4 x_i} + \lambda^4 e^{-\lambda \sum_{i=1}^4 x_i}\left(-\sum_{i=1}^4 x_i\right) \\ &= e^{-\lambda \sum_{i=1}^4 x_i}\lambda^3\left(4 - \lambda \sum_{i=1}^4 x_i\right). \end{aligned}$$

Setting $L'(\lambda) = 0$ and solving for λ yields $\hat{\lambda} = 1/((1/4)\sum_{i=1}^4 x_i)$, the sample mean. We can also find the estimate by taking the log of the likelihood before differentiating. □

Example 6.3 Let $X_1, X_2, \ldots, X_n$ denote a random sample from a distribution with pdf $f(x;\theta) = \theta x^{\theta-1}$ for $0 < x < 1$, where $\theta > 0$. Find the MLE of θ.

Solution

$$\begin{aligned} L(\theta) &= f(X_1;\theta)f(X_2;\theta)\cdots f(X_n;\theta) \\ &= \prod_{i=1}^n \theta X_i^{\theta-1} \\ &= \theta^n \prod_{i=1}^n X_i^{\theta-1}. \end{aligned}$$

Take the log of both sides and then differentiate with respect to θ:

$$\begin{aligned} \ln(L(\theta)) &= n\ln(\theta) + (\theta - 1)\sum_{i=1}^n \ln(X_i) \\ \frac{L'(\theta)}{L(\theta)} &= \frac{n}{\theta} + \sum_{i=1}^n \ln(X_i). \end{aligned}$$

Setting $L'(\theta) = 0$ and solving for θ yields the estimator $\hat{\theta} = -n/\sum_{i=1}^n \ln(X_i)$. Thus, for example, if $X_1 = 0.35$, $X_2 = 0.28$, $X_3 = 0.41$, then the maximum likelihood estimate of θ is $\hat{\theta} = -3/(\ln(0.35) + \ln(0.28) + \ln(0.41)) = 0.9333$. □

Example 6.4 Let $X_1, X_2, \cdots, X_n \overset{\text{i.i.d.}}{\sim} \text{Unif}[0, \beta]$ with pdf $f(x;\beta) = 1/\beta$, where $0 \le x \le \beta$. Find the MLE of β.

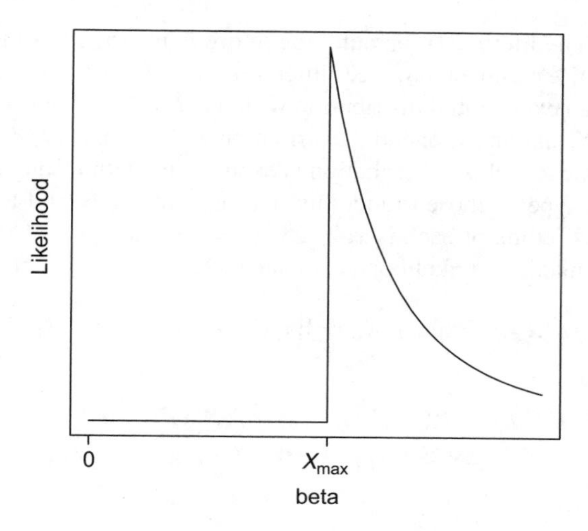

FIGURE 6.3 Likelihood for β.

Solution For $0 \leq X_i \leq \beta$, we have

$$\begin{aligned} L(\beta) &= f(X_1;\beta)f(X_2;\beta)\cdots f(X_n;\beta) \\ &= \frac{1}{\beta}\frac{1}{\beta}\cdots\frac{1}{\beta} \\ &= \left(\frac{1}{\beta}\right)^n \end{aligned} \tag{6.2}$$

(and $L(\beta) = 0$ if any $X_i > \beta$).

This function is positive as long as $X_1, X_2, \ldots, X_n \in [0, \beta]$—in other words, as long as $\beta \geq \max\{X_1, X_2, \ldots, X_n\}$. The function is zero if $\beta < X_{\max}$, jumps at $\beta = X_{\max}$, then is positive. Where the function is positive, the derivative is $L'(\beta) = -n(1/(\beta)^{n-1})$. This negative derivative means the likelihood function is decreasing after the jump, so the maximum of the function occurs right at the jump, $\hat{\beta} = \max\{X_1, X_2, \ldots, X_n\}$ (see Figure 6.3). □

This next example shows that a maximum likelihood estimate may not always exist.

Example 6.5 Suppose instead, we consider $X_1, X_2, \ldots, X_n \overset{\text{i.i.d.}}{\sim} \text{Unif}(0, \beta)$—that is, uniform on the open interval $(0, \beta)$. The likelihood function is nearly the same as Equation 6.2, except we need to assume strict inequalities, $0 < X_i < \beta$ for $i = 1, 2, \ldots, n$.

Again, L is as large as possible when β is as small as possible, so β must be bigger than *but not equal to* the maximum of $X_1, X_2, \ldots, X_n$. Since β can be made arbitrarily close to this maximum, a MLE solution does not exist. □

Example 6.6 The lifetime of an automobile tire is measured in miles rather than time. Suppose a tire company produces three versions of a tire: a standard tire whose lifetime X_s has an exponential distribution with mean $1/\lambda > 0$, an economy version whose lifetime X_e has an exponential distribution with mean $0.77/\lambda$, and a premium tire with a lifetime X_p whose distribution is exponential with mean $1.25/\lambda$. Suppose one tire of each type is chosen randomly and independently and tested to find its lifetime and the lifetime of each is $x_s = 28$, $x_e = 25$, and $x_p = 31$ (in thousands of miles). Find the maximum likelihood estimate of λ.

Solution We have $X_s \sim \text{Exp}(\lambda)$, $X_e \sim \text{Exp}(1.3\lambda)$, and $X_p \sim \text{Exp}(0.8\lambda)$, so the likelihood is

$$\begin{aligned} L(\lambda) &= f_s(X_s;\lambda) f_e(X_e;\lambda) f(X_p;\lambda) \\ &= \lambda e^{-\lambda X_s}(1.3\lambda e^{-1.3\lambda X_e})(0.8\lambda e^{-0.8\lambda X_p}) \\ &= 1.04\lambda^3 e^{-(X_s+1.3X_e+0.8X_p)\lambda}. \end{aligned}$$

The log-likelihood and its derivative with respect to λ are:

$$\begin{aligned} \ln(L(\lambda)) &= \ln(1.04) + 3\ln(\lambda) - (X_s + 1.3X_e + 0.8X_p)\lambda, \\ \frac{L'(\lambda)}{L(\lambda)} &= \frac{3}{\lambda} - (X_s + 1.3X_e + 0.8X_p) \end{aligned}$$

Hence, setting $L'(\lambda) = 0$ yields the estimator $\hat{\lambda} = 3/(X_s + 1.3X_e + 0.8X_p)$; for the given data, we have the estimate $\hat{\lambda} = 0.0395$.

For standard tires, the estimated average lifetime is $\hat{\text{E}}[X_s] = 1/\hat{\lambda} = 25.3$ thousand miles; for economy tires, $\hat{\text{E}}[X_e] = 1/(1.3\,\hat{\lambda}) = 19.5$ thousand miles; for premium tires, $\hat{\text{E}}[X_p] = 1/(0.8\hat{\lambda}) = 31.6$ thousand miles. □

In all examples, so far where a MLE existed, we were able to find a closed form expression. In many situations, this is not possible.

Example 6.7 Suppose $X_1, X_2, \ldots, X_n$ are a random sample from the Cauchy distribution with pdf $f(x;\theta) = 1/(\pi(1 + (x - \theta)^2))$ for $-\infty < x < \infty$, $-\infty < \theta < \infty$. The likelihood function for θ is

$$L(\theta) = \frac{1}{\pi^n \prod_{i=1}^{n} \left[1 + (X_i - \theta)^2\right]}.$$

Thus, $L(\theta)$ will be a maximum when $\prod_{i=1}^{n} \left[1 + (X_i - \theta)^2\right]$ is a minimum. The value of θ that minimizes this expression must usually be determined by numerical methods.

For instance, suppose our observations are $X_1 = 1$, $X_2 = X_3 = 2$, $X_4 = 3$, $X_5 = 4$. Then, to maximize $L(\theta)$, we will need to minimize

$$g(\theta) = (1 + (1 - \theta)^2)(1 + (2 - \theta)^2)^2(1 + (3 - \theta)^2)(1 + (4 - \theta)^2).$$

Setting $g'(\theta) = 0$ gives a ninth degree equation.

$$10\theta^9 - 216\theta^8 + 2072\theta^7 - 11592\theta^6 + 41754\theta^5 - 100680\theta^4 + 163124\theta^3 \\ -182152\theta^2 + 108200\theta - 31200 = 0.$$

R or computer algebra programs with numerical methods such as Mathematica™ can estimate the solution, in this case $\hat{\theta} \approx 2.2212$. □

See Exercises 12 and 13 for some other examples where no closed form solution exists.

6.1.3 Maximum Likelihood for Multiple Parameters

We can find MLE's for distributions with more than one parameter using multivariable calculus.

Theorem 6.1 *Let $X_1 = x_1, X_2 = x_2, \ldots, X_n = x_n$ be a random sample from a normal distribution with mean μ and standard deviation σ. The maximum likelihood estimates of μ and σ are*

$$\hat{\mu} = \frac{1}{n}\sum_{i=1}^{n} x_i = \bar{x}, \tag{6.3}$$

$$\hat{\sigma} = \sqrt{\frac{1}{n}\sum_{i=1}^{n}(x_i - \bar{x})^2}. \tag{6.4}$$

Proof We form the likelihood

$$L(\mu, \sigma^2) = \frac{1}{\sqrt{2\pi}\sigma} e^{-\frac{1}{2}(\frac{x_1-\mu}{\sigma})^2} \frac{1}{\sqrt{2\pi}\sigma} e^{-\frac{1}{2}(\frac{x_2-\mu}{\sigma})^2} \cdots \frac{1}{\sqrt{2\pi}\sigma} e^{-\frac{1}{2}(\frac{x_n-\mu}{\sigma})^2}$$

$$= \left(\frac{1}{\sqrt{2\pi}\sigma}\right)^n e^{-\frac{1}{2}\sum_{i=1}^{n}(\frac{x_i-\mu}{\sigma})^2}. \tag{6.5}$$

Now, take the logarithm of both sides,

$$\ln(L(\mu, \sigma^2)) = -n\ln(\sigma) - \frac{n}{2}\ln(2\pi) - \frac{1}{2\sigma^2}\sum_{i=1}^{n}(x_i - \mu)^2. \tag{6.6}$$

Setting the partial derivatives of Equation 6.6 with respect to μ and σ equal to 0 gives a system of equations:

$$\frac{\partial(\ln(L(\mu, \sigma)))}{\partial\mu} = \frac{1}{\sigma^2}\sum_{i=1}^{n}(x_i - \mu) = 0, \tag{6.7}$$

$$\frac{\partial(\ln(L(\mu, \sigma)))}{\partial\sigma} = -\frac{n}{\sigma} + \frac{1}{2\sigma^4}\sum_{i=1}^{n}(x_i - \mu)^2 = 0. \tag{6.8}$$

Solving for μ in Equation 6.7, we find the maximum likelihood estimate of μ is $(1/n)\sum_{i=1}^{n} x_i = \bar{x}$, the sample mean. We substitute this value into Equation 6.8 to obtain the estimate of σ to be $\sqrt{(1/n)\sum_{i=1}^{n}(x_i - \bar{x})^2}$. □

Case Study: Wind Energy Concerns about climate change and rising costs of fossil fuels and interest in sustainability have made renewable energy from sources such as tides, wind, or the sun more attractive. Wind turbines harness the kinetic energy from the wind to produce electricity.[1] For instance, Carleton College in Northfield Minnesota owns a 1.65 MW wind turbine that has been operational since 2004. In 2008, the turbine produced 3965 MW h of electricity that was then sold to Xcel Energy, a utility company.

Wind speeds are highly variable, affected by the time of day and time of year. Since the amount of energy output from a turbine depends on wind speed, understanding the characteristics of wind speed is important. Engineers use wind speed information to determine suitable locations to build a wind turbine or to optimize the design of a turbine. Utility companies use this information to make predictions on energy availability during peak demand periods (say, during a heat wave) or to estimate yearly revenue.

The Weibull distribution is the most commonly used probability distribution used to model wind speed (Justus et al. (1978); Seguro and Lambert (2000); Weisser (2003); Zhou et al. (2010)). The Weibull distribution has a density function with two parameters, the shape parameter $k > 0$, and the scale parameter $\lambda > 0$,

$$f(x; k, \lambda) = \frac{k\, x^{k-1}}{\lambda^k} e^{-(x/\lambda)^k}, \quad x \geq 0. \tag{6.9}$$

We will use this distribution to model the average wind speeds (m/s) at the site of Carleton's wind turbine for 168 days from February 14 to August 1, 2010 (there were no data for July 2) (Table 6.1).

Given the data $X_1 = x_1, X_2 = x_2, \ldots, X_n = x_n$, the likelihood function is

$$\begin{aligned} L(k, \lambda; x_1, x_2, \ldots, x_n) = L(k, \lambda) &= \prod_{i=1}^{n} \frac{k\, x_i^{k-1}}{\lambda^k} e^{(-x_i/\lambda)^k} \\ &= \frac{k^n}{\lambda^{kn}} \prod_{i=1}^{n} x_i^{k-1} e^{-\sum_{i=1}^{n}(x_i/\lambda)^k}. \end{aligned}$$

Thus, the log-likelihood is

$$\ln(L(k, \lambda)) = n \ln(k) - kn \ln(\lambda) + (k-1)\sum_{i=1}^{n} \ln(x_i) - \sum_{i=1}^{n} \left(\frac{x_i}{\lambda}\right)^k.$$

[1] `http://guidedtour.windpower.org/en/tour/wres/index.htm`

TABLE 6.1 Sample of Wind Speeds (m/s) from Carleton College Turbine

Feb 14	Feb 15	Feb 16	Feb 17	Feb 18	Feb 19
7.8	8.9	9.7	7.7	6.4	3.1

We next compute the partial derivative of $\ln(L(k, \lambda))$ with respect to k and λ and set each result equal to 0:

$$\frac{\partial(\ln(L(k, \lambda)))}{\partial k} = \frac{n}{k} - n\ln(k) + \sum_{i=1}^{n} \ln(x_i) - \sum_{i=1}^{n} \left(\frac{x_i}{\lambda}\right)^k \ln\left(\frac{x_i}{\lambda}\right) = 0, \tag{6.10}$$

$$\frac{\partial(\ln(L(k, \lambda)))}{\partial \lambda} = \frac{-kn}{\lambda} + \frac{k}{\lambda^{k+1}} \sum_{i=1}^{n} x_i^k = 0. \tag{6.11}$$

From Equation 6.11, we find

$$\lambda^k = \frac{1}{n} \sum_{i=1}^{n} x_i^k \tag{6.12}$$

and substituting this into Equation 6.10, we obtain

$$\frac{1}{k} + \sum_{i=1}^{n} \ln(x_i) - \frac{1}{\alpha} \sum_{i=1}^{n} x_i^k \ln(x_i) = 0, \tag{6.13}$$

where $\alpha = \sum_{i=1}^{n} x_i^k$.

Numerical methods must be used to find an approximate value for k in Equation 6.13. Using the wind data from the Carleton turbine, we obtain an estimate of $\hat{k} = 3.169$ and, thus, from Equation 6.12, we get $\lambda^{3.169} = (1/168) \sum_{i=1}^{168} x_i^{3.169}$, which yields $\hat{\lambda} = 7.661$.

From Figure 6.4, it appears that the Weibull distribution models the data quite well, but we will check this with a goodness-of-fit test (see Section 3.8). We find the deciles (the points marking the 10%, 20%, ... probabilities) for the Weibull distribution with the estimated parameters $\hat{k} = 3.169$, $\hat{\lambda} = 7.661$ and count the number of data points that fall into each of the intervals determined by these deciles. Since there are 168 data points, we expect 16.8 points to fall into each of these intervals (Table 6.2).

TABLE 6.2 Distribution of Wind Speed Data into Intervals

Interval	[0.0, 3.77]	(3.77, 4.77]	(4.77, 5.53)	(5.53, 6.20]	(6.20, 6.80]
Observed	17	18	19	13	20
Expected	16.8	16.8	16.8	16.8	16.8
Interval	(6.80, 7.45]	(7.45, 8.12]	(8.12, 8.90]	(8.90, 9.97]	(9.97, ∞)
Observed	14	17	17	14	19
Expected	16.8	16.8	16.8	16.8	16.8

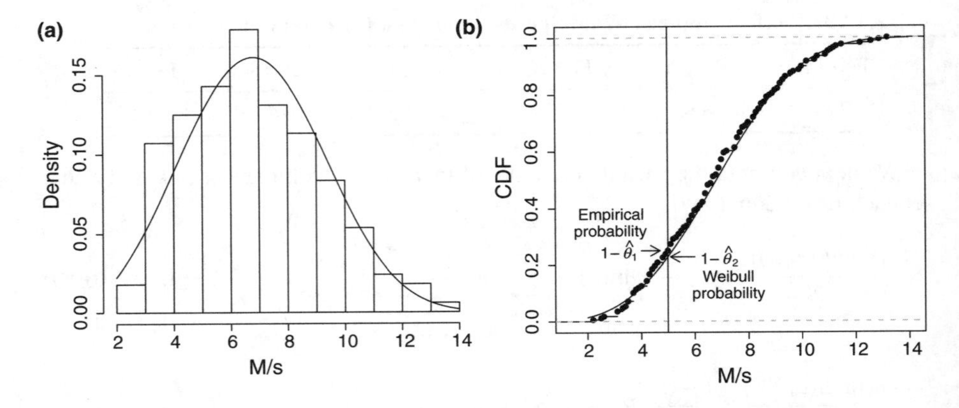

FIGURE 6.4 (a) Histogram of wind speeds (m/s) with the pdf for the Weibull distribution superimposed, $\hat{k} = 3.169$, $\hat{\lambda} = 7.661$. (b) Empirical cumulative distribution function, with cdf for Weibull superimposed. Also shown are the CDF and Weibull CDF values at `wind=5`; $\hat{\theta}_1$ and $\hat{\theta}_2$ are the corresponding estimated probabilities of wind exceeding 5.

The chi-square test statistic is $c = 3.071$ and if the data do come from the proposed Weibull distribution, then c comes from a chi-square distribution with $10 - 2 - 1 = 7$ degrees of freedom. The corresponding P-value is 0.878, so we conclude that the distribution of the wind speed data comes from a Weibull distribution with parameters $\hat{k} = 3.169$ and $\hat{\lambda} = 7.661$.

The `R` code for this analysis is provided on the web site `https://sites.google.com/site/ChiharaHesterberg`.

See Exercise 17 for a problem that uses the Weibull distribution to model time between earthquakes.

6.2 METHOD OF MOMENTS

Another approach for finding estimates for parameters is the *Method of Moments*. Let $f(x)$ denote a probability density function (either continuous or discrete) for a random variable X. For a positive integer r, the rth (theoretical) moment of X is

$$\mathrm{E}\left[X^r\right] = \int_{-\infty}^{\infty} x^r f(x)\, dx \tag{6.14}$$

and the rth sample moment is

$$\frac{1}{n}\sum_{i=1}^{n} x_i^r. \tag{6.15}$$

Let $X_1, X_2, \ldots, X_n$ denote a random sample from a distribution with r unknown parameters, $\theta_1, \theta_2, \ldots, \theta_r$. Let $f(x; \theta_1, \theta_2, \ldots, \theta_r)$ denote the pdf of this distribution. The method of moments procedure equates each of the r theoretical moments with the corresponding rth sample moment to obtain a system of r equations in r unknowns.

$$\int_{-\infty}^{\infty} x\, f(x; \theta_1, \theta_2, \dots, \theta_r) = \frac{1}{n}\sum_{i=1}^{n} X_i,$$

$$\int_{-\infty}^{\infty} x^2\, f(x; \theta_1, \theta_2, \dots, \theta_r) = \frac{1}{n}\sum_{i=1}^{n} (X_i)^2,$$

$$\vdots$$

$$\int_{-\infty}^{\infty} x^r\, f(x; \theta_1, \theta_2, \dots, \theta_r) = \frac{1}{n}\sum_{i=1}^{n} (X_i)^r.$$

The solutions to this system of r equations and r unknowns are then the method of moments estimators of $\theta_1, \theta_2, \dots, \theta_r$.

Remark In the case of a discrete random variable, replace the above integrals with summations. ||

Example 6.8 Let $X_1, X_2, \dots, X_n \overset{\text{i.i.d.}}{\sim} \text{Unif}[0, \beta]$. The first theoretical moment is $\text{E}[X_i] = \beta/2$, while the sample mean $\bar{X}$ is the first sample moment. Thus, setting $\beta/2 = \bar{X}$ yields the method of moments estimator $\hat{\beta} = 2\bar{X}$.

Note that this estimator can give impossible results—for example, if $n = 4$, $x_1 = x_2 = x_3 = 1$, $x_4 = 9$, then $\bar{X} = 3$ and $\hat{\beta} = 6$, which makes $x_4 = 9$ impossible. □

Example 6.9 Suppose $X_1, X_2, \dots, X_n$ are a random sample from the exponential distribution, $f(x; \lambda) = \lambda e^{-\lambda x}$, $x > 0$, $\lambda > 0$. Use the method of moments to find the estimator for λ.

Solution The first theoretical moment is $\text{E}[X] = \int_0^{\infty} x e^{-\lambda x}\, dx = 1/\lambda$. The sample first moment is the sample mean $\bar{x} = (1/n)\sum_{i=1}^{n} X_i$. Equating the two yields $1/\lambda = \bar{X}$; thus, the estimator is $\hat{\lambda} = 1/\bar{X}$. □

Example 6.10 Suppose $x_1 = 1.3$, $x_2 = 1.8$, $x_3 = 2.1$, $x_4 = 2.25$ are a random sample from a distribution with pdf $f(x; \theta) = 2(\theta - x)/\theta$, $0 < x < \theta$. Find the method of moments estimate of θ.

Solution The first theoretical moment is $\text{E}[X] = \int_0^{\theta} x \cdot 2(\theta - x)/\theta\, dx = \theta^2/3$. The first sample moment is the sample mean $\bar{x} = 1.8625$. Thus, $\theta^2/3 = 1.8625$ yields an estimate of $\hat{\theta} = 2.3837$. □

Example 6.11 Suppose $X_1, X_2, \dots, X_n$ are a random sample from a distribution with pdf $f(x; \lambda, \delta) = \lambda e^{-\lambda(x-\delta)}$, $x > \delta$, where $\lambda, \delta > 0$. Find estimators for λ and δ.

Solution Use integration by parts to find

$$\mathrm{E}[X] = \int_{\delta}^{\infty} x\,\lambda e^{-\lambda(x-\delta)}\,dx = \delta + \frac{1}{\lambda}.$$

Then equating this to the first sample moment, we have $\delta + (1/\lambda) = \bar{X}$.

Again, using integration by parts,

$$\mathrm{E}\left[X^2\right] = \int_{\delta}^{\infty} x^2\,\lambda e^{-\lambda(x-\delta)}\,dx = \delta^2 + \frac{2\delta}{\lambda} + \frac{2}{\lambda^2}.$$

Let $m_2 = (1/n)\sum_{i=1}^{n} X_i^2$, the second sample moment. Equating the second sample and theoretical moments and completing the square gives

$$\begin{aligned} m_2 &= \delta^2 + \frac{2\delta}{\lambda} + \frac{2}{\lambda^2} \\ &= \left(\delta + \frac{1}{\lambda}\right)^2 + \frac{1}{\lambda^2} \\ &= \bar{X}^2 + \frac{1}{\lambda^2}. \end{aligned}$$

Thus, $1/\lambda = \sqrt{m_2 - \bar{X}^2}$, and hence $\delta = \bar{X} - \sqrt{m_2 - \bar{X}^2}$.

For instance, suppose $X_1 = 3.5$, $X_2 = 3.9$, $X_3 = 4$, $X_4 = 4.7$. Then, the first and second sample moments are $\bar{x} = 4.025$ and $m_2 = 16.386$, which results in the estimates $\hat{\lambda} = 2.3133$ and $\hat{\delta} = 3.5927$. □

6.3 PROPERTIES OF ESTIMATORS

We now have two very general methods of estimating parameters—maximum likelihood and method of moments—and in any given problem we may be able to think of other more *ad hoc* methods. This raises the question of which method is best. Here, we discuss several criteria for comparing methods and properties that we think good methods should satisfy. The first three of these criteria—unbiasedness, efficiency, and mean square error—are fairly natural. The last two—consistency and transformation invariance—are more mathematical in how we state them, but more visceral in their application; each is a "sniff test," and a procedure that fails these just does not smell right.

6.3.1 Unbiasedness

In Section 5.6, we first introduced the idea of bias: we like an estimator to be, on average, equal to the parameter it is estimating. That is, we want the estimator to be unbiased, or equivalently, bias$\hat{\theta} = \mathrm{E}\left[\hat{\theta}\right] - \theta = 0$.

We have already seen that the sample mean is an unbiased estimator of the population mean μ.

The sample proportion is also an unbiased estimator of the population proportion.

Proposition 6.3 *If $X_1, X_2, \ldots, X_n$ are Bernoulli random variables with parameter p, then* $\mathrm{E}\left[\hat{p}\right] = p$.

Proof Let $X = \sum_{i=1}^{n} X_i$. Then,

$$\mathrm{E}\left[\hat{p}\right] = \mathrm{E}\left[\frac{X}{n}\right] = \frac{np}{n} = p. \qquad \square$$

The case of variance is less straightforward. The maximum likelihood estimate for σ^2 in a normal setting with unknown mean and variance, given by Equation 6.4, is biased. But a minor variation is unbiased.

Theorem 6.2 *Let $X_1, X_2, \ldots, X_n$ be independent random variables from a distribution with unknown* $\mathrm{Var}[X_i] = \sigma^2 < \infty$. *Then an unbiased estimator of σ^2 is* $S^2 = (1/(n-1)) \sum_{i=1}^{n} (X_i - \bar{X})^2$.

Proof We utilize Proposition A.2, Theorem A.5, and the following algebraic fact:

$$\sum_{i=1}^{n} (X_i - \bar{X})^2 = \left(\sum_{i=1}^{n} X_i^2 \right) - n\bar{X}^2. \tag{6.16}$$

Let $\hat{\sigma}^2 = (1/n) \sum_{i=1}^{n} (X_i - \bar{X})^2$ (note that this is the MLE of σ^2 in the normal case, Theorem 6.1). Then,

$$\begin{aligned}
\mathrm{E}\left[\hat{\sigma}^2\right] &= \mathrm{E}\left[\frac{1}{n} \sum_{i=1}^{n} (X_i - \bar{X})^2\right] \\
&= \frac{1}{n} \mathrm{E}\left[\sum_{i=1}^{n} X_i^2 - n\bar{X}^2\right] \\
&= \frac{1}{n} \left[\sum_{i=1}^{n} \mathrm{E}\left[X_i^2\right] - n\mathrm{E}\left[\bar{X}^2\right]\right] \\
&= \frac{1}{n} \left[\sum_{i=1}^{n} (\sigma^2 + \mathrm{E}[X_i]^2) - n\left(\frac{\sigma^2}{n} + \mathrm{E}\left[\bar{X}\right]^2\right)\right] \\
&= \frac{1}{n} \left[\sum_{i=1}^{n} (\sigma^2 + \mu^2) - \sigma^2 - n\mu^2\right] \\
&= \frac{1}{n} \left[n\sigma^2 + n\mu^2 - \sigma^2 - n\mu^2\right] \\
&= \frac{n-1}{n} \sigma^2.
\end{aligned}$$

Thus, $\hat{\sigma}^2$ is a biased estimator of σ^2. However, we can "unbias" it:

$$\frac{n}{n-1}\,\mathrm{E}\left[\hat{\sigma}^2\right] = \sigma^2 \quad \text{or} \quad \mathrm{E}\left[\frac{n}{n-1}\,\hat{\sigma}^2\right] = \sigma^2.$$

Hence, an unbiased estimator of σ^2 is

$$\frac{n}{n-1}\,\hat{\sigma}^2 = \frac{n}{n-1}\cdot\frac{1}{n}\sum_{i=1}^{n}(X_i - \bar{X})^2 = \frac{1}{n-1}\sum_{i=1}^{n}(X_i - \bar{X})^2. \qquad \square$$

Definition 6.3 Let $X_1, X_2, \ldots, X_n$ be independent random variables from a distribution with unknown variance $\sigma^2 < \infty$. The *(sample) variance* of $X_1, X_2, \ldots, X_n$ is

$$S^2 = \frac{1}{n-1}\sum_{i=1}^{n}(X_i - \bar{X})^2,$$

which by Theorem 6.2 is an unbiased estimator of σ^2. ||

In general, if a biased estimator is off by a multiplicative constant, we can "unbias" the estimate by dividing it by that constant. That is, if $\mathrm{E}\left[\hat{\theta}\right] = C \times \theta$ where C does not depend on θ, then $(1/C)\hat{\theta}$ is an unbiased estimator of θ.

Example 6.12 Let $X_1, X_2, \ldots, X_n$ be i.i.d. from Unif$[0, \beta]$. We have already seen that the MLE of β is $\hat{\beta}_{\mathrm{mle}} = X_{\max}$.

With $f_{\max}(x) = (n/\beta^n)x^{n-1}$ (see Corollary 4.1), we have

$$\begin{aligned}\mathrm{E}[X_{\max}] &= \int_0^{\beta} x\,\frac{n}{\beta^n}x^{n-1}\,dx \\ &= \frac{n}{n+1}\beta,\end{aligned}$$

so $\hat{\beta}_{\mathrm{mle}} = X_{\max}$ is a biased estimator of β. However, $((n+1)/n) \times X_{\max}$ is unbiased.

We also have the method of moments estimator $\hat{\beta}_{\mathrm{mom}} = 2\bar{X}$. Computing the expectation, we find $\mathrm{E}\left[\hat{\beta}_{\mathrm{mom}}\right] = \mathrm{E}\left[2\bar{X}\right] = 2(\beta/2) = \beta$. Thus, $2\bar{X}$ is an unbiased estimator of β.

In this case, we have two unbiased estimators of β. To choose between them, we turn to other criteria mentioned below. But first we note some limitations of unbiasedness. $\square$

Limitations of Unbiasedness In practice, we are generally satisfied with estimates that are approximately unbiased; estimates that are exactly unbiased may be impossible to obtain or in some cases are unreasonable. They may also just disagree with our common sense. In Theorem 6.2, we saw that the sample variance $S^2 = (1/(n-1))\sum(X_i - \bar{X})^2$ is an unbiased estimator of the variance σ^2. However, the sample standard deviation S is not an unbiased estimator of the standard deviation σ. For normal distributions, for $n = 10$, $\mathrm{E}[S] \approx 0.973\sigma$. It just seems wrong to use S^2 when estimating σ^2, but to use $S/0.973$ when estimating σ. Furthermore, the correction factor needs to be recomputed for every n. In addition, S^2 is unbiased for the variance for any distribution, not just normal distributions, but the correction factors for S depend on the distribution, which in practice is unknown. So in practice we use S as an estimate of σ. It is approximately unbiased, which is good enough.

If $\hat{\theta}$ is an unbiased estimator of θ and h is a function, then $h(\hat{\theta})$ may not necessarily be an unbiased estimator of $h(\theta)$. A notable exception is when h is a linear transformation, $h(\theta) = a + b\theta$. (Verify!)

For example, consider again the unbiased estimator $\hat{\beta} = 2\bar{X}$ for β in Unif$[0, \beta]$. Is $(\hat{\beta})^2 = 4(\bar{X})^2$ an unbiased estimator of β^2?

$$\begin{aligned}
\mathrm{E}\left[4\bar{X}^2\right] &= 4\mathrm{E}\left[\bar{X}^2\right] \\
&= 4\left(\mathrm{Var}[\bar{X}] + \mathrm{E}\left[\bar{X}\right]^2\right) \\
&= 4\left(\frac{\beta^2}{12n} + \left(\frac{\beta}{2}\right)^2\right) \\
&= \beta^2 + \frac{\beta^2}{3n}.
\end{aligned}$$

Thus, there is a positive bias of $\beta^2/3n$.

Asymptotic Bias In lieu of expecting the bias of an estimator to be zero, we may be satisfied with an estimator whose bias disappears as the sample size increases.

In Theorem 6.2, we saw that the MLE of σ^2 for a sample of size n drawn from $N(\mu, \sigma^2)$ satisfies $\mathrm{E}\left[\hat{\sigma}^2\right] = ((n-1)/n)\sigma^2$, so that $\hat{\sigma}^2$ is a biased estimator of σ^2. But note that

$$\lim_{n\to\infty} \mathrm{E}\left[\hat{\sigma}^2\right] = \lim_{n\to\infty} \frac{n-1}{n}\sigma^2 = \sigma^2,$$

so we say that $\hat{\sigma}^2$ is *asymptotically unbiased*.

Similarly, the MLE, $X_{\max}$ of β for a random sample from the uniform distribution Unif$[0, \beta]$ is also asymptotically unbiased:

$$\lim_{n\to\infty} \mathrm{E}[X_{\max}] = \lim_{n\to\infty} \frac{n}{n+1}\beta = \beta.$$

6.3.2 Efficiency

We have already found two unbiased estimators for β for Unif$[0, \beta]$. Here, we look at one criterion for comparing them—efficiency—that depends on their variance.

We begin with an example comparing two unbiased estimators for a mean. Let X_1, X_2, X_3 be independent random variables from a distribution with mean μ and variance σ^2. Then, $\bar{X}$ is an estimator of μ. Now, consider $Y = (1/6)X_1 + (1/3)X_2 +$

$(1/2)X_3$, a weighted average of X_1, X_2, X_3. Then,

$$\begin{aligned} \mathrm{E}[Y] &= \frac{1}{6}\mathrm{E}[X_1] + \frac{1}{3}\mathrm{E}[X_2] + \frac{1}{2}\mathrm{E}[X_3] \\ &= \frac{1}{6}\mu + \frac{1}{3}\mu + \frac{1}{2}\mu = \mu. \end{aligned}$$

Thus, Y is also an unbiased estimator of μ.

We now have two unbiased estimators of μ, the sample mean $\bar{X}$ and Y. We will compare their variances. Recall that $\mathrm{Var}[\bar{X}] = \sigma^2/3$ (Theorem A.5). On the other hand, since the X_i's are independent,

$$\begin{aligned} \mathrm{Var}[Y] &= \mathrm{Var}\left[\frac{1}{6}X_1\right] + \mathrm{Var}\left[\frac{1}{3}X_2\right] + \mathrm{Var}\left[\frac{1}{2}X_3\right] \\ &= \frac{1}{36}\sigma^2 + \frac{1}{9}\sigma^2 + \frac{1}{4}\sigma^2 = \frac{7}{18}\sigma^2. \end{aligned}$$

Since $\mathrm{Var}[\bar{X}] < \mathrm{Var}[Y]$, we see that $\bar{X}$ is less variable than Y.

Definition 6.4 If $\hat{\theta}_1$ and $\hat{\theta}_2$ are both estimators of θ and $\mathrm{Var}[\hat{\theta}_1] < \mathrm{Var}[\hat{\theta}_2]$, then $\hat{\theta}_1$ is said to be more *efficient* than $\hat{\theta}_2$. ||

Example 6.13 We saw earlier that if $X_1, X_2, \ldots, X_n$ are a random sample from the uniform distribution Unif$[0, \beta]$, then $((n+1)/n)X_{\max}$ and $2\bar{X}$ are two unbiased estimators of β.

We will run a simulation to see how these two estimators perform. We draw random samples of size 25 from Unif[0, 12]. For each sample, we compute $2\bar{x}$ and $25/24 \times \max\{x_1, x_2, \ldots, x_{25}\}$ and record these values. We repeat this 1000 times.

We know that the true β is 12. Which estimator exhibits the smallest amount of variability?

R Note:

```
my.mean <- numeric(1000)
my.max <- numeric(1000)
for (i in 1:1000)
{
  x <- runif(25, 0, 12)        # sample n=25 from Unif[0, 12]
  my.mean[i] <- 2 * mean(x)
  my.max[i] <- 25/24 * max(x)
}

mean(my.mean)
sd(my.mean)
mean(my.max)
sd(my.max)
```

Here we scale the axes to be the same. You may need to adjust this setting for your simulation.

```
hist(my.mean, xlim=c(8,16), ylim=c(0,650), xlab="2*mean")
dev.new()          # open new graphics device
hist(my.max,  xlim=c(8,16), ylim=c(0,650), xlab="(25/24)*max")
```

In one run of this simulation, we obtained a mean of 12.0054 and standard deviation of 1.3721 for the estimates of β based on the method of moments $2\bar{X}$ compared to a mean of 12.0003 and standard deviation of 0.46224 for the estimates based on $((n+1)/n)X_{\max}$. Histograms are shown in Figure 6.5.

The simulation shows that the estimates based on $((n+1)/n)X_{\max}$ have less variability. We can also compute the variances exactly.

$$\mathrm{Var}[2\bar{X}] = 2^2\mathrm{Var}[\bar{X}] = 2^2\frac{\beta^2}{12n} = \frac{\beta^2}{3n}.$$

On the other hand,

$$\begin{aligned}
\mathrm{Var}\left[\frac{n+1}{n}X_{\max}\right] &= \frac{(n+1)^2}{n^2}\mathrm{Var}[X_{\max}] \\
&= \frac{(n+1)^2}{n^2}\left\{\mathrm{E}\left[X_{\max}^2\right] - \mathrm{E}[X_{\max}]^2\right\} \\
&= \frac{(n+1)^2}{n^2}\left\{\int_0^\beta x^2\frac{n}{\beta^n}x^{n-1}\,dx - \left(\int_0^\beta x\frac{n}{\beta^n}x^{n-1}\,dx\right)^2\right\} \\
&= \frac{\beta^2}{n(n+2)}.
\end{aligned}$$

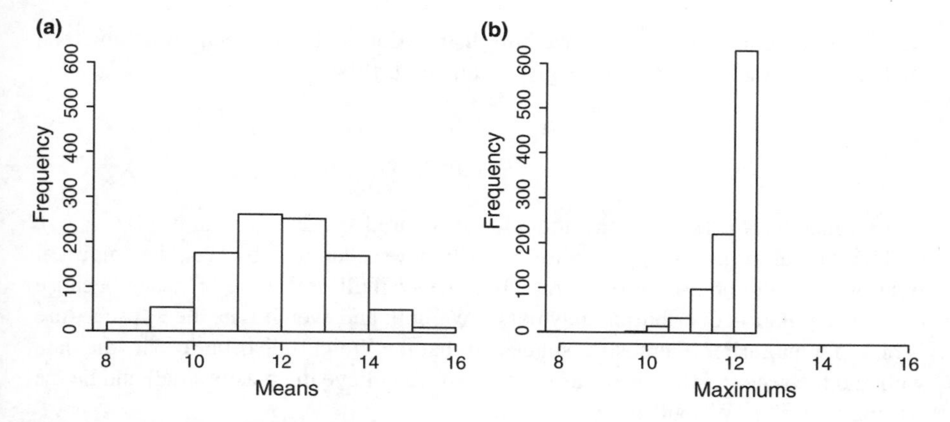

FIGURE 6.5 Sampling distribution of estimators for β. (a) Distribution of $2\bar{X}$. (b) Distribution of $(25/24)X_{\max}$.

Since $\beta^2/(n(n+2)) < \beta^2/3n$, for $n > 1$, $((n+1)/n)X_{\max}$ is more efficient than $2\bar{X}$. □

Remark Suppose we have an unbiased estimators $\hat{\theta}$ of a parameter θ. Are there other unbiased estimators of θ that are more efficient than $\hat{\theta}$? There is a theorem that provides a lower bound for the variance of any unbiased estimator of θ.

Cramer–Rao Inequality If $X_1, X_2, \ldots X_n$ are a random sample from a distribution with continuous pdf $f(x;\theta)$ and f satisfies certain smoothness criteria, then any unbiased estimator $\hat{\theta}$ of θ satisfies

$$\text{Var}[\hat{\theta}] \geq \frac{1}{n\,\text{E}\left[\left((\partial/\partial\theta)\ln(f(X;\theta))\right)^2\right]}.$$

Thus, if the variance of an unbiased estimator θ achieves this lower bound, then in some sense it is a "best" estimator. However, there are cases where *no* unbiased estimator of a parameter θ achieves the lower bound in the Cramer–Rao inequality. ||

In practice, we may use efficiency to compare estimators, even if we do not know if the estimators are unbiased.

Example 6.14 We continue the Wind Energy Case Study in Section 6.1.3. We are interested in the fraction of time that the wind speed exceeds 5 m/s. Two ways to estimate this are the empirical fraction of the 168 measurements that exceed 5 and the probability based on the Weibull distribution with parameters estimated using maximum likelihood.

$$\hat{\theta}_1 = 126/168 = 0.75,$$
$$\hat{\theta}_2 = 1 - F(5;\alpha = 3.169, \lambda = 7.661) = 0.772,$$

where $F(\cdot;\alpha,\lambda)$ is the CDF of a Weibull distribution with the specified parameters. Bootstrapping these two estimates gives standard errors:

$$s_{\hat{\theta}_1} = 0.033,$$
$$s_{\hat{\theta}_2} = 0.023.$$

The estimated *relative efficiency*, the ratio of squared standard errors, is 2.02.

The Weibull procedure appears to be much more accurate. However, the empirical fraction θ_1 is an unbiased procedure, while the Weibull procedure is biased because the true population distribution may not be Weibull; and even if it is, the θ_2 procedure is biased. Our earlier diagnostics suggested that the Weibull distribution fits the data well and the second effect is usually small, so we believe the bias is small and hence willing to use the Weibull-based estimate.

The relative efficiency is amazingly large. In practice, statisticians may put great effort into developing estimation procedures with a relative efficiency gain of 1%, because data may be very expensive to collect, and this would cut data requirements by 1%.

For this example, the code for the bootstrap procedure is provided on the web site `https://sites.google.com/site/ChiharaHesterberg`. □

6.3.3 Mean Square Error

We now turn to a criterion that combined bias and variance. This is useful for comparing estimators that are not both unbiased. We may prefer an estimator with small bias and small variance over one that is unbiased but with large variances. This criterion provides a way to quantify the preference.

Definition 6.5 The *Mean Square Error* of an estimator is $\text{MSE}[\hat{\theta}] = \text{E}\left[(\hat{\theta} - \theta)^2\right]$. ||

The mean square error measures the average squared distance between the estimator and the parameter. It combines bias and variance in a particular way.

Proposition 6.4 $\text{MSE}[\hat{\theta}] = \text{Var}[\hat{\theta}] + \text{Bias}[\hat{\theta}]^2$.

Proof

$$\begin{aligned}
\text{MSE}[\hat{\theta}] &= \text{E}\left[(\hat{\theta} - \theta)^2\right] \\
&= \text{E}\left[(\hat{\theta} - \text{E}\left[\hat{\theta}\right] + \text{E}\left[\hat{\theta}\right] - \theta)^2\right] \\
&= \text{E}\left[((\hat{\theta} - \text{E}\left[\hat{\theta}\right]) + (\text{E}\left[\hat{\theta}\right] - \theta))^2\right] \\
&= \text{E}\left[(\hat{\theta} - \text{E}\left[\hat{\theta}\right])^2\right] + 2\text{E}\left[\hat{\theta} - \text{E}\left[\hat{\theta}\right]\right](\text{E}\left[\hat{\theta}\right] - \theta) + (\text{E}\left[\hat{\theta}\right] - \theta)^2 \\
&= \text{Var}[\hat{\theta}] + \left(\text{Bias}[\hat{\theta}]\right)^2.
\end{aligned}$$

□

Also, if $\hat{\theta}$ is unbiased, then $\text{MSE}[\theta] = \text{Var}[\hat{\theta}]$. So, the unbiased estimator $\hat{\theta}_1$ of θ is more efficient than the unbiased estimator $\hat{\theta}_2$ if and only if $\text{MSE}[\hat{\theta}_1] < \text{MSE}[\hat{\theta}_2]$.

MEAN SQUARE ERROR

The mean square error takes into account the variability of the estimator as well as the bias.

In general, when comparing two estimators $\hat{\theta}_1$ and $\hat{\theta}_2$ of θ, we are often faced with a trade-off between small variability or small bias.

Example 6.15 Let $X \sim \text{Bin}(n, p)$, n known and p unknown. The sample proportion $\hat{p}_1 = X/n$ is an unbiased estimator of p (see Exercises 1 and 21) with $\text{E}\left[\hat{p}_1\right] = np$ and $\text{Var}[\hat{p}_1] = p(1 - p)/n$, and mean square error $\text{MSE}[\hat{p}_1] = p(1 - p)/n$.

An alternative estimator of p is $\hat{p}_2 = ((X+1)/(n+2))$; this adds one artificial success and one failure to the real data. Then,

$$\mathrm{E}\left[\hat{p}_2\right] = \frac{1}{n+2}(\mathrm{E}[X] + \mathrm{E}[1]) = \frac{np+1}{n+2}.$$

This is a case where we will not be able to "unbias" $\hat{p}_2$. The bias is

$$\mathrm{Bias}[\hat{p}_2] = \frac{np+1}{n+2} - p = \frac{1-2p}{n+2}$$

and the variance is

$$\begin{aligned}\mathrm{Var}[\hat{p}_2] &= \mathrm{Var}\left[\frac{X+1}{n+2}\right] = \frac{1}{(n+2)^2}(\mathrm{Var}[X] + \mathrm{Var}[1]) = \frac{1}{(n+2)^2} np(1-p)\\ &= \frac{np(1-p)}{(n+2)^2}\end{aligned}$$

resulting in mean square error

$$\mathrm{MSE}[\hat{p}_2] = \frac{np(1-p)}{(n+2)^2} + \left(\frac{1-2p}{n+2}\right)^2 = \frac{np(1-p) + (1-2p)^2}{(n+2)^2}.$$

Figure 6.6 shows the mean square error for both estimators as a function of p. Thus, while $\hat{p}_2$ is biased (except for $p = 0.5$), it has smaller mean square error than $\hat{p}_1$ except for p near 0 or 1.

R Note:

```
n <- 16
curve(x*(1-x)/n, from=0, to=1, xlab= "p", ylab="MSE")
curve(n*(1-x)*x/(n+2)^2 + (1-2*x)^2/(n+2)^2, add=TRUE, col="blue", lty=2)
```

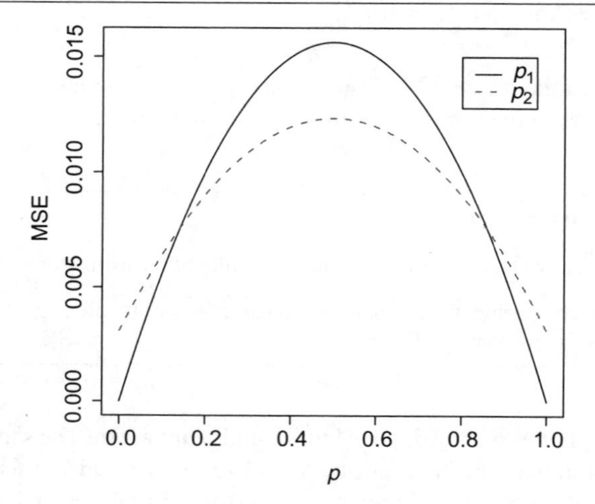

FIGURE 6.6 Mean square error against p, $n = 16$.

In Exercise 31, you will investigate what happens to the mean square error when you adjust the sample size n. □

6.3.4 Consistency

As we get more data, we expect an estimator to become more accurate. Our next criterion, *consistency*, says roughly that an estimator gives the right answer in the long run, as the sample size goes to infinity. This is a "sniff test" criterion—it is rare to use an estimator that fails this test.

Definition 6.6 For a random sample of size n, let $\hat{\theta}_n$ denote an estimator of θ and let $\{\hat{\theta}_n\}_{n=0}^{\infty}$ be a sequence of estimators. The estimators are *consistent* for θ if and only if

$$\lim_{n\to\infty} P(|\hat{\theta}_n - \theta| < \epsilon) = 1 \tag{6.17}$$

for every $\epsilon > 0$. ||

Since Equation 6.17 can also be expressed as $\lim_{n\to\infty} P(|\hat{\theta}_n - \theta| \geq \epsilon) = 0$, this says that for any acceptable amount of error $\epsilon > 0$, the probability of an actual error worse than ϵ goes to zero.

Remark

- The limit in Equation 6.17 is referred to as *convergence in probability*.
- The "sequence of estimators" referred to in the definition generally refers to a single estimation procedure, such as $\bar{X}$, S^2, or $((n+1)/n)X_{\max}$, being recomputed every time an observation is added to the sample. ||

Example 6.16 Let $X_1, X_2, \ldots, X_n$ denote a random sample from $N(\mu, 1)$ and let $\bar{X}_n$ denote the sample mean.

From Theorem A.5, $\bar{X}_n$ is normal with mean μ and variance $1/n$. Thus,

$$\begin{aligned}\lim_{n\to\infty} P(|\bar{X}_n - \mu| < \epsilon) &= \lim_{n\to\infty} P(\mu - \epsilon < \bar{X}_n < \mu + \epsilon)\\ &= \lim_{n\to\infty} \frac{1}{\sqrt{2\pi/n}} \int_{\mu-\epsilon}^{\mu+\epsilon} e^{-(t-\mu)^2/(2/\sqrt{n})}\, dt\\ &= \lim_{n\to\infty} \frac{1}{\sqrt{2\pi}} \int_{-\epsilon\sqrt{n}}^{\epsilon\sqrt{n}} e^{-z^2/2}\, dz, \qquad \text{where } z = \sqrt{n}(t-\mu).\end{aligned}$$

Since we are integrating the standard normal density over an interval approaching $(-\infty, \infty)$, the limit is 1.

Thus, the sample means are a consistent sequence of estimators of μ. □

Rather than integrating pdf's, we can use the mean square error of an estimator to determine consistency. Before we prove this, we need some preliminary results.

Theorem 6.3 (**Markov's Inequality**) *Let X be a random variable, $X \geq 0$. For any $c > 0$,*

$$P(X \geq c) \leq \frac{\mathrm{E}[X]}{c}.$$

Proof Let f denote the pdf of X and $\mathrm{Im}(X)$ denote the range of X. Set $A = \{X \geq c\}$. Then,

$$\mathrm{E}[X] = \int_{t \in \mathrm{Im}(X)} t\, f(t)\, dt \geq \int_{t \in A} t\, f(t)\, dt = \int_{c}^{\infty} t\, f(t)\, dt \geq c\, P(X \geq c).$$

□

Proposition 6.5 (**Chebyshev's Inequality**) *Let X be a random variable with finite mean μ and variance σ^2. Then, for any $k > 0$,*

$$P(|X - \mu| \geq k) \leq \frac{\sigma^2}{k^2}.$$

Proof We apply Markov's inequality to $(X - \mu)^2$ setting $c = k^2$:

$$P((X - \mu)^2 \geq k^2) \leq \frac{\mathrm{E}\left[(X - \mu)^2\right]}{k^2}.$$

But $P((X - \mu)^2 \geq k^2) = P(|X - \mu| \geq k)$, so the result follows. □

There is a close connection between consistency and mean square error—an estimator is consistent if its MSE goes to zero.

Proposition 6.6 *Let $\{\hat{\theta}_n\}$ be a sequence of estimators for θ. If*

$$\lim_{n \to \infty} \mathrm{E}\left[\hat{\theta}_n\right] = \theta \quad \textit{and} \quad \lim_{n \to \infty} \mathrm{Var}[\hat{\theta}_n] = 0,$$

then θ_n is consistent for θ.

Proof Recall from Proposition 6.4 that

$$\mathrm{E}\left[(\hat{\theta}_n - \theta)^2\right] = \mathrm{Var}[\hat{\theta}_n] + \mathrm{Bias}[\hat{\theta}_n]^2.$$

In addition, Chebyshev's inequality gives

$$P(|\hat{\theta}_n - \theta| \geq \epsilon) \leq \frac{\mathrm{E}\left[(\hat{\theta}_n - \theta)^2\right]}{\epsilon^2}.$$

Putting these two results together gives

$$P(|\hat{\theta}_n - \theta| \geq \epsilon) \leq \frac{\mathrm{Var}[\hat{\theta}_n] + \mathrm{Bias}[\hat{\theta}_n]^2}{\epsilon^2}.$$

Thus, $\lim_{n\to\infty} \mathrm{E}\left[\hat{\theta}_n\right] = \theta$ and $\lim_{n\to\infty} \mathrm{Var}[\hat{\theta}_n] = 0$ give $\lim_{n\to\infty} P(|\hat{\theta}_n - \theta| \geq \epsilon) = 0$. This is equivalent to Equation 6.17. □

Example 6.17 Returning to Example 6.16, we have that $\mathrm{E}\left[\bar{X}_n\right] = \mu$ and $\mathrm{Var}[\bar{X}_n] = 1/n$. So by Proposition 6.6, the sample means are a consistent sequence of estimators of μ.

More generally, let $X_1, X_2, \ldots, X_n$ be i.i.d. from a distribution with mean μ and variance σ^2. If $\bar{X}_n$ denotes the sample mean, then $\mathrm{E}\left[\bar{X}\right] = \mu$ and $\mathrm{Var}[\bar{X}_n] = \sigma^2/n$, so again by Proposition 6.6, the sample means are a consistent sequence of estimators for μ. □

Example 6.18 Let $X_1, X_2, \ldots, X_n$ be a random sample from the Cauchy distribution (refer to Example 6.7). It is a fact that neither the mean nor any of its moments exist, although we can still compute the sample mean $\bar{X}_n$. Is this sample mean a consistent estimator of θ? It turns out this sample mean has the same distribution as a single observation (Stuart and Ord (2009))! Hence, $\bar{X}_n$ is not a consistent estimator of θ.

On the other hand, the sample medians are a consistent estimator of θ. We illustrate this with a simulation: we draw random samples of sizes n from a Cauchy distribution with $\theta = 5$, compute the sample medians, and plot the distribution. Taking $\epsilon = 0.05$, we note the proportion of sample medians that fall within $\theta \pm \epsilon$. Figure 6.7 gives the results for runs of this simulation for different sample sizes. We can see that as sample sizes increase, a larger proportion of the sample medians fall within $\theta \pm \epsilon$. For $n = 250$, only 38.1% of the sample medians fall within this interval. On the other hand, for $n = 5000$, 97.8% of the sample medians are within ϵ of θ. □

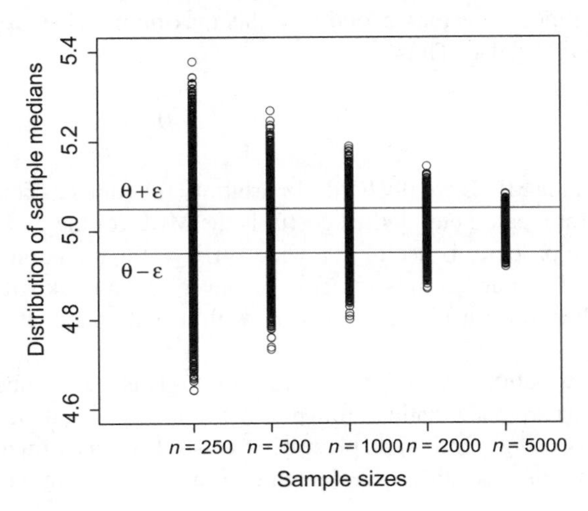

FIGURE 6.7 Comparison of distributions of sample medians from a Cauchy distribution with $\theta = 5$. For $\epsilon = 0.05$, the fraction of medians that fall in 5 ± 0.05 are 0.381, 0.521, 0.683, 0.847, and 0.978 for sample sizes 250, 500, 1000, 2000, and 5000, respectively.

6.3.5 Transformation Invariance

In Section 6.3.1, page 148, we declared that it seems wrong to use S^2 when estimating σ^2, but to use $S/0.973$ when estimating σ. This discrepancy is ugly. Our next criterion is one that, if satisfied, prevents such discrepancies.

An estimation procedure is *invariant to parameter transformations* if it yields equivalent answers for transformations of parameters.

For example, the exponential distribution (Section B.8) can be given in terms of a rate parameter λ or a scale parameter $\rho = 1/\lambda$,

$$f(x) = \lambda e^{-\lambda x} = \frac{1}{\rho} e^{-x/\rho}$$

for $x \geq 0$. Both maximum likelihood and method of moments are invariant under this parameter transformation, with $\hat{\rho} = 1/\hat{\lambda}$.

Proposition 6.7 *Let $f(x;\theta) = g(x;\rho)$, where f and g are discrete or continuous densities with parameters θ and ρ related by $\rho = h(\theta)$ for some invertible function h. If $\hat{\theta}$ and $\hat{\rho}$ are the maximum likelihood estimators for θ and ρ, then $\hat{\rho} = h(\hat{\theta})$.*

Proof Let L_{f} denote the likelihood function for the θ parameterization and L_{g} the likelihood function for the ρ parameterization. We have

$$L_{\mathrm{f}}(\theta) = \prod f(x_i;\theta) = \prod g(x_i;\rho) = L_{\mathrm{g}}(\rho)$$

when $\rho = h(\theta)$. Since L_{f} is maximized at $\hat{\theta}$, this maximum value $L_{\mathrm{f}}(\hat{\theta})$ must be the maximum value of L_{g} also. Thus,

$$L_{\mathrm{f}}(\hat{\theta}) = L_{\mathrm{g}}(h(\theta)) = L_{\mathrm{g}}(\hat{\rho}).$$

□

The invariance of the MLE actually holds for arbitrary functions g. Thus, for instance, if $\hat{\theta}$ is the MLE for a parameter θ, then $\cos(\hat{\theta})$ is the MLE for the parameter $\cos(\theta)$.

The invariance property is useful in a wide variety of applications. For example, with exponential distributions, it is sometimes convenient to work with a rate parameter λ, while other times it is useful to work with a scale parameter $\rho = 1/\lambda$ (see Section B.8).

For normal distributions, we may work with σ or σ^2. This case is subtle—it involves thinking of σ^2 as a parameter distinct from σ. Let $\rho = h(\sigma) = \sigma^2$, then we may rewrite the normal density using ρ and $\sqrt{\rho}$ in place of σ^2 and σ. If we maximize the likelihood with respect to ρ, this gives the equivalent answer as maximizing with respect to σ, with $\hat{\rho} = \hat{\sigma}^2$.

Similarly, if p is the probability for a binomial distribution and $\hat{p}$ is its MLE (see Exercise 1), then the MLE for $\sqrt{p(1-p)}$ is $\sqrt{\hat{p}(1-\hat{p})}$ (for $p < 0.5$).

Methods of moments estimators are also invariant under parameter transformations.

6.4 EXERCISES

1. Let X be a binomial random variable, $X \sim \text{Binom}(n, p)$. Show that the MLE of p is $\hat{p} = X/n$.
2. Prove Proposition 6.2.
3. Suppose a random sample with $X_1 = 5, X_2 = 9, X_3 = 9, X_4 = 10$ is drawn from a distribution with pdf $f(x; \theta) = (\theta/(2\sqrt{x}))e^{-\theta\sqrt{x}}$, where $x > 0$. Use maximum likelihood to find an estimate for θ.
4. Let $X_1 = x_1, X_2 = x_2, \ldots X_n = x_n$ be a random sample from the distribution with pdf $f(x; \theta) = (x^3 e^{-x/\theta}/6\theta^4)$, $x \geq 0$. Calculate the maximum likelihood estimate of θ.
5. Recall Theorem 6.1 where we found the maximum likelihood estimates for μ and σ for a random sample $X_1, X_2, \ldots, X_n \sim N(\mu, \sigma^2)$.
   (a) Suppose instead, μ is unknown but σ is *known*. Find the maximum likelihood estimate of μ.
   (b) Now suppose σ is unknown and μ is known. Find the MLE of σ.
6. Let $X_1 = x_1, X_2 = x_2, \ldots, X_n = x_n$ be a random sample from a distribution with pdf $f(x; \theta) = e^{\theta - x}$ for $x > \theta > 0$.
   (a) Show that the MLE of θ does not exist.
   (b) What if we change the domain of x to $x \geq \theta > 0$?
7. Let $X_1, X_2, \ldots, X_n$ be a random sample from a distribution with pdf $f(x; \theta) = e^{-x}/(1 - e^{-\theta})$, $0 < x < \theta$. Find the MLE of θ.
8. The Maxwell–Boltzmann distribution is used to model the speed of particles in gases. The pdf is $f(x; \theta) = \sqrt{2/\pi}\, x^2 e^{-x^2/(2\theta^2)}/\theta^3$. If $X_1, X_2, \cdots, X_n$ are a random sample from this distribution, find the MLE of θ.
9. Katie has N keys in her purse of which one opens the door to her house. When she arrives home at the end of a long day, she puts her hand into her purse, randomly grabs a key and tries to unlock the door. If she fails, she puts the key back into her purse and then randomly draws another key. Suppose over the course of 5 days, she manages to unlock her front door on the 8th, 12th, 7th, 6th, and 12th attempts. Find the MLE of N, the number of keys in her purse. (Assume the keys are similar, so each has the same chance to be picked.)
10. Suppose the weight X (in pounds) of a girl in a certain town has distribution $N(\mu, 15^2)$, while the weight Y of a boy in this town has distribution $N(1.3\mu, 20^2)$. The weights of a randomly chosen girl and boy are $x = 95$, $y = 130$ pounds, respectively. Find the MLE of μ.
11. Let $X_1, X_2, \ldots, X_n \overset{\text{i.i.d.}}{\sim} \text{Exp}(\lambda)$ and $Y_1, Y_2, \ldots, Y_m \overset{\text{i.i.d.}}{\sim} \text{Exp}(2\lambda)$, and assume the X_i's are independent of the Y_j's. Find the MLE of λ.
12. Let $X_1, X_2, \ldots, X_n$ denote a random sample from the gamma distribution Gamma(r, λ). Assume r and λ are unknown. Write down the equations that would be solved simultaneously to find the maximum likelihood estimators of r and λ.

13. Suppose the random variable X has pdf $f(x; \alpha, \beta) = \alpha\beta x^{\beta-1} e^{-\alpha x^{\beta}}$ for $x \geq 0; \alpha, \beta > 0$.
    (a) Find the maximum likelihood estimator for α, assuming that β is known.
    (b) Suppose α and β are both unknown. Write down the equations that would be solved simultaneously to find the maximum likelihood estimators of α and β.
14. Let the five numbers 2, 3, 5, 9, 10 come from the uniform distribution on $[\alpha, \beta]$. Find the method of moments estimates of α and β.
15. Let $x_1 = 7.13, x_2 = 5.26, x_3 = 9.93, x_4 = 6.62, x_5 = 7.52$ be five random numbers from the gamma distribution Gamma(r, λ). Use the method of moments to find estimates for r and λ.
16. Nobody likes to stand in line to order a hamburger or buy groceries, so understanding how to model service times (time to be served) is one area of research in queuing theory. Four students collected data on customers waiting in line at a college snack bar (Haynor et al. (2010)). The data set `Service` contains service times (in minutes) for 174 customers. We will model the distribution of service times using the gamma distribution.
    (a) Use the method of moments to estimate the parameters of the gamma distribution.
    (b) Use a goodness-of-fit test to see whether the gamma distribution is an adequate model for these data.
    (c) Provide a histogram and ecdf of the data with the gamma distribution superimposed.
17. The Weibull distribution has been used to model the time between successive earthquakes (Hasumi et al. (2009); Tiampo et al. (2008)). The data set `Quakes` contains the time between earthquakes (in days) for all earthquakes of magnitude 6 or greater from 1970 through 2009 (from `http://earthquake.usgs.gov/earthquakes/eqarchives/`). Modify the `R` scripts from the Wind Speeds Case Study to see if the times between earthquakes can be modeled with the Weibull distribution. Include graphs like those in Figure 6.4. *Hint:* When searching for the shape parameter with the `uniroot` command, use `lower=0.8, upper=1`.
18. Let $X_1 = 2, X_2 = 2, X_3 = 8, X_4 = 12$ be a random sample from the distribution with pdf $f(x) = \lambda e^{-\lambda(x-a)}$ where $x \geq a > 0$ and $\lambda > 0$. Use the method of moments to find estimates of λ and a.
19. Let $X_1 = x_1, X_2 = x_2, \ldots, X_n = x_n$ denote a random sample from a distribution with pdf $f(x; \theta) = \theta 2^{\theta}/x^{\theta+1}, x \geq 2, \theta > 1$.
    (a) Use the method of moments to estimate θ.
    (b) Use maximum likelihood to estimate θ.
20. Let $X_1 = 0.4, X_2 = 0.5, X_3 = 0.25, X_4 = 0.9, X_5 = 0.92$ be a random sample from a distribution with pdf $f(x; \theta) = \theta x^{\theta-1}, 0 \leq x \leq 1, \theta > 0$.
    (a) Find the MLE of θ.
    (b) Find the method of moments estimate of θ.

21. Let X be a binomial random variable $X \sim \text{Binom}(n, p)$. Show that $\hat{p} = X/n$ is an unbiased estimator of p.
22. Verify Equation 6.16.
23. Return to the setting in Exercise 6. We will use $\bar{X}$ as an estimator of θ. Determine whether or not $\bar{X}$ is an unbiased estimator. If it is biased, calculate the bias and determine whether or not it is possible to obtain an unbiased estimator from it.
24. Let $X_1, X_2, \ldots, X_n$ be i.i.d. from the negative binomial distribution with $P(X = x) = \binom{x-1}{r-1} p^r (1-p)^{x-r}$, $x = r, r+1, \ldots$ (see Section B.4). Show that $\hat{p} = (r-1)/(X-1)$ is an unbiased estimator of p. *Hint*: $1/(1-w)^m = \sum_{k=0}^{\infty} \binom{k+m-1}{m-1} w^k$.
25. Let $X_1, X_2, \ldots, X_n$ be random variables with $\text{E}[X_i] = \mu, i = 1, 2, \ldots, n$. Under what condition on the constants $a_1, a_2, \ldots, a_n$ is $X = a_1X_1 + a_2X_2 + \cdots + a_nX_n$ an unbiased estimator of μ?
26. Let $X_1, X_2, \ldots, X_n$ be independent random variables from a distribution with mean $\mu = 0$ and variance σ^2. Let $W = (1/n) \sum_{i=1}^{n} X_i^2$. Show that W is an unbiased estimator of σ^2.
27. Let $X_1, X_2, \ldots, X_n \overset{\text{i.i.d.}}{\sim} N(\mu, \sigma^2)$ and $\hat{\sigma}^2$ denote the MLE of σ^2 (Theorem 6.1).
    (a) Find the bias of $\hat{\sigma}^2$.
    (b) Find the variance of $\hat{\sigma}^2$. *Hint*: Theorem B.16.
    (c) Find the mean square error of $\hat{\sigma}^2$.
28. Let $\hat{\theta}_1$ and $\hat{\theta}_2$ be two estimators of θ with $\text{E}\left[\hat{\theta}_1\right] = 0.9\theta$ and $\text{E}\left[\hat{\theta}_2\right] = 1.2\theta$. Also, suppose $\text{Var}[\hat{\theta}_1] = 3$ and $\text{Var}[\hat{\theta}_2] = 2$. Find two unbiased estimators of θ and determine which one is more efficient.
29. Suppose X_1, X_2, X_3 are a random sample of size 3 from a distribution with pdf $f(x) = (1/\theta)e^{-x/\theta}$ for $x > 0, \theta > 0$. Let $\hat{\theta}_1 = X_1$, $\hat{\theta}_2 = (X_1 + X_2)/2$ and $\hat{\theta}_3 = (X_1 + 2X_2)/3$. Show that these estimators of θ are all unbiased, and determine the relative efficiencies between them.
30. Let $\hat{\theta}_1$ and $\hat{\theta}_2$ be two different estimators for a parameter θ. Suppose $\text{Var}[\hat{\theta}_1] = 25$ and $\text{Var}[\hat{\theta}_2] = 4$.
    (a) If $\text{E}\left[\hat{\theta}_1\right] = \theta$ and $\text{E}\left[\hat{\theta}_2\right] = \theta + 3$, which estimator has the smaller mean square error?
    (b) Suppose $\text{E}\left[\hat{\theta}_1\right] = \theta$ and $\text{E}\left[\hat{\theta}_2\right] = \theta + b$, for some positive number b. For what values of b (if any) does $\hat{\theta}_2$ have a smaller mean square error than $\hat{\theta}_1$?
31. Consider the MSE of the two different estimators for the binomial proportion in Section 6.3.3. The `R` code shows how to graph the MSE as a function of p. Recreate these graphs but with sample sizes $n = 30$, $n = 50$, $n = 100$, $n = 200$. What is the effect of increasing the sample size?
32. Let $X_1, X_2, \ldots, X_n \overset{\text{i.i.d.}}{\sim} \text{Unif}[0, \beta]$. Let $\hat{\beta}_1 = (n+1)X_{\text{min}}$.
    (a) Is $\hat{\beta}_1$ an unbiased estimator of β?

(b) Let $\hat{\beta}_2$ be $((n+1)/n)X_{\max}$. (This is an unbiased estimator that we considered earlier in this chapter.) Compute $\text{Var}[\hat{\beta}_1]/\text{Var}[\hat{\beta}_2]$. What do you conclude?

33. Let X_1, X_2, X_3 be randomly drawn from a distribution with pdf $f(x;\theta) = 2\theta^2 x$ for $0 < x < 1/\theta$.
    (a) Find the expected value of X_i.
    (b) Let $T = X_1/9 + X_2/9 + X_3/3$ be an estimator of $1/\theta$. Find the bias and mean square error.
    (c) If possible, use T to get an unbiased estimator for $1/\theta$.
34. Let $X_1, X_2, \ldots, X_n$ be a random sample from a distribution with pdf $f(x) = (6/\theta^6)x^5$, $0 \le x \le \theta$. We will use $X_{\max}$ as an estimator of θ.
    (a) Find the pdf for $X_{\max}$, the maximum of the sample.
    (b) Compute $\text{E}[X_{\max}]$.
    (c) Compute the bias of $X_{\max}$.
    (d) Compute the mean square error of $X_{\max}$.
35. Let $X_1, X_2, \ldots, X_n$ be i.i.d. from a distribution with pdf $f(x;\beta) = \alpha x^{\alpha-1}/\beta^{\alpha}$ for $0 \le x \le \beta$, where $\alpha > 0$ is a known constant but β is unknown.
    (a) Find the pdf for $X_{\max}$, the maximum of the sample.
    (b) Compute $\text{E}[X_{\max}]$.
    (c) Compute the bias of $X_{\max}$.
    (d) Compute the mean square error of $X_{\max}$.
36. Let $X_1, X_2, \ldots, X_n$ be independent random variables with mean μ and variance σ_1^2. Let $Y_1, Y_2, \ldots, Y_m$ be independent random variables with mean μ and variance σ_2^2. Let $W = a\bar{X} + (1-a)\bar{Y}$, where $0 < a < 1$.
    (a) Compute the expected value of W.
    (b) For what value of a is the variance of W a minimum?
37. Let X_1, X_2 be independent exponential random variables with parameter λ. Let $\bar{X} = (X_1 + X_2)/2$ be an estimator of $1/\lambda$.
    (a) Show that $\bar{X}$ is an unbiased estimator of $1/\lambda$.
    (b) Show that $\text{Var}[\bar{X}] = 1/(2\lambda^2)$.
    (c) Show that $\text{E}\left[\sqrt{X_1 X_2}\right] = \pi/(4\lambda)$. *Fact:* $\text{E}\left[\sqrt{X_i}\right] = \sqrt{\pi}/(2\sqrt{\lambda})$.
    (d) Compute the bias of the estimator $\sqrt{X_1 X_2}$ of $1/\lambda$.
38. In Chapter 5, we claimed that "the ratio of sample means is not generally an unbiased estimator of the ratio of population means." Prove this under the additional assumptions that all random variables are strictly positive, that the denominator has nonzero variance, and that the numerator and denominator are independent.
    (a) Show that $1/\bar{X}$ is not unbiased for $1/\mu$. *Hint:* Approximate $f(x) = 1/x$ with a Taylor series expansion about μ and then evaluate at $x = \bar{X}$.
    (b) Show that if the two samples are independent, $\bar{Y}/\bar{X}$ is not unbiased for μ_Y/μ_X.

39. Let $X_1, X_2, \ldots, X_n \overset{\text{i.i.d.}}{\sim}$ Unif$[0, \beta]$ and let $\hat{\beta}_n = X_{\max}$. Show that the sequence $\{\hat{\beta}_n\}$ is consistent for β.
40. Let $X_1, X_2, \ldots, X_n$ be i.i.d. from a normal distribution with mean μ and variance σ^2. Show that $\hat{\sigma}_n^2 = (1/n)\sum_{i=1}^{n}(X_i - \bar{X})^2$ is a consistent estimator of σ^2. *Hint:* Theorem B.16.
41. Let $X_1, X_2, \cdots, X_n$ be a random sample from the exponential distribution with parameter $\lambda > 0$. Determine whether or not $\hat{\lambda} = \sum_{i=1}^{n} X_i$ is a consistent estimator of λ.

7

CLASSICAL INFERENCE: CONFIDENCE INTERVALS

In 2010, according to an AP-Gfk Poll conducted on October 13–18, 59% of 846 likely voters responded that they felt things in this country were heading in the wrong direction (`http://www.ap-gfkpoll.com/poll_archive.html`). We learned in Chapter 6 that $\hat{p} = 0.59$ is an unbiased point estimate of the true proportion p of those who think the country is headed in the wrong direction, but we do not have any indication of how far off $\hat{p}$ is from the true p. In Chapter 5, we used bootstrap percentile confidence intervals to give a range of plausible values for a parameter. In this chapter, we learn some other ways to obtain confidence intervals.

7.1 CONFIDENCE INTERVALS FOR MEANS

We begin with confidence intervals for a mean, or a difference in means. These are important in their own right and are instructive for other situations.

7.1.1 Confidence Intervals for a Mean, σ Known

Example 7.1 The Centers for Disease Control maintains growth charts for infants and children (`http://www.cdc.gov/growthcharts/zscore.htm`). For 13-year-old girls, the mean weight is 101 pounds with a standard deviation of 24.6 pounds. We assume that weights are normally distributed. The public health officials in Sodor are interested in the weights of the teens in their town: they suspect that the

Mathematical Statistics with Resampling and R, First Edition. By Laura Chihara and Tim Hesterberg.

mean weight of their girls might be different from the mean weight in the growth chart but are willing to assume that the variation is the same. If they survey a random sample of 150 thirteen-year-old girls and find that their mean weight—an estimate of the population mean weight—is 95 pounds, how accurate will this estimate be?

We assume the 150 sample values are from a normal distribution, $N(\mu, 24.6^2)$. Then the sampling distribution of mean weights is $N(\mu, 24.6^2/150)$ by Corollary A.2. Let $\bar{X}$ denote the mean of the 150 weights, so standardizing gives $Z = (\bar{X} - \mu)/(24.6/\sqrt{150}) \sim N(0, 1)$. For the standard normal random variable Z, we have $P(-1.96 < Z < 1.96) = 0.95$. Thus, we compute

$$
\begin{aligned}
0.95 &= P\left(-1.96 < \frac{\bar{X} - \mu}{24.6/\sqrt{510}} < 1.96\right) && (7.1)\\
&= P\left(-1.96(24.6/\sqrt{150}) < \bar{X} - \mu < 1.96(24.6/\sqrt{150})\right) && (7.2)\\
&= P\left(-\bar{X} - 1.96(24.6/\sqrt{150}) < -\mu < -\bar{X} + 1.96(24.6/\sqrt{150})\right) && (7.3)\\
&= P\left(\bar{X} + 1.96(24.6/\sqrt{150}) > \mu > \bar{X} - 1.96(24.6/\sqrt{150})\right) && (7.4)\\
&= P\left(\bar{X} - 3.937 < \mu < \bar{X} + 3.937\right). && (7.5)
\end{aligned}
$$

The random interval $(\bar{X} - 3.937, \bar{X} + 3.937)$ has a probability of 0.95 of containing the mean μ. Now, once you have drawn your sample, the random variable $\bar{X}$ is replaced by the (observed) sample mean weight of $\bar{x} = 95$, and the interval $(91.1, 98.9)$ is no longer a random interval. We interpret this interval by stating that we are 95% confident that the population mean weight of 13-year-old girls in Sodor is between 91.1 and 98.9 pounds. □

Remark

- The trick of doing algebra on equations that are inside a probability is often handy.
- Be careful when reading an equation such as $0.95 = P(\bar{X} - 3.937 < \mu < \bar{X} + 3.937)$. This does not mean that μ is random, with a 95% probability of falling between two values. The parameter μ is an unknown *constant*. Instead, it is the interval that is random, with a 95% probability of including μ.

  In the previous example, we computed a confidence interval of $(91.1, 98.9)$. We should not attribute a probability to this interval: either the true mean is in this interval or it is not! The statement "we are 95% confident" means that if we repeated the same process of drawing samples and computing intervals many times, then in the long run, 95% of the intervals would include μ. ||

More generally, for a sample of size n drawn from a normal distribution with unknown μ and known σ^2, a 95% confidence interval for the mean μ is

$$\left(\bar{X} - 1.96\frac{\sigma}{\sqrt{n}}, \bar{X} + 1.96\frac{\sigma}{\sqrt{n}}\right). \tag{7.6}$$

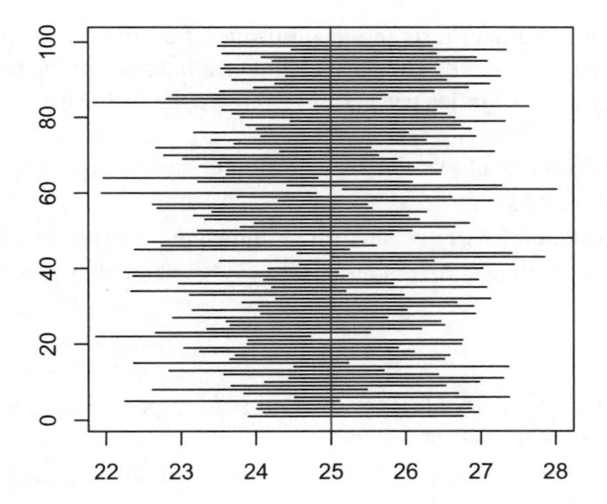

FIGURE 7.1 Random confidence intervals, $X_i \sim N(25, 4^2), n = 30$. Notice that several miss the mean 25.

If we draw thousands of random samples from a normal distribution with parameters μ, σ^2 and compute the 95% confidence interval from each sample, then about 95% of the intervals would contain μ.

We illustrate this with a simulation by drawing random samples of size 30 from $N(25, 4^2)$. For each sample, we construct the 95% confidence interval and check to see whether it contains $\mu = 25$. We do this 1000 times and keep track of the number of times that the interval contains μ. For good measure, we will graph some of the random intervals (Figure 7.1).

R Note:

```
counter <- 0                            # set counter to 0
plot(x = c(22, 28), y = c(1, 100), type = "n", xlab = "",
     ylab = "")         # set up a blank plot with specified ranges
for (i in 1:1000)
{
  x <- rnorm(30, 25, 4)                 # draw a random sample of size 30
  L <- mean(x) - 1.96*4/sqrt(30) # lower limit
  U <- mean(x) + 1.96*4/sqrt(30) # upper limit
  if (L < 25 && 25 < U)                 # check if 25 is in interval
    counter <- counter + 1              # if yes, increase counter by 1
  if (i <= 100)                         # plot first 100 intervals
    segments(L, i, U, i)
}

abline(v = 25, col = "red")             # vertical line at mu
counter/1000            # proportion of times interval contains mu.
```

Example 7.2 An engineer tests the gas mileage of a random sample of 30 of his company cars ready to be sold. The 95% confidence interval for the mean mileage of all the cars is (29.5, 33.4) miles per gallon. Evaluate the following statements:

1. We are 95% confident that the gas mileage for cars in this company is between 25.5 and 33.4 mpg.
2. 95% of all samples will give an average mileage between 29.5 and 33.4 mpg.
3. There is a 95% chance that the true mean is between 29.5 and 33.4 mpg.

Solution

1. This is not correct: a confidence interval is for a population parameter, and in this case the mean, not for individuals.
2. This is not correct: each sample will give rise to a *different* confidence interval and 95% of these intervals will contain the true mean.
3. This is not correct: μ is not random. The probability that it is between 29.5 and 33.4 is 0 or 1. □

In our first example, we constructed a 95% confidence interval, but we can use other levels of confidence. More generally, let q denote the $(1-\alpha/2)$ quantile that satisfies $P(Z < q) = 1 - \alpha/2$ (see Figure 7.2). Then, by symmetry, $P(-q < Z < q) = 1 - \alpha$.

Let $\bar{X}$ denote the mean of a random sample of size n from a normal distribution $N(\mu, \sigma^2)$. Then, since $\bar{X} \sim N(\mu, \sigma^2/n)$, by mimicking the algebra steps on page 168,

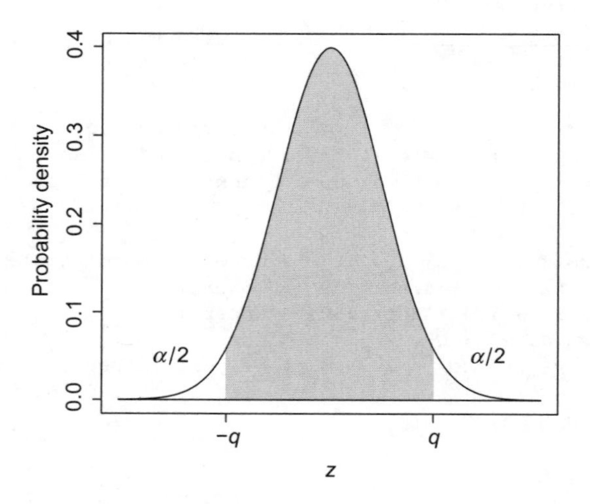

FIGURE 7.2 Standard normal density with shaded area $1 - \alpha$.

we have

$$1 - \alpha = P\left(-q < \frac{\bar{X} - \mu}{\sigma/\sqrt{n}} < q\right)$$
$$= P\left(\bar{X} - q \cdot \frac{\sigma}{\sqrt{n}} < \mu < \bar{X} + q \cdot \frac{\sigma}{\sqrt{n}}\right).$$

In summary,

Z CONFIDENCE INTERVAL FOR NORMAL MEAN WITH KNOWN STANDARD DEVIATION

If $X_i \sim N(\mu, \sigma^2)$, $i = 1, \ldots, n$, with known σ, then a $(1 - \alpha) \times 100\%$ *confidence interval for* μ is given by

$$\left(\bar{X} - q \cdot \frac{\sigma}{\sqrt{n}},\ \bar{X} + q \cdot \frac{\sigma}{\sqrt{n}}\right), \tag{7.7}$$

where q denotes the $(1 - \alpha/2)$ quantile of $N(0, 1)$.

Example 7.3 Suppose the sample 3.4, 2.9, 2.8, 5.1, 6.3, 3.9 is drawn from the normal distribution with unknown mean μ and known $\sigma = 2.5$. Find a 90% confidence interval for μ.

Solution The mean of the six numbers is 4.067. Here, $1 - \alpha = 0.90$, so $\alpha/2 = 0.05$. Thus, the 95th quantile is $q = q_{0.95} = 1.645$ (i.e., $P(Z < 1.645) = 0.95$). The 90% confidence interval is

$$\left(4.067 - 1.645\frac{2.5}{\sqrt{6}},\ 4.067 + 1.645\frac{2.5}{\sqrt{6}}\right).$$

We are 90% confident that the population mean lies in the interval (2.389, 5.746). □

The term $q(\sigma/\sqrt{n})$ is called the *margin of error* (we abbreviate this as ME).

MARGIN OF ERROR

The *margin of error* for a symmetric confidence interval is the distance from the estimate to either end. The confidence interval is of the following form: estimate ± ME.

Example 7.4 Suppose researchers want to estimate the mean weight of girls in Sodor. They assume that the distribution of weights is normal with unknown mean μ, but known standard deviation $\sigma = 24.6$. How many girls should they sample if they want, with 95% confidence, their margin of error to be at most 5 pounds?

Solution Since $q_{0.975} = 1.96$, we set $1.96(24.6/\sqrt{n}) \leq 5$. This leads to $n \geq 92.99$, so there should be at least 93 girls in the sample. □

Remark Note that the width of the interval, given by $q(\sigma/\sqrt{n})$, depends on the level of confidence (which determines q), the standard deviation, and the sample size. Analysts cannot control σ, but can adjust q or n. To make the confidence interval narrower, they can either increase the sample size n or decrease the size of the quantile q, which amounts to decreasing the confidence level. ||

7.1.2 Confidence Intervals for a Mean, σ Unknown

In most real-life settings, a data analyst will not know the mean or the standard deviation of the population of interest. How then would we get an interval estimate of the mean μ? We have used the sample mean $\bar{X}$ as an estimate of μ, so it seems natural to consider the sample standard deviation S as an estimate of the σ.

However, in deriving the confidence interval for μ, we used the fact that $(\bar{X} - \mu)/(\sigma/\sqrt{n})$ follows a standard normal distribution. Does changing σ to S, the sample standard deviation, change the distribution? We will use a simulation to investigate the distribution of $(\bar{X} - \mu)/(S/\sqrt{n})$ for random samples drawn from $N(\mu, \sigma^2)$.

R Note:

```
N <- 10^4
w <- numeric(N)
n <- 15                    # sample size
for (i in 1:N)
{
  x <- rnorm(n, 25, 7)   # draw a size 15 sample from N(25, 7^2)
  xbar <- mean(x)
  s <- sd(x)
  w[i] <- (xbar-25) / (s/sqrt(n))
}

hist(w)
dev.new()                  # open new graphics device
qqnorm(w, pch = ".")
abline(0, 1)               # y=x line (y=0+1x)
```

This distribution does have slightly longer tails than the normal distribution; you could never tell this from a histogram, but it is apparent in the normal quantile plot (Figure 7.3). Sometimes S is smaller than σ, and when the denominator is small, the ratio is large. In effect, having to estimate σ using S adds variability.

It turns out that $T = (\bar{X} - \mu)/(S/\sqrt{n})$ has a Students t distribution with $n - 1$ degrees of freedom.

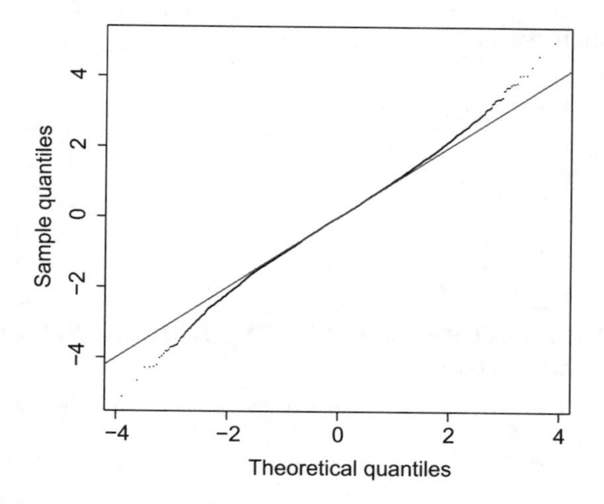

FIGURE 7.3 qq plot for sampling distribution of $(\bar{X} - \mu)/(S/\sqrt{n})$.

The density of a t distribution with k degrees of freedom is bell shaped and symmetric about 0, with heavier (longer) tails than that of the standard normal. As k tends toward infinity, the density of the t distribution tends toward the density of the standard normal. For more details on this distribution, refer to Section B.11. Figure 7.4 shows densities for the standard normal and two t distributions; the shapes are similar, though the t distributions are far greater in the tails.

We derive the confidence interval for μ when σ is unknown in the same way as when σ is known. Let $q = q_{(1-\alpha/2)}$ denote the $(1 - \alpha/2)$ quantile of the t distribution with $n - 1$ degrees of freedom, $P(T < q) = 1 - \alpha/2,\ 0 < \alpha < 1$. Then using symmetry

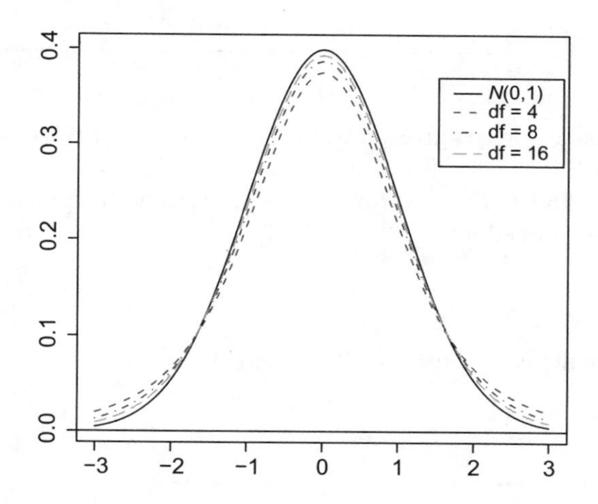

FIGURE 7.4 Density for standard normal and Students t distributions with 4 and 8 degrees of freedom.

of the t distribution, we have

$$
\begin{aligned}
1-\alpha &= P\left(-q < \frac{\bar{X}-\mu}{S/\sqrt{n}} < q\right) \\
&\vdots \\
&= P\left(\bar{X} - q\cdot\frac{S}{\sqrt{n}} < \mu < \bar{X} + q\cdot\frac{S}{\sqrt{n}}\right).
\end{aligned}
$$

T CONFIDENCE INTERVAL FOR NORMAL MEAN WITH UNKNOWN STANDARD DEVIATION

If $X_i \sim N(\mu, \sigma^2)$, $i = 1, \ldots, n$, with σ unknown, then a $(1-\alpha)\times 100\%$ confidence interval for μ is given by

$$
\left(\bar{X} - q\cdot\frac{S}{\sqrt{n}},\ \bar{X} + q\cdot\frac{S}{\sqrt{n}}\right), \tag{7.8}
$$

where q denotes the $(1-\alpha/2)$ quantile of the t distribution with $(n-1)$ degrees of freedom.

Example 7.5 The distribution of weights of boys in Sodor is normal with unknown mean μ. From a random sample of 28 boys, we find a sample mean of 110 pounds and a sample standard deviation of 7.5 pounds. To compute a 90% confidence interval, find the 0.95 quantile of the t distribution with 27 degrees of freedom, which is $q = 1.7033$ using software. The interval is $(110 - 1.7033(7.5/\sqrt{28}), 110 + 1.7033 \times (7.5/\sqrt{28}))$; thus, we are 90% confident that the true mean weight is between 107.6 and 112.4 pounds. □

R Note:

The commands `pt` or `qt` give probabilities or quantiles, respectively, for the t distribution.

For instance, to find $P(T < 2.8)$ for the random variable T from a t distribution with 27 degrees of freedom,

```
> pt(2.8, 27)
[1] 0.9953376
```

To find the quantile $q_{0.95}$ satisfying $P(T < q_{0.95}) = 0.95$,

```
> qt(0.95, 27)
[1] 1.703288
```

Compare the 0.95 quantile for a t distribution with 27 degrees of freedom with that of the standard normal: 1.703 versus 1.645. Thus, the t interval is slightly wider than

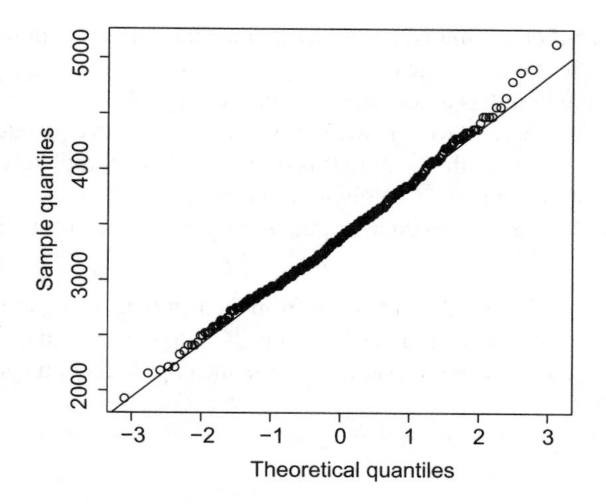

FIGURE 7.5 Normal quantile plot of weights of baby girls.

the z interval, reflecting, as we noted previously, the extra uncertainty in not knowing the true σ.

Example 7.6 Find a 99% confidence interval for the mean weight of baby girls born in North Carolina in 2004 (Case Study in Section 1.2).

Solution The mean and standard deviation of the weights of $n = 521$ girls is 3398.317 and 485.691 g, respectively. A normal quantile plot shows that the weights are approximately normally distributed (Figure 7.5), so a t interval is reasonable.

$1 - \alpha = 0.99$, so $\alpha/2 = 0.005$. The 0.995 quantile for the t distribution with 520 degrees of freedom is $q_{.995} = 2.585$. Thus, the 99% confidence interval is $3398.317 \pm 2.585(485.691/\sqrt{521}) = (3343.30, 3453.33)$ g.

R Note:

Use the command `t.test` to find confidence intervals.

```
> girls <- subset(NCBirths2004, select = Weight,
                  subset = Gender == "Female", drop = T)
> t.test(girls, conf.level = .99)$conf
[1] 3343.305 3453.328
attr(,"conf.level")
[1] 0.99
```

□

Assumptions With any statistical procedure, one of the first questions to ask is, How robust is it? That is, what happens if the assumptions underlying the procedure are violated? The t confidence interval assumes that the underlying population is normal, so what happens if that is not the case?

When the population has a normal distribution, the t confidence interval is exact: a $(1-\alpha) \times 100\%$ interval covers μ with probability $1-\alpha$ or, equivalently, misses μ on either side with probability $\alpha/2$; that is, the interval is completely above μ with probability $\alpha/2$ or is completely below with probability $\alpha/2$.

Let us check this for a nonnormal population by running a simulation.

Example 7.7 We draw random samples from the right-skewed gamma distribution with $r = 5$ and $\lambda = 2$ (see the graph of the density on page 85) and count the number of times the 95% confidence interval misses the mean $\mu = 5/2$ on each side.

R Note:

```
tooLow <- 0                          # set counter to 0
tooHigh <- 0                         # set counter to 0
n <- 20   # sample size
q <- qt(0.975, n-1)                  # quantile
N <- 10^5
for (i in 1:N)
{
   x <- rgamma(n, shape = 5, rate = 2)
   xbar <- mean(x)
   s <- sd(x)
   L <- xbar - q*s/sqrt(n)
   U <- xbar + q*s/sqrt(n)
   if (U < 5/2)                      # Does right endpt miss 5/2?
     tooLow <- tooLow + 1            # If yes, increases counter
   if (5/2 < L)                      # Does left endpt miss 5/2?
     tooHigh <- tooHigh + 1          # If yes, increase counter
}
tooLow/N
tooHigh/N
```

What proportion of times did the confidence intervals miss the true mean $5/2$? In one run of this simulation, about 4.5% of the time, the interval was too low and was below $5/2$, and about 1.3% of the time, the interval was too high and was above $5/2$. In Exercise 8, you will examine the effect of changing the sample size. □

When the population is nonnormal but symmetric and the sample size is moderate or large, the t interval is very accurate. The main weakness of the t confidence interval occurs when the population is skewed. The simulation illustrated this problem. To see this from another point of view, we will look at the distributions of the t statistics, $T = (\bar{X} - \mu)/(S/\sqrt{n})$, since the accuracy of t intervals depends on how close the t statistic is to having a t distribution.

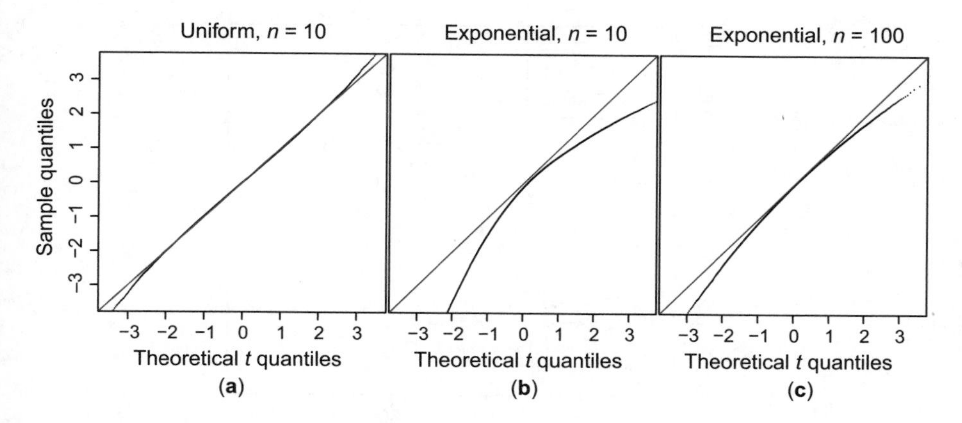

FIGURE 7.6 Quantile–quantile plot of t statistics versus a t distribution with $n-1$ degrees of freedom for samples from (a) a uniform distribution with $n=10$, (b) an exponential distribution with $n=10$, and (c) an exponential distribution with $n=100$.

Figure 7.6 compares the distribution of t statistics for samples of size $n=10$ from a uniform distribution, size $n=10$ from an exponential distribution, and size $n=100$ from an exponential distribution to the t distribution.[1] The range on all plots is truncated so that we can focus on the range of values important for confidence intervals. Notice that for the uniform population, the distribution of the t statistic is close to the t distribution, except in the tails. For exponential populations, the discrepancy is much larger, and the discrepancy decreases only slowly as the sample size increases. To reduce the discrepancy (the difference between actual and nominal probabilities) by a factor of 10 requires a sample size 100 times larger. For an exponential population, we must have $n > 5000$ before the actual probabilities of a 95% t interval missing the true mean in either tail are within 10% of the desired probability of 2.5%; that is, the actual tail probabilities are between 2.25% and 2.75%.

Before using a t confidence interval, you should create a normal quantile plot to see whether the data are skewed. The larger the sample size, the more skew can be tolerated. There are skewness-adjusted versions of t intervals. One of these is a formula-based version, given in Exercise 29. Another is bootstrap t intervals that we cover later (page 195.)

However, be particularly careful with outliers: since $\bar{x}$ is sensitive to extreme values, outliers can have a big impact on confidence intervals—a bigger impact than skewness. If you have outliers in your data, you should investigate them: are these recording errors or observations that are not representative of the population? If the former, correct them; and if the latter, remove them. If the outliers cannot be removed, then advanced, more robust techniques may be required.

[1]The principle here is the same as in the normal quantile plot. The quantiles of a sample are compared to the quantiles of the t distribution. If the distributions are the same, the points should roughly fall on a line.

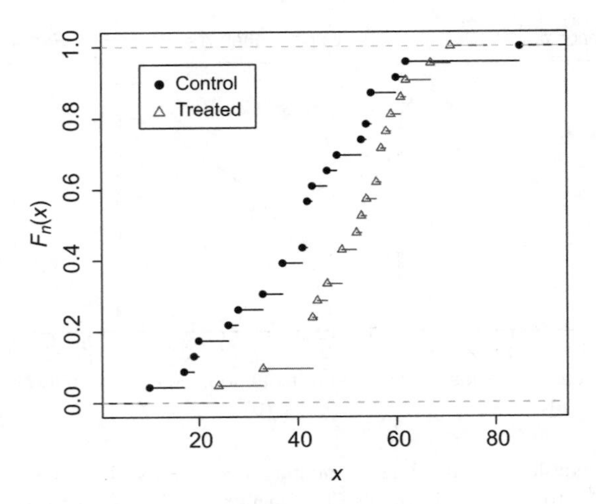

FIGURE 7.7 Comparison of the empirical cdf's of reading scores for control and treatment groups. One student in the control group had an unusually high score (in the 80s) compared to everybody else in his/her group.

7.1.3 Confidence Intervals for a Difference in Means

Will new directed reading activities improve certain aspects of a child's reading ability? An educator conducted an experiment to test this (DASL). Twenty-one third graders took part in these directed reading activities for 8 weeks, while another class of 23 third graders did not. At the end of the study, all children took the Degree of Reading Power (DRP) test, a standard test that measures various aspects of reading ability. The scores are shown in Figure 7.7.

The mean DRP score for the students who received the treatment was 51.48, while the mean for the control group was 41.52. Though the difference seems large, it may well be owing to sampling variability. To check this, we need to understand the sampling distribution of $\bar{X}_1 - \bar{X}_2$.

Let X and Y be random variables with $X \sim N(\mu_1, \sigma_1^2)$ and $Y \sim N(\mu_2, \sigma_2^2)$. Then from Theorem A.10, $X - Y \sim N(\mu_1 - \mu_2, \sigma_2^2 + \sigma_2^2)$. For sample sizes n_1 and n_2,

$$\bar{X} - \bar{Y} \sim N\left(\mu_1 - \mu_2, \frac{\sigma_1^2}{n_1} + \frac{\sigma_1^2}{n_2}\right). \tag{7.9}$$

Of course, in practice, we usually do not know the population variances, so we will plug in the sample variances. As in the single-sample case, we call this a t statistic:

$$T = \frac{(\bar{X} - \bar{Y}) - (\mu_1 - \mu_2)}{\sqrt{S_1^2/n_1 + S_2^2/n_2}}. \tag{7.10}$$

The exact distribution of this statistic is an unsolved problem. It does, however, have approximately a t distribution if the populations are normal. The difficult part is the degrees of freedom. A quick rule is to set the degrees of freedom equal to the smaller of $n_1 - 1$ and $n_2 - 1$. A more accurate rule, which gives larger degrees of freedom (and hence shorter intervals), is Welch's approximation:

$$\nu = \frac{(s_1^2/n_1 + s_2^2/n_2)^2}{(s_1^2/n_1)^2/(n_1-1) + (s_2^2/n_2)^2/(n_2-1)}. \tag{7.11}$$

This is based on how accurately $s_1^2/n_1 + s_2^2/n_2$ estimates $\sigma_1^2/n_1 + \sigma_2^2/n_2$ (see Exercise 40).

Let q denote the $(1-\alpha/2)$ quantile for the t distribution with ν degrees of freedom:

$$P\left(-q < \frac{(\bar{X}-\bar{Y}) - (\mu_1 - \mu_2)}{\sqrt{S_1^2/n_1 + S_2^2/n_2}} < q\right) \approx 1 - \alpha,$$

and solve for $\mu_1 - \mu_2$ to obtain a $(1-\alpha) \times 100\%$ confidence interval for $\mu_1 - \mu_2$.

T CONFIDENCE INTERVAL FOR DIFFERENCE IN MEANS

If $X_i \sim N(\mu_1, \sigma_1^2)$, $i = 1, 2, \ldots, n_1$, and $Y_j \sim N(\mu_2, \sigma_2^2)$, $j = 1, 2, \ldots, n_2$, then an approximate $(1-\alpha) \times 100\%$ confidence interval for $\mu_1 - \mu_2$ is

$$(\bar{X} - \bar{Y}) \pm q\sqrt{\frac{S_1^2}{n_1} + \frac{S_2^2}{n_2}}, \tag{7.12}$$

where the degree of freedom ν is given by Equation 7.11 and q denotes the $(1-\alpha/2)$ quantile of the t distribution with ν degree of freedom.

Example 7.8 Reading scores in directed reading study.

For the $n_1 = 21$ third graders who participated in directed reading activities, the mean and the standard deviation of their DRP scores are $\bar{x} = 51.48$, $s_1 = 11.01$, respectively, and for the $n_2 = 23$ third graders in the control group, they are $\bar{y} = 41.52$, $s_2 = 17.14$ g, respectively. The mean difference is $\bar{x} - \bar{y} = 9.96$ with the standard error of the difference and degrees of freedom

$$\sqrt{\frac{11.01^2}{21} + \frac{17.14^2}{23}} = 4.31 \quad \text{and} \quad \frac{(11.01^2/21 + 17.4^2/23)^2}{(11.01^2/21)^2/20 + (17.4^2/23)^2/22} = 37.577,$$

respectively. The 0.975 quantile of the t distribution is 2.0251. Thus, the 95% confidence interval is $9.96 \pm 2.0251 \times 4.31 = (1.23, 18.69)$. With 95% confidence, third graders who participated in directed reading activities score, on average, from 1.23 to 18.69 points higher than third graders who did not participate.

Now, one child in the control group received an unusually high score (85). If we remove this child, then the 95% confidence interval for the difference in mean scores becomes (3.98, 19.89). It does not appear that this child is influential: removing this outlier does not change the confidence interval much.

The directed reading activities may improve DRP scores, on average, by as much as 19 points, or it may have little practical effect, improving by only about a few points.

R Note:

```
> treated <- subset(Reading, select = Response,
                    subset = Treatment == "Treated", drop = T)
> control <- subset(Reading, select = Response,
                    subset = Treatment == "Control", drop = T)
> t.test(treated, control)$conf
[1]  1.23302 18.67588
attr(, "conf.level"):
[1] 0.95
```

□

Remark

- If the confidence interval for the difference in means contains 0, then we cannot rule out the possibility that the means might be the same, $\mu_1 - \mu_2 = 0$ or, equivalently, $\mu_1 = \mu_2$. See exercise 30.
- If we were to construct separate 95% confidence intervals for the mean reading score for each of the treatment group and control group, then we would find an interval of $(46.47, 56.49)$ for the treatment group and $(34.11, 48.94)$ for the control group. The confidence intervals overlap, so we would not be able say whether or not there was a difference in the true mean scores. Looking for overlap in individual confidence intervals is not a good way to test for significant differences.
- Skewness is less of an issue for two-sample t confidence intervals than for one-sample intervals, because the skewness from the two samples tends to cancel out. In particular, if the populations have the same skewness and variance and the sample sizes are equal, then the skewness cancels out exactly, and the distribution of t statistics can be very close to a t distribution even for quite small samples.

Figure 7.8 shows the results of three simulations in which two samples are drawn repeatedly from a right-skewed exponential distribution with $\lambda = 1$. In Figure 7.8a, the two samples are the same size, and we can see that the distribution of the two-sample t statistic (Equation 7.10) is very close to the t distribution. Figure 7.8b and c shows the results when the two samples are different sizes. If the sample sizes are unbalanced, the skewness partially cancels out. The extreme case, as one sample size goes to infinity, reduces to a one-sample problem. ||

Example 7.9 Consider the weights of boy and girl babies born in Texas in 2004 (see page 2 for the description of the data set). Construct a 95% t confidence interval for the mean difference in weights (boys–girls).

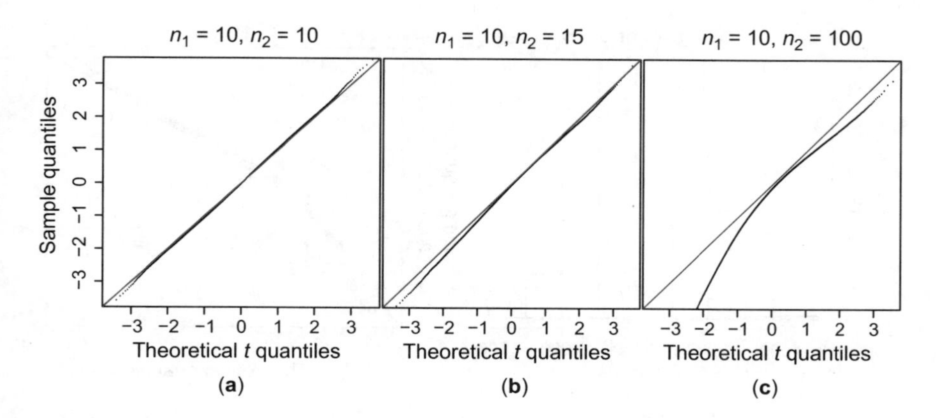

FIGURE 7.8 Quantile–quantile plot of t statistics versus a t distribution when both samples are exponential and when the sample sizes are (a) $n_1 = 10$, $n_2 = 10$, (b) $n_1 = 10$, $n_2 = 15$, and (c) $n_1 = 10$, $n_2 = 100$. The degrees of freedom for the theoretical t quantiles are obtained using known variances in Welch's approximation.

Solution The weights are shown in Figure 7.9. The distribution of weights is left skewed for both boys and girls. While this may make t intervals for the individual means inaccurate, the skewness largely cancels out when computing a t confidence interval for the difference in means if the sample sizes are similar.

The mean and standard deviation for the $n_1 = 848$ boys are 3336.84 and 547.53. g, respectively, while for the $n_2 = 739$ girls, they are, respectively, 3220.94 and 550.82 g. Thus, the mean difference weights is 115.90 with standard error 27.64. The degree of freedom is approximately 1552.91, so the corresponding 0.975 quantile is $q_{0.975} = 1.96$. The confidence interval is $115.90 \pm 1.96 \times 27.64 = (61.68, 170.12)$ g. Thus, we are 95% confident that boy babies born in Texas in 2004 were, on average, from 61.68 to 170.12 g heavier than girl babies.

In this example, separate 95% t confidence intervals for the mean weights are (3299.94, 3373.75) for the boys and (3181.16, 3260.72) for the girls. The intervals do not overlap, providing evidence that the mean weights are different between boys and girls. □

Pooling the Variances If $\sigma_1^2 = \sigma_2^2$, then we can "pool the variances," estimating the common variance using all the data. The *pooled sample variance* is

$$S_{\mathrm{p}}^2 = \frac{(n_1 - 1)S_1^2 + (n_2 - 1)S_2^2}{n_1 + n_2 - 2}.$$

We have the following:

Theorem 7.1 *Let $X_1, X_2, \ldots, X_{n_1} \sim N(\mu_1, \sigma^2)$ and $Y_1, Y_2, \ldots, Y_{n_2} \sim N(\mu_2, \sigma^2)$ be two independent random samples with sample means and variances $\bar{X}, S_1^2, \bar{Y}, S_2^2$, respectively.*

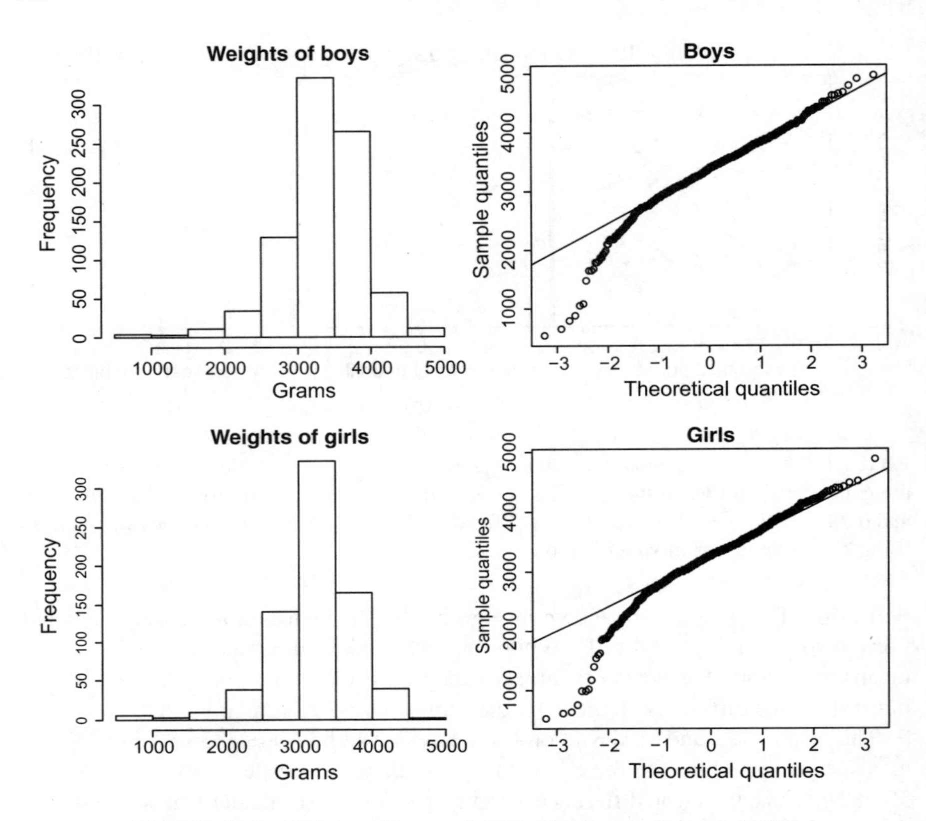

FIGURE 7.9 Weights of boy and girl babies born in Texas in 2004.

Then

$$T = \frac{(\bar{X} - \bar{Y}) - (\mu_1 - \mu_2)}{S_p\sqrt{1/n_1 + 1/n_2}} \tag{7.13}$$

has a Student's t distribution with $(n_1 + n_2 - 2)$ degrees of freedom.

This leads to the following $(1 - \alpha) \times 100\%$ pooled variance two-sample t interval for $\mu_1 - \mu_2$:

$$\left(\bar{X} - \bar{Y} - q \cdot S_p\sqrt{\frac{1}{n_1} + \frac{1}{n_2}},\ \bar{X} - \bar{Y} + q \cdot S_p\sqrt{\frac{1}{n_1} + \frac{1}{n_2}}\right), \tag{7.14}$$

where q is the $(1 - \alpha/2)$ quantile of the t distribution with $(n_1 + n_2 - 2)$ degrees of freedom.

In general, we advise against using this confidence interval, especially if the two sample sizes are unequal. Pooling the variances usually provides only a small gain when the variances are the same, but can be badly off when the variances are different. In practice, it is hard to test whether two variances are the same. An eminent statistician, George Box, wrote about pooling: "To make the preliminary test on variances is rather like putting to sea in a rowing boat to find out whether conditions are sufficiently calm for an ocean liner to leave port." (Box (1953)).

An exception is when one sample size is very small, so that the sample provides little information about the variance of its population. In that case, it may be better to assume equal variances, in spite of the bias this causes when the variances are unequal, to avoid the extra variability caused by a wild variance estimate. For more discussion about pooling variances, see Moser and Stevens (1992), Miao and Chiou (2008), or Scheffe (1970).

7.2 CONFIDENCE INTERVALS IN GENERAL

In the previous section, we derived confidence intervals for means and difference of means by considering the sampling distribution of a statistic that depended on both μ and $\bar{x}$. We were able to solve for μ to obtain lower and upper limits that did not involve μ. We can use this same idea for finding confidence intervals for other parameters.

Suppose $X_1, X_2, \ldots, X_n$ are a random sample from a distribution F with parameter θ. Suppose for all θ,

$$P(L < \theta < U) = 1 - \alpha, \qquad (7.15)$$

where $L = g_1(X_1, X_2, \ldots, X_n)$ and $U = g_2(X_1, X_2, \ldots, X_n)$ are functions of the n random variables, but not θ. Then the interval (L, U) is a $(1 - \alpha) \times 100\%$ confidence interval for θ.

Example 7.10 In the 95% confidence interval for a mean with σ known (Equation 7.6), we have $L = \bar{X} - 1.96(\sigma/n)$ and $U = \bar{X} + 1.96(\sigma/n)$. □

One approach to forming a confidence interval is to find a *pivotal statistic*, a statistic that depends on the sample $X_1, X_2, \ldots, X_n$ and the parameter θ, say $h(X_1, X_2, \ldots, X_n, \theta)$, but whose distribution *does not* depend on θ or on any unknown parameters. For example, for a normal population with σ known, $(\bar{X} - \mu)/\sigma$ and $(\bar{X} - \mu)/(\sigma/\sqrt{n})$ are both pivotal since their distributions, $N(0, n)$ and $N(0, 1)$, respectively, do not depend on μ.

If q_1 and q_2 denote the $\alpha/2$ and $1 - \alpha/2$ quantiles of the distribution of the pivotal statistic, respectively, then we have

$$P(q_1 < h(X_1, X_2, \ldots, X_n, \theta) < q_2) = 1 - \alpha.$$

Set $q_1 = h(X_1, X_2, \ldots, X_n, \theta)$ and $q_2 = h(X_1, X_2, \ldots, X_n, \theta)$ and solve for θ to find the limits of the interval.

Example 7.11 Let $X_1, X_2, \ldots, X_n$ be a random sample from an exponential distribution with unknown parameter λ. Then $X_1 + X_2 + \cdots + X_n$ has a gamma

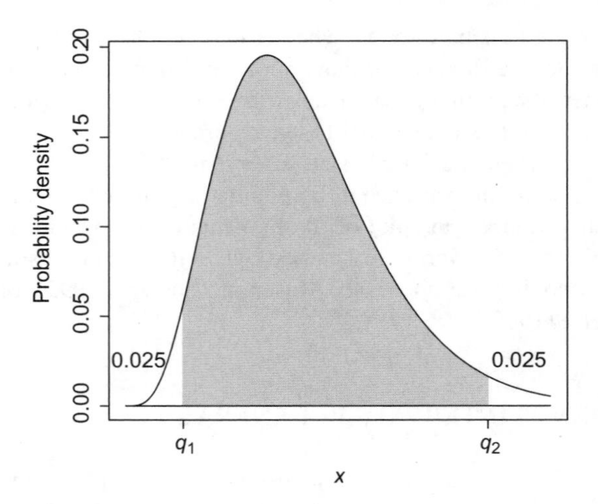

FIGURE 7.10 Density for Gamma(5, 1). Shaded region represents an area of 0.95.

distribution with parameters n and λ by Theorem B.11, and $\lambda(X_1 + X_2 + \cdots + X_n)$ has a gamma distribution with parameters n and 1.

Let q_1 and q_2 mark the 0.025 and 0.975 quantiles, respectively, for Gamma(n, 1) (see Figure 7.10). Then,

$$\begin{aligned} 0.95 &= P(q_1 < \lambda(X_1 + X_2 + \cdots + X_n) < q_2) \\ &= P\left(\frac{q_1}{\sum_{i=1}^n X_i} < \lambda < \frac{q_2}{\sum_{i=1}^n X_i}\right). \end{aligned}$$

Thus, the 95% confidence interval for λ is

$$\left(\frac{q_1}{\sum_{i=1}^n X_i}, \frac{q_2}{\sum_{i=1}^n X_i}\right).$$

For instance, suppose we observe 2, 2.5, 3, 4, 9. Then $n = 5$, $\sum_{i=1}^5 x_i = 20.5$, and the 0.025 and 0.975 quantiles for Gamma(5,1) are 1.6235 and 10.2415, respectively. Now, we are 95% confident that the parameter λ lies in the interval $(1.6235/20.5, 10.2415/20.5) = (0.0792, 0.4996)$. The estimate of λ is $\hat{\lambda} = 1/\bar{x} = 0.244$, so the interval can be written as $(\hat{\lambda} - 0.165, \hat{\lambda} + 0.256)$. In particular, note that in contrast to the z and t confidence intervals, this interval is not symmetric about the estimate. □

Example 7.12 Let X be a single observation from the exponential distribution $f_X(x) = \lambda e^{-\lambda x}$ for $x > 0$, $\lambda > 0$. We will find a 95% confidence interval for λ. By Proposition B.2, $Y = \lambda X$ has a standard exponential distribution with pdf $f_Y(y) = e^{-y}$

for $y > 0$. Hence, Y is pivotal. First, we find the 0.025 and 0.975 quantiles of Y using

$$\int_0^{q_1} e^{-y} dy = F_Y(q_1) = 0.025 \quad \text{and} \quad \int_0^{q_2} e^{-y} dy = F_Y(q_2) = 0.975,$$

where $F_Y(y) = 1 - \exp(-y)$ for $y > 0$. Thus, $q_1 = 0.0253$ and $q_2 = 3.689$. Therefore,

$$\begin{aligned} 0.95 &= P(q_1 \le Y \le q_2) \\ &= P(0.0253 \le \lambda X \le 3.689) \\ &= P\Big(\frac{0.0253}{X} \le \lambda \le \frac{3.689}{X}\Big). \end{aligned}$$

The confidence interval for λ is $(0.0253/X, 3.689/X)$. Thus, for example, if the single observation was $X = 0.03$, then the confidence interval is (0.843, 122.97), a rather wide interval. □

Example 7.13 During World War II, the Western Allies estimated German tank production from the serial numbers on tanks, in particular on gearboxes. There is a nice discussion of this at `http://en.wikipedia.org/wiki/German_tank_problem`. This statistical approach was sometimes much more accurate than other approaches, for example, (from Wikipedia):

> The Allied conventional intelligence estimates believed the number of tanks the Germans were producing between June 1940 and September 1942 was around 1,400 a month. Using the above formula on the serial numbers of captured German tanks, (both serviceable and destroyed) the number was calculated to be 256 a month. After the war captured German production figures from the ministry of Albert Speer show the actual number to be 255.

We will consider a simplified version of the problem—that the serial numbers began at 1 and continued up to θ, where θ is the total number produced, and that any tank is equally likely to be captured, independent of all others. (In practice, neither of these assumptions is true. For example, the newest tanks have had less opportunity to be captured, and tanks produced around the same time are more likely to be shipped to the same region, be in the same battles, and be captured together.)

We will simplify further by considering a continuous distribution. Suppose that the serial numbers are $X_1, \ldots, X_n \sim \text{Unif}[0, \theta]$. In Section 6.1.2, we saw that the maximum of the observations $X_{\max}$ is the maximum likelihood estimate, and we saw in Section 6.3.2 that a rescaled version $((n+1)/n)X_{\max}$ was substantially more efficient than another unbiased estimate $2\bar{X}$. Since estimation based on $X_{\max}$ worked so well, it should also work well for confidence intervals.

The pdf for $X_{\max}$ is $f_{\max}(x) = (n/\theta^n)x^{n-1}$ for $0 \le x \le \theta$ (Equation 4.4). By Proposition B.1, dividing every observation X by θ makes every observation standard uniform, Unif[0, 1], so $Y = X_{\max}/\theta$ has pdf $f_{\max}(y) = ny^{n-1}$ for $0 \le y \le 1$. This does not depend on θ, so Y is pivotal. The corresponding cdf is $F_{\max}(y) = y^n$

for $0 \le y \le 1$. We solve the two equations

$$F_{\max}(q_1) = \alpha/2,$$
$$F_{\max}(q_2) = 1 - \alpha/2$$

to obtain $q_1 = (\alpha/2)^{1/n}$ and $q_2 = (1-\alpha/2)^{1/n}$.

We obtain the endpoints of the confidence interval by solving these equations for θ:

$$\begin{aligned} 1-\alpha &= P(q_1 < Y < q_2) \\ &= P\left(q_1 \le \frac{X_{\max}}{\theta} \le q_2\right) \\ &= P\left(\frac{X_{\max}}{q_2} \le \theta \le \frac{X_{\max}}{q_1}\right). \end{aligned}$$

For a 95% interval with $n = 400$ captured tanks, $q_1 = 0.025^{1/400} = 0.9908202$ and $q_2 = 0.975^{1/400} = 0.9999367$, so the confidence interval is $(1.000063X_{\max}, 1.009265X_{\max})$. This is a remarkably narrow confidence interval; if the largest serial number observed to date is 10^4, we would be 95% confident that θ, the true number of tanks produced, would be between 10,001 and 10,093. For a more in-depth analysis of serial numbers, see Goodman (1952). □

7.2.1 Location and Scale Parameters

The examples are all examples of either *location* or *scale* parameters. It is useful to recognize these situations because it governs what confidence intervals should look like.

Definition 7.1 Let $f(\cdot;\theta)$ be a family of densities, with $-\infty < \theta < \infty$. θ is a *location parameter* if and only if the density can be written as a function of $x-\theta$, that is, $f(x;\theta) = h(x-\theta)$ for some function h. Equivalently, θ is a location parameter if and only if the distribution of $X - \theta$ does not depend on θ. ||

In other words, changing a location parameter shifts a distribution sideways. The standard deviation does not change as the mean (or median, etc.) changes. For example, the mean μ of a normal distribution is a location parameter. For location parameters, good estimates have the property that if you add c to every observation, then $\hat\theta$ increases by c. Then $\hat\theta - \theta$ is pivotal, $1-\alpha = P(q_1 \le \hat\theta - \theta \le q_2)$, and confidence intervals are of the form $(L = \hat\theta - q_2, U = \hat\theta - q_1)$.

This may seem backward and indeed it is. With confidence intervals, we need to think backward, to think like detectives—given the $\hat\theta$ we observed, what values of θ could have produced it? Figure 7.11 shows this. Figure 7.11a shows the sampling distribution of $\hat\theta$ for three different values of θ. This estimator is positively biased and skewed, so $\hat\theta$ tends to be greater than θ, and sometimes much greater. Hence, thinking backward, it is likely that θ is less than $\hat\theta$. Figure 7.11b shows the two sampling

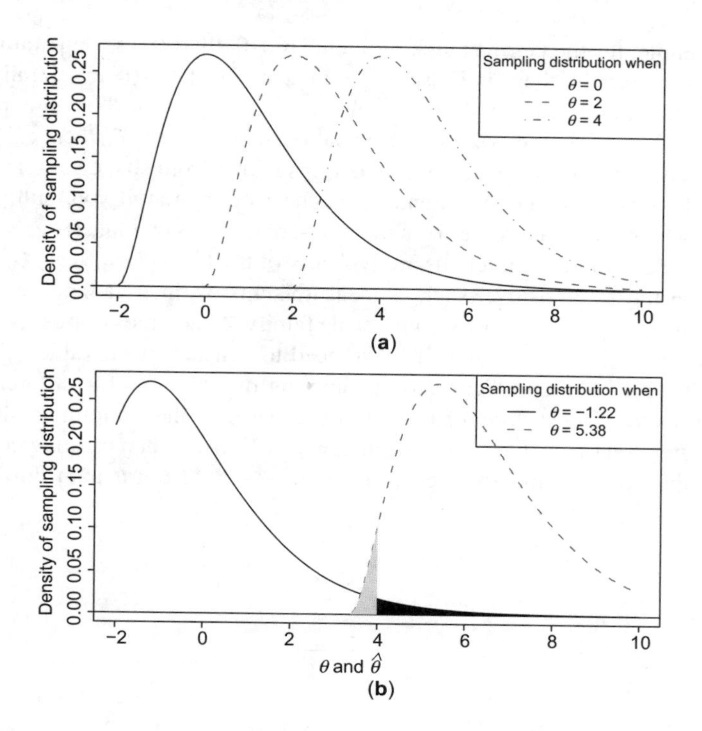

FIGURE 7.11 Sampling distributions for a location parameter. (a) Sampling distributions of $\hat{\theta}$ for three values of θ; this estimator is positively biased and skewed. (b) The two sampling distributions with 2.5% probability of being above and below 4, respectively; the corresponding θ values are the lower and upper endpoints of a confidence interval for θ, when $\hat{\theta} = 4$.

distributions that have 2.5% probability of being above and below 4, respectively; if $\hat{\theta} = 4$, then (1) when $\theta = -1.22$, the sampling distribution for $\hat{\theta}$ has 2.5% probability above 4, and (2) when $\theta = 5.38$, the sampling distribution for $\hat{\theta}$ has 2.5% probability below 4. These are the endpoints of an exact confidence interval. The interval $(-1.22, 5.38) = (4 - 5.22, 4 + 1.38)$ reaches farther to the left to adjust the positive bias and skewness of the estimator. Any θ inside this range has the observed value of 4 in the middle of its sampling distribution, so these are the θ values that are consistent with $\hat{\theta} = 4$.

Definition 7.2 Let $f(\cdot;\theta)$ be a family of distributions, with $0 < \theta < \infty$. θ is a *scale parameter* if and only if the density can be written as $f(x;\theta) = (1/\theta)h(x/\theta)$ for some function h. Equivalently, θ is a scale parameter if and only if the distribution of X/θ does not depend on θ. ||

In other words, changing a scale parameter multiplies all values in a distribution, for example, making the density wider and shorter or narrower and taller (Figure 7.12). The standard deviation is proportional to the mean (or median, etc.).

For example, for the German tank problem Unif$[0, \theta]$, θ is a scale parameter. For the Unif$[\alpha, \beta]$ family, there is no scale parameter. For the $N(0, \sigma^2)$ family, σ is a scale parameter, and also for $N(a\sigma, \sigma^2)$ for a fixed constant a. For an exponential distribution, if we parameterize the distribution using $\rho = 1/\lambda$, then ρ is a scale parameter and similarly for gamma distributions with a fixed shape parameter r.

For scale parameters, good estimates have the property that if you multiply every observation by c, then $\hat{\theta}$ increases by a factor c. Then, $\hat{\theta}/\theta$ is pivotal, $1 - \alpha = P(q_1 \leq \hat{\theta}/\theta \leq q_2)$, and confidence intervals are typically of the form $(L = \hat{\theta}/q_2, U = \hat{\theta}/q_1)$.

Data that must be positive, such as measurements of time, cannot come from a location family. They often come from a scale family. Again, the confidence intervals seem backward. Given an observed $\hat{\theta}$, we need to consider what values of θ could have produced that. Figure 7.12 shows sampling distributions for different values of the parameter. The striking characteristic is that here the sampling distributions corresponding to large values of θ are much wider. Hence, when creating confidence intervals, the upper endpoints need to be farther away from θ than do the lower endpoints.

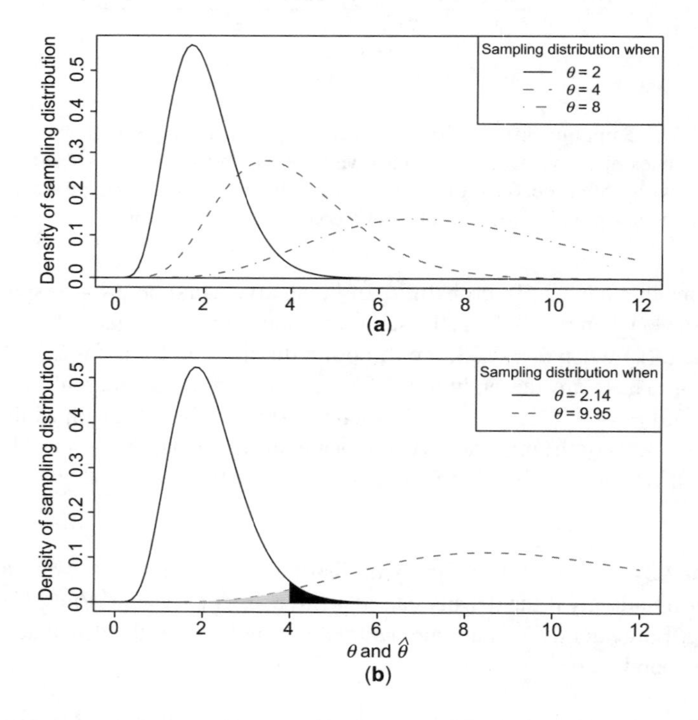

FIGURE 7.12 Sampling distributions for a scale parameter. (a) Sampling distributions of $\hat{\theta}$ for three values of θ; this estimator is unbiased and skewed. (b) The two sampling distributions with 2.5% probability of being above and below 4, respectively; the corresponding θ values are the lower and upper endpoints of a confidence interval for θ, when $\hat{\theta} = 4$.

Remark Indeed "thinking backward" yields a very general way to create confidence intervals, to *invert a significance test*—to let the confidence interval include all values of θ for which a two-sided significance test would not reject the null hypothesis for that θ. ||

7.3 ONE-SIDED CONFIDENCE INTERVALS

We have been discussing confidence intervals of the form (L, U) that gives both lower bound and upper bounds for the parameter θ being estimated. In practice, we are often interested in only an upper bound U or only a lower bound L. For example, engineers at an automobile company might want a lower bound on the lifetime of a muffler, or financial analysts may want an upper bound on the proportion of mortgages in a portfolio that will default.

Thus, if for all θ we have $P(L \leq \theta) = 1 - \alpha$, then L is a $(1 - \alpha) \times 100\%$ lower confidence bound for θ and $[L, \infty)$ is a $(1 - \alpha) \times 100\%$ one-sided confidence interval for θ. Similarly, if $P(\theta \leq U) = 1 - \alpha$, then U is a one-sided $(1 - \alpha) \times 100\%$ upper confidence bound for θ. For instance, let $X_1, X_2, \ldots, X_n$ be a random sample from a normal distribution with unknown μ and let q denote the $(1 - \alpha)$ quantile of the t distribution with $n - 1$ degrees of freedom. Then

$$1 - \alpha = P\left(\frac{\bar{X} - \mu}{S/\sqrt{n}} < q\right) = P\left(\bar{X} - q \cdot \frac{S}{\sqrt{n}} < \mu\right),$$

so a one-sided $(1 - \alpha) \times 100\%$ confidence interval for μ is $[\bar{X} - q \cdot \frac{S}{\sqrt{n}}, \infty)$. A similar argument can be used to find the lower confidence interval for μ. In summary,

ONE-SIDED CONFIDENCE INTERVAL FOR THE MEAN

If $X_1, X_2, \ldots, X_n$ are a random sample from a normal distribution with unknown mean μ, and q is the $(1 - \alpha)$ quantile of the t distribution with $n - 1$ degrees of freedom, then a $(1 - \alpha) \times 100\%$ lower confidence bound is

$$\bar{X} - q\frac{S}{\sqrt{n}} \leq \mu,$$

with corresponding upper confidence interval

$$\left[\bar{X} - q\frac{S}{\sqrt{n}}, \infty\right),$$

and a $(1 - \alpha) \times 100\%$ upper confidence bound is

$$\mu \leq \bar{X} + q\frac{S}{\sqrt{n}},$$

with corresponding lower confidence interval

$$\left(-\infty, \bar{X} + q\frac{S}{\sqrt{n}}\right].$$

Example 7.14 Chemists at a state pollution control agency are concerned about lead levels in a certain lake. They take 15 samples of lake water and find an average lead level of 7 μg/dL with standard deviation of 2 μ/dL. Find a 95% lower confidence bound for μ.

Solution We have $\bar{x} = 7$, $s = 2$, and $q = 1.761$ is the 0.95 quantile of a t distribution with 14 degrees of freedom. Thus, $7 - 1.761(2/\sqrt{15}) \approx 6.091$, so we are 95% confident that the mean lead level in this lake is at least 6.09 μ/dL (the corresponding confidence interval is $[6.09, \infty)$). □

R Note:

To compute one-sided confidence intervals using `t.test`, add the argument `alt="less"` for the lower confidence interval and `alt="greater"` for the upper confidence interval.

```
> t.test(NCBirths2004$Weight, alt = "greater")$conf
[1] 3422.98    Inf
attr(, "conf.level"):
[1] 0.95
```

With 95% confidence, the mean weight of babies born in North Carolina in 2004 is at least 3422.98 grams (the interval is $[3422.08, \infty)$).

Example 7.15 Let X be a single observation from the exponential distribution $f(x) = \lambda e^{-\lambda x}$ for $x > 0$, $\lambda > 0$. Find a 95% lower confidence interval, that is, an upper bound, for λ.

Solution We already noted (page 184) that $Y = \lambda X$ is exponential with parameter $\lambda = 1$. Thus, $F_Y(q) = 0.95$ gives $q = 2.996$ for a 0.95 quantile,

$$0.95 = P(Y \leq 2.996) = P(\lambda X \leq 2.996) = P(\lambda \leq 2.996/X),$$

so the interval is $(-\infty, 2.996/X)$. □

Example 7.16 (This example is for those who covered Section 7.2.1.) We revisit Example 7.14. In the original solution, we treated the problem of estimating the mean as a location problem but this time, we treat this as a scale problem. This is more realistic—the data must be positive, and it is more likely that the standard deviation is proportional to the mean rather than being constant. We will treat $\bar{X}/\mu$ as a pivotal statistic and create an interval of the form $[\bar{X}/q, \infty)$, where q is the 95% quantile of the distribution of $\bar{X}/\mu$.

For a quick-and-dirty interval, we approximate the sampling distribution of $\bar{X}/\mu$ by estimating the distribution of $\bar{X} \sim N(7, 2^2/15)$, and then plugging in $\mu = 7$ gives the distribution of $\bar{X}/\mu = \bar{X}/7 \sim N(1, 2^2/(15 \times 7^2))$, which has 95% quantile, $1 + 1.64(2/(7\sqrt{15})) = 1.12$. The interval is $[7/1.12, \infty)$ or $[6.24, \infty)$.

We use the t distribution quantiles with the scale approach to provide a bit more insurance due to parameters being estimated. Thus, the 95% quantile is $1 + 1.761(2/(7\sqrt{15})) = 1.13$, so the lower endpoint would be $7/1.13 = 6.19$. Neither multiplicative endpoint is as low as the subtractive endpoint. Multiplicative is probably better. Note what happens in the extreme as the confidence level approaches 100%—the original approach would eventually make the lower endpoint negative, while a multiplicative endpoint would approach zero. □

7.4 CONFIDENCE INTERVALS FOR PROPORTIONS

We return to the poll results given at the beginning of this chapter: 59% of 846 likely voters believe that the country is headed in the wrong direction (page 167). Let X denote the number of likely voters in a sample of size n who think the country is headed in the wrong direction. We assume X is binomial, $X \sim \text{Binom}(n, p)$. From Chapter 6, we know that the proportion of likely voters, $\hat{p} = X/n$, is an unbiased estimator of p and from Corollary 4.2, for large n, $Z = (\hat{p} - p)/\sqrt{p(1-p)/n}$ is approximately standard normal.

Thus,

$$P\left(-1.96 < \frac{\hat{p} - p}{\sqrt{p(1-p)/n}} < 1.96\right) \approx 0.95. \tag{7.16}$$

Isolating the p in this expression requires a bit more algebra than the earlier problems. We set

$$-1.96 = \frac{\hat{p} - p}{\sqrt{p(1-p)/n}}$$

and solve for p (we get the same answer if we had set the right-hand side of the above to 1.96). This leads to the quadratic equation

$$\left(\frac{n}{1.96^2} + 1\right) p^2 - \left(\frac{2n\hat{p}}{1.96}\right) p + \frac{n\hat{p}^2}{1.96} = 0.$$

Using the quadratic formula to solve for p gives a 95% confidence interval (L, U), where

$$L = \frac{\hat{p} + 1.96^2/(2n) - 1.96\sqrt{\hat{p}(1-\hat{p})/n + 1.96^2/(4n^2)}}{1 + 1.96^2/n}, \tag{7.17}$$

$$U = \frac{\hat{p} + 1.96^2/(2n) + 1.96\sqrt{\hat{p}(1-\hat{p})/n + 1.96^2/(4n^2)}}{1 + 1.96^2/n}. \tag{7.18}$$

Thus, for the example at the beginning of the section, using $\hat{p} = 0.62$ and $n = 300$, we have

$$\frac{0.59 + (1.96^2/2 \cdot 846) \pm 1.96\sqrt{(0.59 \cdot 0.41/846) + 1.96^2/(4 \cdot 846^2)}}{1 + (1.96^2/846)}$$
$$= 0.589 \pm 0.033 = (0.556, 0.622).$$

Thus, we are 95% confident that between 55.6% and 62.2% of likely voters believe the country is headed in the wrong direction.

More generally, we have

SCORE CONFIDENCE INTERVAL FOR A PROPORTION

Theorem 7.2 *Consider a random sample of size n from a population where the parameter p indicates the true proportion with a certain binary characteristic,* $0 < p < 1$. *Let X denote the number in the sample with this characteristic and* $\hat{p} = X/n$ *the sample proportion. Then an approximate* $(1-\alpha) \times 100\%$ *confidence interval* (L, U) *for p is* (L, U) *with*

$$L = \frac{\hat{p} + q^2/(2n) - q \cdot \sqrt{\hat{p}(1-\hat{p})/n + q^2/(4n^2)}}{1 + q^2/n},$$

$$U = \frac{\hat{p} + q^2/(2n) + q \cdot \sqrt{\hat{p}(1-\hat{p})/n + q^2/(4n^2)}}{1 + q^2/n}$$

where q denotes the $(1 - \alpha/2)$ *quantile of* $N(0, 1)$.

This confidence interval is called the $(1-\alpha) \times 100\%$ *score confidence interval* for the proportion p.

Example 7.17 In the 2002 General Social Survey (Case Study in Section 1.6), 1308 participants chose to answer the question about whether they favor the death penalty for murder. Of the respondents, 899 favor the death penalty. Find a 90% confidence interval for the proportion of the population that favor the death penalty.

Solution With $X = 899$, $n = 1308$, we have $\hat{p} = 0.6873$ and $q = 1.645$. Thus, the 90% confidence interval is (0.6658, 0.7079), so we are 90% confident that between 66.6% and 70.8% of the population favors the death penalty for murder.

R Note:

The command `prop.test` computes score confidence intervals.

```
> prop.test(899, 1308, conf.level = .9, correct = FALSE)$conf
[1] 0.6658563 0.7079882
attr(,"conf.level")
[1] 0.9
```

If you omit the `conf.level` argument, then by default, a 95% confidence interval is calculated. The `correct=FALSE` argument is needed because by default, `prop.test` uses a different algorithm (one that uses a continuity correction) for the calculations.

`prop.test` also takes an argument of `alt="greater"` or `alt="less"` for one-sided confidence intervals.

```
> prop.test(899, 1308, conf.level = .9, correct = FALSE,
    alt = "greater")$conf
[1] 0.6706553 1.0000000
...
```

We are 90% confident that at least 67.1% of the population favor the death penalty.

□

Remark

- This interval is also called a Wilson or Wilson score interval (Wilson (1927)).
- The center of the score interval is $(\hat{p} + q^2)/(2n/1 + q^2/n)$. If we set $\kappa = q^2$, then this center can be written as $\hat{p}(n/(n+\kappa)) + (1/2)(\kappa/(n+\kappa))$, a weighted average of the observed proportion and $1/2$. As n increases, more weight is given to $\hat{p}$.

||

7.4.1 The Agresti–Coull Interval for a Proportion

Now, the limits of the interval given by Theorem 7.2 are pretty messy, so in general, we would want to use software to do the calculations. However, in the event that we must resort to hand calculations, it would be nice to find a simpler expression for the confidence interval. Agresti and Coull (1998) considered the 95% confidence interval for which the 0.975 quantile is $q \approx 1.96$, and hence $q^2 \approx 4$.

The AGRESTI–COULL 95% CONFIDENCE INTERVAL FOR A PROPORTION

If X denotes the number of successes in a sample of size n, let $\tilde{X} = X + 2$, $\tilde{n} = n + 4$, and $\tilde{p} = \tilde{X}/\tilde{n}$. Then an approximate 95% confidence interval for p is

$$\left(\tilde{p} - 1.96\sqrt{\frac{\tilde{p}(1-\tilde{p})}{\tilde{n}}},\ \tilde{p} + 1.96\sqrt{\frac{\tilde{p}(1-\tilde{p})}{\tilde{n}}}\right). \quad (7.19)$$

Example 7.18 Suppose the sample size is $n = 210$ with $x = 130$. Then $\tilde{x} = 132$, $\tilde{n} = 214$, and $\tilde{p} = 132/214 = 0.6168$ Thus, an approximate 95% confidence interval is given by

$$0.6168 \pm 1.96\sqrt{\frac{0.6168(1-0.6168)}{214}} = 0.6168 \pm 0.0651 = (0.5517, 0.6819),$$

which gives almost the same result as Theorem 7.2. □

Remark The above interval is similar to an interval that is often taught in introductory statistics courses, sometimes called the *Wald confidence interval for a binomial proportion.* Suppose that in Equation 7.16 we ignore the fact that the standard error $\sqrt{p(1-p)/n}$ depends on p and simply substitute $\hat{p}$ for p. We obtain the interval $\hat{p} \pm 1.96\sqrt{\hat{p}(1-\hat{p})/n}$. Many papers have been published recently cautioning against the use of this interval: it is not very accurate, particularly for $\hat{p}$ near zero or one (Agresti and Coull (1998), Brown et al. (2001), Newcombe (1998b)), (Pan (2009)). ||

Example 7.19 A political candidate prepares to conduct a survey to gauge voter support for his candidacy for senator. He would like a confidence interval with an error of at most 4%, with 95% confidence. How large should the sample size be for the survey?

Solution Since the interval given by Equation 7.19 is symmetric, the margin of error is $1.96\sqrt{\tilde{p}(1-\tilde{p})/\tilde{n}}$. Thus, we want to solve for $\tilde{n}$ in

$$1.96\sqrt{\frac{\tilde{p}(1-\tilde{p})}{\tilde{n}}} \leq 0.04.$$

Unfortunately, we do not know $\tilde{p}$—if we did, the candidate would not need to conduct the survey! We will use $\tilde{p} = 0.5$ since this will maximize the expression under the radical sign (see Exercise 22).

$$\begin{aligned} 1.96\sqrt{\frac{0.05(1-0.05)}{\tilde{n}}} &\leq 0.04 \\ \left(\frac{1.96(0.5)}{0.04}\right)^2 &\leq \tilde{n} \\ 596.25 &\leq n. \end{aligned}$$

Thus, he should survey at least 597 people. In some instances, based on prior knowledge, the analyst may substitute another estimate for $\tilde{p}$ (e.g., the proportion from a previous poll). □

7.4.2 Confidence Interval for the Difference of Proportions

For the difference of two proportions, $p_1 - p_2$, an interval extending the score interval can be constructed, though there is no closed form version (Wilson (1927)). Statistical software should be employed for constructing confidence intervals.

Example 7.20 From the 2002 General Social Survey (Case Study in Section 1.6), 475 out of 631 males favored the death penalty compared to 424 out of 677 females. Find a 95% confidence interval for the true difference in proportions (male–female).

Let p_1 and p_2 denote the proportions of males and females who support the death penalty. We want a confidence interval for $p_1 - p_2$ based on our sample estimates of $\hat{p}_1 = 475/631 = 0.753$ and $\hat{p}_2 = 424/677 = 0.626$. In R, the command `prop.test` computes the score confidence interval for the difference in proportions.

R Note:

```
> prop.test(c(475, 424), c(631, 677), correct = FALSE)$conf
[1] 0.07687197 0.17608985
...
```

Thus, we are 95% confident that between 7.7% and 17.6% more men than women favor the death penalty for murder. □

Remark Agresti and Caffo (2000) derive an approximate interval for the difference in proportions,

$$\tilde{p}_1 - \tilde{p}_2 \pm q\sqrt{\frac{\tilde{p}_1(1-\tilde{p}_1)}{\tilde{n}_1} + \frac{\tilde{p}_2(1-\tilde{p}_2)}{\tilde{n}_2}}, \qquad (7.20)$$

where $\tilde{p}_i = (X_i + 1)/(n_i + 2)$ and $\tilde{n}_i = n_i + 2$ for $i = 1, 2$, and q is the $(1 - \alpha/2)$ quantile of $N(0, 1)$.

Using the Agresti–Caffo equation on the previous death penalty survey example, we let $\tilde{p}_1 = (475 + 1)/(631 + 2) = 0.7520$ and $\tilde{p}_2 = (424 + 1)/(677 + 2) = 0.6259$. Then

$$(0.7520 - 0.6259) \pm 1.96\sqrt{\frac{0.7520 \times 0.248}{633} + \frac{0.6259 \times 0.3741}{679}}$$
$$= 0.1261 \pm 0.04912 = (0.07698, 0.1752).$$

With 95% confidence, between 7.7% and 17.5% more men than women favor the death penalty for murder. ||

7.5 BOOTSTRAP *t* CONFIDENCE INTERVALS

In Chapter 5, we introduced the bootstrap percentile confidence interval for giving a range of plausible values for a parameter. Another confidence interval is the bootstrap *t* interval that is based on estimating the actual distribution of the *t* statistic from the data, rather than just assuming that the *t* statistic has a Student's *t* distribution.

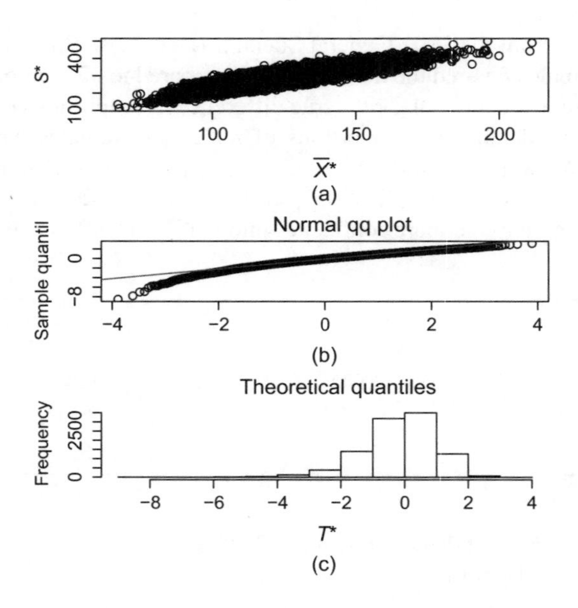

FIGURE 7.13 (a) Plot of S^* against $\bar{X}^*$. (b) A normal quantile plot of the bootstrap T statistics. (c) A histogram of the bootstrap T statistics.

Recall the Bangladesh arsenic levels data in Section 5.2.2 (page 110). The distribution of arsenic levels was right skewed (Figure 5.6). To use the formula derived earlier (Section 7.1.2) for the confidence interval of the mean μ, we would have to assume that the statistic $T = (\bar{x} - \mu)/(S/\sqrt{n})$ follows a t distribution. That seems unlikely for data this skewed. Instead, we bootstrap the t statistic: for each of 10^5 resamples, we compute the resample mean $\bar{X}^*$, resample standard deviation S^*, and then compute the resample T statistic $T^* = (\bar{X}^* - \bar{x})/(S^*/\sqrt{n})$.

In Figure 7.13b, we can see that the bootstrap distribution for the t statistic is left skewed; in fact, it is more left skewed than the bootstrap distribution of the mean is right skewed (Figure 5.7)! The reason for this is the strong positive relationship between $\bar{X}^*$ and S^* (Figure 7.13a). If a bootstrap resample contains a large number of the big values from the right tail of the original data, then $\bar{X}^*$ is large and hence S^* is especially large (standard deviations are computed by squaring distances from the mean, so they are affected even more by large observations than a mean is). The large denominator thus keeps T^* from being particularly large. Conversely, when there are relatively few of the big observations in the resample, then $\bar{X}^* - \bar{x}$ is negative and the denominator can be especially small, thus resulting in a T ratio that is large negative (Figure 7.13c).

The 2.5% and 97.5% percentiles of the bootstrap t distribution are -2.644 and 1.632, compared to ± 1.97 for the Student's t distribution. This is a reflection of the skewed nature of the bootstrap t distribution compared to the symmetric Student's t distribution.

Before proceeding with the bootstrap t, let us think about what skewness implies for the accuracy of the formula-based t confidence intervals $(\bar{X} \pm q(S/\sqrt{n}))$, where q is the $(1-\alpha/2)$ quantile. For right-skewed data, when $\bar{X} < \mu$, typically $S < \sigma$, so the confidence intervals tend to be narrow, and the interval falls below μ more often than $\alpha/2 \times 100\%$ of the time. This is bad. Conversely, when $\bar{X} > \mu$, typically $S > \sigma$, so the intervals tend to be wide and do not miss μ often enough; this is also bad. Overall, the intervals tend to be to the left of where they should be and give a biased picture of where the mean is likely to be.

We return to the bootstrap t and consider samples from nonnormal populations.

Let $T = (\bar{X} - \mu)/(S/\sqrt{n})$ and F be the cdf for the T statistic (the cdf of the sampling distribution). Let Q_1 and Q_2 denote the $\alpha/2$ and $(1-\alpha/2)$ quantiles of this distribution; that is, $Q_1 = F^{-1}(\alpha/2)$ and $Q_2 = F^{-1}(1-\alpha/2)$.

Then,

$$
\begin{aligned}
1-\alpha = P(Q_1 < T < Q_2) &= P\left(Q_1 < \frac{\bar{X}-\mu}{S/\sqrt{n}} < Q_2\right) \\
&= P\left(Q_1 \frac{S}{\sqrt{n}} < \bar{X}-\mu < Q_2 \frac{S}{\sqrt{n}}\right) \\
&= P\left(-\bar{X} + Q_1 \frac{S}{\sqrt{n}} < -\mu < -\bar{X} + Q_2 \frac{S}{\sqrt{n}}\right) \\
&= P\left(\bar{X} - Q_2 \frac{S}{\sqrt{n}} < \mu < \bar{X} - Q_1 \frac{S}{\sqrt{n}}\right).
\end{aligned}
$$

This suggests the confidence interval

$$\left(\bar{X} - Q_2 \frac{S}{\sqrt{n}},\ \bar{X} - Q_1 \frac{S}{\sqrt{n}}\right).$$

The quantiles Q_1 and Q_2 are unknown, but they can be estimated using quantiles of the bootstrap distribution of the t statistic:

$$T^* = \frac{\bar{X}^* - \bar{x}}{(S^*/\sqrt{n})},$$

where $\bar{X}^*$ and S^* are the mean and standard deviation of a bootstrap resample.

We use the standard error formula for every bootstrap sample because the bootstrap statistic should mimic $T = (\bar{X} - \mu)/(S/\sqrt{n})$.

BOOTSTRAP *t* CONFIDENCE INTERVAL FOR μ

For each of many resamples, calculate the bootstrap t statistic, $T^* = (\bar{X}^* - \bar{x})/(S^*/\sqrt{n})$. Let Q_1^* and Q_2^* be the empirical $\alpha/2$ and $(1-\alpha/2)$ quantiles, respectively (i.e. , the interval between them contains the middle $(1-\alpha) \times 100\%$ of all the bootstrap t statistics). The bootstrap t confidence interval is

$$\left(\bar{x} - Q_2^* \frac{s}{\sqrt{n}},\ \bar{x} - Q_1^* \frac{s}{\sqrt{n}}\right).$$

Remark For α near 0, Q_1^* is negative and Q_2^* is positive. So a positive number is subtracted from $\bar{X}$ to get the left endpoint, and a negative number is subtracted from $\bar{X}$ to get the right endpoint. ||

Example 7.21 We return to the arsenic example. We have already seen $n = 271$, $\bar{x} = 125.32$, and $s = 297.98$. So the 95% bootstrap t interval is $(125.32 - Q_2^*(297.98/\sqrt{271}), 125.32 - Q_1^*(297.98/\sqrt{271}))$, where Q_1^* and Q_2^* are the 0.025 and 0.975 quantiles from the bootstrapped t statistics.

R Note:

The following script bootstraps the t statistics and stores it in the vector `Tstar`.

```
Arsenic <- Bangladesh$Arsenic
xbar <- mean(Arsenic)
N <- 10^4
n <- length(Arsenic)
Tstar <- numeric(N)
for (i in 1:N)
{
  x <- sample(Arsenic, size = n, replace = T)
  Tstar[i] <- (mean(x)-xbar) / (sd(x)/sqrt(n))
}
```

For bootstrap t confidence intervals, we need the empirical quantiles.

```
> quantile(Tstar, c(.025, .975))
-2.644125  1.631868
```

Thus, $Q_1^* = -2.644125$ and $Q_2^* = 1.631868$, so we compute

$$\left(125.32 - 1.632\frac{297.98}{\sqrt{271}}, 125.32 - (-2.644)\frac{297.98}{\sqrt{271}}\right) = (95.78, 173.18)\ \mu\text{g/dL}.$$

The 95% bootstrap percentile interval computed earlier (Section 5.3) is (92.95, 164.44), while the formula t confidence interval is (89.68, 160.956). The bootstrap t interval is stretched further to the right, reflecting the right-skewed distribution of the data. Because of the large sample size, we report the 95% bootstrap t confidence interval (95.53, 173.04) μ g/dL. □

Example 7.22 In Section 5.3, we found the 95% bootstrap percentile interval for the mean weight of all babies born in North Carolina in 2004 to be (3419, 3478) g. The 95% t bootstrap confidence interval is (3418.8, 3478.4) g, and the 95% t formula confidence interval is (3419.3, 3478.4). All three intervals are quite close, so we may use any one of them. A check of a normal quantile plot shows that the distribution of weights is approximately normal. □

The bootstrap intervals for a difference in means follows the same idea.

BOOTSTRAP t CONFIDENCE INTERVAL FOR $\mu_1 - \mu_2$

For each of many resamples, calculate the bootstrap t statistic

$$T^* = \frac{\bar{X}_1^* - \bar{X}_2^* - (\bar{x}_1 - \bar{x}_2)}{\sqrt{S_1^{2*}/n_1 + S_2^{2*}/n_2}}.$$

Let Q_1^* and Q_2^* be the empirical $\alpha/2$ and $(1 - \alpha/2)$ quantiles of the bootstrap t distribution, respectively. The bootstrap t confidence interval is

$$\left(\bar{x}_1 - \bar{x}_2 - Q_2^* \sqrt{s_1^2/n_1 + s_2^2/n_2},\quad \bar{x}_1 - \bar{x}_2 - Q_1^* \sqrt{s_1^2/n_1 + s_2^2/n_2}\right).$$

Recall the Verizon example (page 117), where we considered the difference in means of two very skewed distributions of repair times for two very unbalanced samples ($n_1 = 23$ versus $n_2 = 1664$). The 95% bootstrap t interval for the difference in means is $(-22.1, -2.1)$. For comparison, the formula t interval is $(-16.6, 0.4)$ and the bootstrap percentile interval is $(-17.2, -1.6)$. The more accurate bootstrap t interval stretches farther in the negative direction, even more than the bootstrap percentile interval.

R Note:

```
Time.ILEC <- subset(Verizon, select = Time,
                    subset = Group == "ILEC", drop = T)
Time.CLEC <- subset(Verizon, select = Time,
                    subset = Group == "CLEC", drop = T)
thetahat <- mean(Time.ILEC) - mean(Time.CLEC)
nx <- length(Time.ILEC)  # nx=1664
ny <- length(Time.CLEC)  # ny=23
SE <- sqrt(var(Time.ILEC)/nx + var(Time.CLEC)/ny)

N <- 10000
Tstar <- numeric(N)
for (i in 1:N)
{
  bootx <- sample(Time.ILEC, nx, replace = TRUE)
  booty <- sample(Time.CLEC, ny, replace = TRUE)
  Tstar[i] <- (mean(bootx) - mean(booty) - thetahat) /
   sqrt(var(bootx)/nx + var(booty)/ny)
}
```

To find the 95% bootstrap t confidence interval

```
> thetahat - quantile(Tstar, c(.975, .025)) * SE
    97.5%        2.5%
-22.073888  -2.096639
> t.test(Time.ILEC, Time.CLEC)$conf   # compare to t-test
[1]  -16.5568985   0.3618588
...
```

The same basic procedure can also be used for confidence intervals for statistics other than one or two means—to compute a t statistic for each of many resamples, using the appropriate standard error for $\hat{\theta}$, find the quantiles of that bootstrap t distribution and create an estimate of the form $(\hat{\theta} - Q_2^* \times SE, \hat{\theta} - Q_1^* \times SE)$.

7.5.1 Comparing Bootstrap t and Formula t Confidence Intervals

It is useful to compare the bootstrap distributions to classical statistical inferences. With classical t intervals of the form $\bar{x} \pm q \times s/\sqrt{n}$, the confidence interval width varies substantially in small samples as the sample standard deviation s varies. Correspondingly, the classical standard error $s/\sqrt{n}$ varies as s varies. The bootstrap is no different in this regard—bootstrap standard errors and widths of confidence intervals for the mean are proportional to s.

Where the bootstrap does differ from classical inference is in how it handles skewness. The bootstrap percentile interval and bootstrap t interval are in general asymmetrical, with asymmetry depending on the sample. These intervals estimate the skewness of a population from the skewness of the sample (see Section A.7 for the definition of skewness.) In contrast, classical t intervals assume that the population has no underlying skewness (skewness is 0).

Which is preferred? Frankly, neither, but rather something in between. This is an area that needs attention from statistical researchers. Until then, we will recommend the formula t if $n \leq 10$, and the bootstrap t otherwise; the reason being in large samples, we should put more trust in the data—in that case, the bootstrap t is preferred. In small samples, the classical procedure is probably better—if the sample size is small, then skewness cannot be estimated accurately from the sample, and it may be better to assume that there is no skewness (skewness is 0) in spite of the bias, rather than to use an estimate that has high variability. Something between classical intervals and the bootstrap procedures would be best—something that makes a trade-off between bias and variance and transitions smoothly from being like the formula t for small n and bootstrap t for large n.

The bootstrap percentile makes less of a skewness correction than does the bootstrap t. Hence, for smaller samples, it is less variable than the bootstrap t. For larger samples, the bootstrap t is preferred. In the long run, increasing the sample size by a factor of 10 reduces the coverage errors of the bootstrap t intervals by a factor of 10 but reduces the errors of symmetric formula intervals and the bootstrap percentile interval only by a factor of $\sqrt{10}$.

For comparing two samples, if the sample sizes are equal or nearly equal, you may use the formula t for every sample size unless there is reason to believe that the skewness differs between the two populations. However, it is good to also do the bootstrap t interval for a check. If the sample sizes are unbalanced and skewed, then it is best to do a bootstrap t interval except for small samples.

7.6 EXERCISES

1. A researcher hired by a farming organization obtains a random sample of 50 cows in a state and finds a 95% confidence interval for the mean

milk production for cows μ to be (22, 30) kg/day. Critique the following interpretations:

(a) There is a 95% chance that a cow produces, on average, between 22 and 30 kg milk per day.

(b) We are 95% confident that $\bar{x}$ is between 22 and 30 kg of milk per day.

(c) The mean milk production μ for cows in the state will be 22–30 kg/day 95% of the time.

(d) We are 95% confident that cows in this state produce, on average, between 22 and 30 kg milk per day.

(e) In 95% of samples, the mean milk production will be between 22 to 30 kg/day.

2. For high school seniors in 2009 who took the SAT exam, the mean math SAT score was $\mu = 515$ with a standard deviation of $\sigma = 116$. From a random sample of 34 students at your university, you find the average SAT score to be 538. You forgot to compute the standard deviation, so you decide to assume that the standard deviation of scores at your university is the same as the national standard deviation of 116. Compute a 95% confidence interval for the mean SAT score at your university. (Assume that the university is large enough that we can ignore the fact that the mean of the 34 students is known.)

3. Suppose that 20 years ago, the mean cholesterol level of adult men in a certain town was 185 mg/dL with a standard deviation of 50 mg/dL.

   (a) Suppose you obtain a sample of size 100 and find the mean cholesterol level to be $\bar{x} = 210$. Assuming that σ has not changed, find a 90% confidence interval for the mean cholesterol level of the population (of adult men in this town).

   (b) Suppose you decide to conduct a new study to determine the mean cholesterol levels of adult men in this town. Assuming that the standard deviation has not changed, how many people should you include in your sample if you want the margin of error to be at most 10 mg/dL, using 95% confidence?

   (c) If you want to be 99% confident, then how large should your sample size be?

4. In R, the `qt` command computes quantiles of the t distribution.

   (a) Find the quantile q used in a 90% t confidence interval for a sample of size $n = 5$, $n = 15$, $n = 30$, and $n = 100$. Compare to the corresponding quantile for the standard normal.

   (b) For a fixed sample size n, say $n = 15$, find the quantile q used in a $(1 - \alpha) \times 100\%$ t confidence interval for 90% confidence, 95% confidence, 99% confidence. Which confidence level results in the narrowest interval and which the widest interval (assuming a fixed n)?

5. Suppose you draw a random sample of size n from a normal distribution with unknown mean μ and known standard deviation σ and construct a 95% confidence interval for μ. If you want to halve the margin of error, how much larger would the sample size have to be?

6. Julie is interested in the sugar content of vanilla ice cream. She obtains a random sample of $n = 20$ brands and finds an average of 18.05 g with standard deviation 5 g (per half cup serving). Assuming that the data come from a normal distribution, find a 90% confidence interval for the mean amount of sugar in a half cup serving of vanilla ice cream.
7. An engineer is studying the length of time his company's rechargeable batteries will work before needing to be recharged. He tests a random sample of 100 batteries and finds the average time that these batteries hold a charge is 120 h with standard deviation 12 h. Assume that the data exhibit only moderate skewness and find a 95% one-sided lower t confidence bound for the true mean length of battery life and give an interpretation of the bound.
8. In the simulation on page 176, we drew random samples of size 20 from Gamma(5, 2) to see how often the confidence interval for the mean missed the true mean. Repeat this simulation by changing the sample size, say $n = 30$, $n = 60$, $n = 100$, and $n = 250$. How does the sample size affect the frequency of missing the μ?
9. Import the data set `Spruce` (Case Study in Section 1.9) into `R`.
   (a) Create exploratory plots to check the distribution of the variable `Ht.change`.
   (b) Find a 95% t confidence interval for the mean height change over the 5-year period of the study and give a sentence interpreting your interval.
10. Researchers conducted a small study to determine the effects of diet on severely obese patients (Samaha et al. (2009)). After 6 months, the 43 patients who were randomly assigned to eat a low-carbohydrate diet lost an average of 5.8 kg with standard deviation 8.6 kg. The 36 patients on the low-fat diet lost an average of 1.9 kg with standard deviation 4.2 kg. Find a 95% t confidence interval for the mean difference in weight loss between the two groups (low carbohydrate versus low fat) and state the interpretation of the interval.
11. Consider the data set `Girls2004` with birth weights of baby girls born in Wyoming or Alaska (Case Study in Section 1.2).
    (a) Create exploratory plots and compare the distribution of weights between the babies born in the two states.
    (b) Find a 95% t confidence interval for the mean difference in weights for girls born in these two states. Give a sentence interpreting this interval.
12. Consider the data set `Girls2004` (see Case Study in Section 1.2).
    (a) Create exploratory plots and compare the distribution of weights between babies born to nonsmokers and babies born to smokers.
    (b) Find a 95% one-sided lower t confidence bound for the mean difference in weights between babies born to nonsmokers and smokers. Give a sentence interpreting the interval.

13. Import the data set `Spruce` (Case Study in Section 1.9) into `R`. We will compare the mean height change of the seedlings planted in a fertilized plot with those planted in a nonfertilized plot.
    (a) Create exploratory plots to compare the distributions of the variable `Ht.change` for the seedlings in the fertilized and nonfertilized plots.
    (b) Find a 95% one-sided lower t confidence bound for the mean difference in height change (F–NF) over the 5-year period of the study and give a sentence interpreting your interval.
14. Import the `FlightDelays` data set (see Case Study in Section 1.1) into `R`. Although the data represent all flights for United Airlines and American Airlines in May and June 2009, assume for this exercise that these flights are a sample from all flights flown by the two airlines under similar conditions. We will compare the lengths of flight delays between the two airlines.
    (a) Create exploratory plots of the lengths of delays for the two airlines.
    (b) Find a 95% t confidence interval for the difference in mean flight delays between the two airlines and interpret this interval.
15. Suppose the heights (inches) of 12 men randomly chosen from town A are

60.3	62.0	65.0	54.7	65.6	66.5	60.7	53.2	68.7	63.2	72.9	85.5

    (a) Begin with an exploratory plot of the data. What do you observe?
    (b) Find a 95% t confidence interval for the mean population height μ.
    (c) Remove the outlier and find the 95% t confidence interval for the mean population height μ. Did this change much?
16. (Exercise 15 continued) In town B, 11 men were randomly chosen and their heights (inches) measured.

68.0	65.4	68.0	66.0	66.6	69.8	68.0	70.0	70.3	65.7	75.9

    (a) Begin with an exploratory plot of this set and describe the distribution.
    (b) Create a 95% t confidence interval for the difference in mean height between men from town A and B. State your interpretation.
    (c) There is an outlier in town A. Remove this measurement and re-create the 95% t confidence interval. Does your conclusion change?
17. Run a simulation to see if the t ratio $T = (\bar{X} - \mu)/(S/\sqrt{n})$ has a t distribution or even an approximate t distribution when the samples are drawn from a nonnormal distribution. Be sure to superpose the appropriate t density curve onto your histograms. Try two different nonnormal distributions and remember to see if sample size makes a difference.
18. In the remark on page 181, we stated a result for the confidence interval for the difference in means if we knew that the population variances are the same. We also cautioned against using this result since, in general, it is difficult to

determine whether the population variances are indeed the same. Run a simulation to see how well the confidence interval for the difference in means compare in the pooled and unpooled variance cases when, in fact, the population variances are not the same.

The R code below will draw random samples of size m and n from $N(8, 10^2)$ and $N(3, 15^2)$. We will count the number of times the two 95% confidence intervals capture the true difference in mean of 5.

```
pooled.count <- 0             # set counter to 0
unpooled.count <- 0           # set counter to 0

m <- 20                       # sample size
n <- 10                       # sample size

N <- 10000                    # number of runs
for (i in 1:N)
{
  x <- rnorm(m, 8, 10)        # random sample from N(8,10^2)
  y <- rnorm(n, 3, 15)        # random sample from N(3,15^2)
  CI.pooled <- t.test(x, y, var.equal = T)$conf  # CI, pooled variance
  CI.unpooled <- t.test(x, y)$conf               # CI, unpooled variance

  if (CI.pooled[1] <= 5 && 5 <= CI.pooled[2])    # Is 5 in interval?
    pooled.count <- pooled.count + 1             # If yes, increment counter.

  if (CI.unpooled[1] <= 5 && 5 <= CI.unpooled[2])# Is 5 in interval?
    unpooled.count <- unpooled.count + 1         # If yes, increment counter.
}

pooled.count/N                                   # Prop. of time 0 in CI.
unpooled.count/N                                 # Prop. of time 0 in CI.
```

(a) Compare the performance of the two versions of the confidence interval for the difference in means.

(b) Repeat the simulation with different sample sizes, for example, $m = 80, n = 40; m = 120, n = 80; m = 80, n = 80$. Discuss.

19. As many states, Tennessee conducts audits of stores to determine whether or not proper sales tax was assessed. The Department of Revenue obtains a random sample of transactions at the audited store and for each transaction looks at the tax error defined to be the amount of tax owed minus the amount of tax paid. The auditors examine the lower bound of a 75% one-sided upper t confidence interval and if it larger than 0, the store owes the state money (`www.tn.gov/revenue/tntaxes/sales/statsamplingapr06.pdf`). Suppose the Department of Revenue samples 500 transactions of a certain store and finds the average tax error to be \$5.29 with a standard deviation of \$3.52. Compute a 75% one-sided upper t confidence interval for the true mean tax error.

20. One question in the 2002 General Social Survey asked participants whom they voted for in the 2000 election. Of the 980 women who voted, 459 voted for Bush. Of the 759 men who voted, 426 voted for Bush.
    (a) Find a 95% confidence interval for the proportion of women who voted for Bush.
    (b) Find a 95% confidence interval for the proportion of men who voted for Bush. Do the intervals for the men and women overlap? What, if anything, can you conclude about gender difference in voter preference?
    (c) Find a 95% confidence interval for the difference in proportions and interpret your interval.
21. A retail store wishes to conduct a marketing survey of its customers to see if customers would favor longer store hours. How many people should be in their sample if the marketers want their margin of error to be at most 3% with 95% confidence, assuming
    (a) they have no preconceived idea of how customers will respond, and
    (b) a previous survey indicated that about 65% of customers favor longer store hours.
22. Verify that $\sqrt{\tilde{p}(1-\tilde{p})}$ is a maximum when $\tilde{p} = 0.5$.
23. Suppose researchers wish to study the effectiveness of a new drug to alleviate hives due to math anxiety. Seven hundred math students are randomly assigned to take either this drug or a placebo. Suppose 34 of the 350 students who took the drug break out in hives compared to 56 of the 350 students who took the placebo.
    (a) Compute a 95% confidence interval for the proportion of students taking the drug who break out in hives.
    (b) Compute a 95% confidence interval for the proportion of students on the placebo who break out in hives.
    (c) Do the intervals overlap? What, if anything, can you conclude about the effectiveness of the drug?
    (d) Compute a 95% confidence interval for the difference in proportions of students who break out in hives by using or not using this drug and give a sentence interpreting this interval.
24. An article in the March 2003 *New England Journal of Medicine* describes a study to see if aspirin is effective in reducing the incidence of colorectal adenomas, a precursor to most colorectal cancers (Sandler et al. (2003)). Of 517 patients in the study, 259 were randomly assigned to receive aspirin and the remaining 258 received a placebo. One or more adenomas were found in 44 of the aspirin group and 70 in the placebo group. Find a 95% one-sided upper bound for the difference in proportions ($p_A - p_P$) and interpret your interval.
25. The data set `Bangladesh` has measurements on water quality from 271 wells in Bangladesh (Example 5.3).

(a) Compute the numeric summaries of the chlorine levels and create a plot and comment on the distribution.

(b) Find a 95% t confidence interval for the mean μ of chlorine levels in Bangladesh wells.

(c) Find the 95% bootstrap percentile and bootstrap t confidence intervals for the mean chlorine level and compare results. Which confidence interval will you report?

There are two missing values in the chlorine variable. Use the following R code to remove these two observations.

```
chlorine <- with(Bangladesh, Chlorine[!is.na(Chlorine)])
```

26. The data set MnGroundwater has measurements on water quality of 895 randomly selected wells in Minnesota.
    (a) Create a histogram or normal quantile plot of the alkalinity and comment on the distribution.
    (b) Find a 95% t confidence interval for the mean μ of alkalinity levels in Minnesota wells.
    (c) Find the 95% bootstrap percentile and bootstrap t confidence intervals for the mean alkalinity level and compare results. Which confidence interval will you report?
27. Consider the babies born in Texas in 2004 (TXBirths2004, Case Study in Section 1.2). We will compare the weights of babies born to nonsmokers and smokers.
    (a) How many nonsmokers and smokers are there in this data set?
    (b) Create exploratory plots of the weights for the two groups and comment on the distributions.
    (c) Compute the 95% confidence interval for the difference in means using the formula t, bootstrap percentile, and bootstrap t methods and compare your results. Which interval would your report?
    (d) Modify your result from above to obtain a one-sided 95% t confidence interval (hypothesizing that babies born to nonsmokers weigh more than babies born to smokers).
28. As we have seen, we can create confidence intervals for many different parameters besides a mean or proportion. One statistic common in epidemiology or medical research is relative risk, p_1/p_2, where p_1 and p_2 are the proportions of people in two groups who have a disease or condition. Suppose in a study of college students, 23% of sophomores reported binge drinking during homecoming compared to 12% of seniors. The sample relative risk is $0.23/0.12 = 1.9$, so sophomores are 1.9 times more likely to binge drink than seniors.
    (a) If in the study the researchers compute a 95% confidence interval for the true relative risk to be (1.6, 2.3), give a sentence interpreting this interval.

(b) Under what circumstance would a researcher conclude, based on a confidence interval for the relative risk, that there is no difference between the two groups?

29. *Johnson's t confidence interval* adjust for skewness by shifting endpoints right or left for positive or negative skewness, respectively. The interval is $\bar{X} + \hat{\kappa}_3/(6\sqrt{n})(1 + 2q^2) \pm q(S/\sqrt{n})$, where $\hat{\kappa}_3$ is a sample estimate of the population skewness $E(X - \mu)^3)/\sigma^3$ and q denotes the α quantile for a t distribution with $n - 1$ degrees of freedom. Calculate Johnson's t interval for the arsenic data (in `Bangladesh`) and compare with the formula t and bootstrap t intervals.

30. In this exercise, we compare two different ways to judge whether the parameters θ_1 and θ_2 for two populations differ. We assume here that methods based on the CLT are reasonable. We calculate estimates $\hat{\theta}_1$ and $\hat{\theta}_2$ and the corresponding standard errors $\hat{\text{SE}}_1$ and $\hat{\text{SE}}_2$, and calculate intervals:

$$\hat{\theta}_1 \pm 1.96\hat{\text{SE}}_1, \tag{7.21}$$

$$\hat{\theta}_2 \pm 1.96\hat{\text{SE}}_2, \tag{7.22}$$

One approach to determining whether or not $\theta_1 = \theta_2$ is to note whether or not these confidence intervals overlap. Alternately, we may compute the confidence interval for the difference $\theta_1 - \theta_2$

$$(\hat{\theta}_1 - \hat{\theta}_2) \pm 1.96\sqrt{\hat{\text{SE}}_1^2 + \hat{\text{SE}}_2^2} \tag{7.23}$$

and note to see whether or not the interval contains 0.

(a) Explain why the intervals given by Equations 7.21 and 7.22 overlap if and only if $(\hat{\theta}_1 - \hat{\theta}_2) \pm 1.96(\hat{\text{SE}}_1 + \hat{\text{SE}}_2)$ contains 0.

(b) Explain why the ratio of the width of the interval in (a) to the width of interval (7.23),
$(\hat{\text{SE}}_1 + \hat{\text{SE}}_2)/\sqrt{\hat{\text{SE}}_1^2 + \hat{\text{SE}}_2^2}$, is greater than 1.

(c) What does this imply about the two methods for gauging whether or not $\theta_1 = \theta_2$?

See Schenker and Gentleman (2001) for more discussion of this issue.

31. Let $X_i \sim N(\mu_1, \sigma^2), i = 1, 2, \ldots, n_1$ and $Y_j \sim N(\mu_2, \sigma^2), j = 1, 2, \ldots, n_2$ be independent samples with sample means and variances $\bar{X}, S_1^2, \bar{Y}, S_2^2$, respectively. Use Theorem 7.1 to verify that the confidence interval for $\bar{X} - \bar{Y}$ is given by Equation 7.14.

32. Suppose $X \sim N(0, \sigma^2)$. It is a fact that X^2/σ^2 has a chi-square distribution with 1 degree of freedom (see Theorem B.14). Use this fact to find a 95% confidence interval for σ^2. (In `R`, the command `qchisq` computes quantiles for the chi-square distribution.)

33. Let $X_1, X_2, \ldots, X_n$ be a random sample from a Poisson distribution with $\lambda > 0$. From Proposition 6.2 and Theorem A.5, we know that $\bar{X}$ is an unbiased estimator of λ.
    (a) Use the CLT approximation to find a 95% confidence interval for λ.
    (b) Compute the 95% confidence interval for the sample 4, 6, 7, 9, 10, 13.
34. Let $X \sim \text{Gamma}(2, \lambda)$. It is a fact that $2\lambda X$ has a chi-square distribution with 4 degrees of freedom (see Exercise 11). Use this fact to find a 95% confidence interval for λ. (In R, the command `qchisq` computes quantiles for the chi-square distribution.)
35. For the German Tank example (Example 7.13), derive a confidence interval based on using $\bar{X}/\theta$ as a pivotal statistic. Assume that $\bar{X}$ has an approximate normal distribution (by the CLT). Find the 95% confidence interval, for $n = 400$ when $\bar{X} = 5000$. Compare the width of this interval with the interval based on $X_{\max}$.
36. In this exercise, you will derive and use a "bootstrap Z" interval.
    (a) Following the steps in the derivation of the bootstrap t interval in Section 7.5, derive a bootstrap Z interval for μ, for cases when σ is known.
    (b) Calculate this interval for the Verizon CLEC data; for σ, use the sample variance of the Verizon ILEC data. (In practice, we may use methods for known σ when we can estimate σ from a large related data set.)
    (c) Compare that interval with a formula z interval. How does the bootstrap Z interval adjust for skewness?
37. Let $X_1, X_2, \ldots, X_n$ denote a random sample from $N(\mu, \sigma^2)$ with sample mean $\bar{X}$ and variance S^2. Let q_1 and q_2 denote the $\alpha/2$ and $(1 - \alpha/2)$ quantiles, respectively, of the chi-square distribution with $n - 1$ degrees of freedom. Show that $(1 - \alpha) \times 100\%$ confidence interval for σ^2 is given by $((n-1)S^2/q_2, (n-1)S^2/q_1)$. (*Hint:* Theorem B.16).
38. Robin wonders how much variation there is in a box of his favorite cereal. He buys eight boxes from eight different stores and finds the weights of the contents to be (in grams)

    560 568 580 550 581 581 562 550

    Assuming the data are from a normal distribution, find a 90% confidence interval for the variance σ^2. (In R, the command `qchisq` computes quantiles for the chi-square distribution.)
39. Let f be the pdf for some random variable and suppose μ and $\sigma > 0$ are two constants. Let $h(x) = \frac{1}{\sigma} f((x - \mu)/\sigma)$. Show that $h(x)$ is also a pdf.
40. Welch's approximation is based on the following idea: If (a) $Y \sim N(0, \sigma^2)$ and σ^2 is estimated by $\hat{\sigma}^2$, with $\hat{\sigma}^2$ independent of Y, (b) $\text{E}\left[\hat{\sigma}^2/\sigma^2\right] = 1$, and (c) $\text{Var}[\hat{\sigma}^2/\sigma^2] = 2/c$, then $Y/\hat{\sigma}$ has approximately a t distribution with $\nu = c$. Hence, an accurate estimate gives high degrees of freedom and a poor estimate lower degrees of freedom (and a correspondingly wide confidence interval).

In the one-sample case where the sample is from a normal distribution, S^2/σ^2 is a gamma distribution with mean 1 and variance $2/(n-1)$ and is independent of the sample mean. Hence, the degrees of freedom is $\nu = n - 1$ in the one-sample case.

In the two-sample case, we use $S_1^2/n_1 + S_2^2/n_2$ as an estimate for $\sigma_1^2/n_1 + \sigma_2^2/n_2$. Show that if both populations are normal, then this estimate satisfies the three conditions for the idea. In particular, show that the value of c equals Welch's approximation. *Hint:* See Exercise 12.

8

CLASSICAL INFERENCE: HYPOTHESIS TESTING

In Section 3.3, we introduced hypothesis testing, a formal procedure to evaluate a statement about a population or populations. In particular, we tested hypotheses involving two population means using a permutation test. This test makes no assumptions about the distributions underlying the two populations. We now consider hypothesis testing in situations where we can make some assumptions about the distribution of a population or populations.

The underlying procedure for analyzing a hypothesis test remains the same as before: we compute a test statistic from the data, and then compute a P-value, measuring how often chance alone would give a test statistic as extreme as the observed statistic, assuming the null hypothesis is true. If the P-value is small, then the results cannot easily be explained by chance alone, and we reject the null hypothesis.

To compute the P-value, we need to find a reference distribution, the null distribution—the distribution that the test statistic follows if the null hypothesis is true. We then compute a one-sided P-value using that distribution. For a two-sided test, we multiply the smaller of the one-sided P-values by two.

8.1 HYPOTHESIS TESTS FOR MEANS AND PROPORTIONS

8.1.1 One Population

Example 8.1 For college-bound seniors in 2009, SAT math scores are normally distributed with a mean of 515 and a standard deviation of 116. You suspect that

Mathematical Statistics with Resampling and R, First Edition. By Laura Chihara and Tim Hesterberg.

seniors in the town of Sodor are much brighter than the country as a whole, so you decide to conduct a test. In a random sample of 25 seniors, you find the mean SAT math score to be 555. You assume that the standard deviation of math scores in Sodor are the same as the national standard deviation of $\sigma = 116$. Is this sufficient evidence to conclude that Sodor seniors are smarter, or could a mean score of 555 be attributable to random variability?

The parameter of interest is μ, the mean math SAT score in Sodor. The hypotheses are as follows:

$$H_0\colon \mu = 515 \quad \text{versus} \quad H_A\colon \mu > 515.$$

As in Section 3.3, to determine the null distribution, we assume that the null hypothesis is true; here we assume that the data are from a normal distribution with mean 515 and standard deviation 116. Thus, the sample mean $\bar{x}$ comes from $N(515, 116^2/25)$. We standardize

$$z = \frac{555 - 515}{116/5} = 1.724 \sim N(0, 1).$$

How likely is this? We compute the P-value, $P(Z \geq 1.724) = 0.043$. So if the null hypothesis is true, then only 4.3% of samples of size 25 give rise to a mean as extreme as 555. This is (mild) evidence against the null hypothesis, so we conclude that Sodor seniors are better than the national pool of seniors (at least as measured by the SAT!) □

Now, suppose in the previous example, we are not willing to assume that the variability of scores in Sodor is the same as the variability in the national scores. We test the same hypotheses, but now estimate the standard error from the data, using $s/\sqrt{n}$ in place of $\sigma/\sqrt{n}$. Suppose $s = 120$. Then the test statistic is

$$t = \frac{\bar{x} - \mu}{s/\sqrt{n}} = \frac{555 - 515}{120/5} = 1.67.$$

We saw previously that $t = (\bar{x} - \mu)/(s/\sqrt{n})$ has a Student's t distribution with $(n - 1)$ degrees of freedom (page 172), so that is the null distribution. The one-sided P-value is $P(T \geq 1.67) = 0.0543$, so about 5.4% of samples of size 25 from $N(515, \sigma^2)$ would give a t statistic this large or larger. Thus, at a 5% significance level, we would conclude that we do not have enough evidence to support the claim that Sodor seniors score, on average, higher than the national pool of seniors.

Thus, in these two examples, we see a common theme: we calculate a test statistic from the data and then find or estimate the distribution for this test statistic assuming the null hypothesis is true. In permutation testing, the reference distribution is the permutation distribution obtained by permuting the data. Here, the reference distributions are parametric distributions: the normal distribution and the t distribution.

ONE SAMPLE MEANS T TEST

Let $X_1, X_2, \ldots, X_n$ be a random sample from a normal distribution with unknown μ and σ. Let $\bar{X}$ and S denote the sample mean and the standard deviation. To test

$$H_0\colon \mu = \mu_0 \quad \text{versus} \quad H_a\colon \mu \neq \mu_0,$$

we form the t test statistic

$$T = \frac{\bar{X} - \mu_0}{S/\sqrt{n}}.$$

Under the null hypothesis, T has a t distribution with $(n-1)$ degrees of freedom. The P-value is the probability that chance alone would produce a test statistic as extreme as or more extreme than the observed value $t = (\bar{x} - \mu_0)/(s/\sqrt{n})$, if the null hypothesis is true.

Example 8.2 The coffee vending machine at work dispenses 7 ounces of coffee in paper cups. The staff suspects that the machine is under-filling the cups. From a sample of $n = 15$ cups, they compute a mean of 6.6 ounces and a standard deviation of 0.8 ounces. Does this evidence support their suspicions? Assume that the coffee amounts dispensed by the machine are normally distributed.

Solution Let μ denote the true amount of coffee dispensed by the machine. We test $H_0\colon \mu = 7$ versus $H_A\colon \mu < 7$.

We assume the null hypothesis is true, so the amount of coffee dispensed follows a normal distribution with mean 7 and an unknown standard deviation. The test statistic is $t = (6.6 - 7.0)/(0.8/\sqrt{15}) = -1.9365$. We compare this to a t distribution with 14 degrees of freedom and find a P value of $P(t \leq -1.9365) = 0.0366$. Thus, we conclude that the vending machine is under-filling the cups. □

We consider another one-population setting, where the parameter of interest is now a proportion rather than a mean.

Example 8.3 About 13% of the population is left-handed. A biologist suspects that the scientific community is not like the general population in terms of handedness. He queries 200 scientists and finds that 36, or 18%, are left-handed. Does this evidence support the biologist's theory?

Solution Let p denote the true proportion of left-handed scientists. Then

$$H_0\colon p = 0.13 \quad \text{versus} \quad H_A\colon p \neq 0.13.$$

To calculate the P-value, we assume the null hypothesis is true. Then X, the number of left-handed scientists, is a binomial random variable, $X \sim \text{Binom}(200, 0.13)$. How unusual is it to get 36 or more left-handers in a sample of size 200? The one-sided P-value is

$$\begin{aligned} P(X \geq 36) &= P(X = 36) + P(X = 37) + \cdots + P(X = 200) \\ &= \sum_{i=36}^{200} \binom{200}{i} 0.13^i (1 - 0.13)^{200-i} \\ &= 0.0267. \end{aligned}$$

We multiply by 2 for a two-sided alternative hypothesis, so the P-value is 0.0533. Thus, our evidence does not support the hypothesis that scientists are different from the general population in terms of handedness. □

Remark Instead of the exact calculation, we may do quick-and-dirty calculations using the CLT. Under the null hypothesis, the distribution of $\hat{p}$ is approximately normal with mean 0.13 and variance $0.13(1 - 0.13)/200 = (0.02378)^2$. Using the continuity correction, the one-sided P-value is

$$P(X \geq 36) = P(X > 35.5) = P(\hat{p} > 35.5/200) = P(\hat{p} > 0.1775).$$

Standardizing, we obtain

$$z = (0.1775 - 0.13)/0.02378 = 1.997.$$

Thus, the one-sided P-value is $P(z \geq 1.997) = 0.022887$, and the two-sided P-value is 0.0458. This indicates mild evidence against the null hypothesis.

The exact calculation and the CLT yielded dramatically different results, reaching opposite conclusions using a 5% significance level, with P-values differing by 0.8%, a fraction of about 1/6 of the true value.

Here, $np = 200 \times 0.13 = 26$, well above a common rule of thumb that says if $np > 10$ and $n(1 - p) > 10$, then the CLT approximation is appropriate. Indeed, we saw in Section 4.3.3 that requiring $np > 384$ and $n(1 - p) > 384$ was a much better condition when accuracy is critical. So, in the case of significance testing, when the CLT yields a borderline answer, it is best to use exact calculations.

In contrast to confidence intervals, we do not add anything to numerator or denominator (Section 7.4.1). This is because in hypothesis testing we do calculations assuming the null hypothesis is true, whereas in confidence intervals we allow for uncertainty in the true parameter. ||

One Sample Proportion Test

Let X be a binomial random variable, $X \sim \text{Binom}(n, p)$. Suppose $X = x$ is observed. To test

$$H_0\colon p = p_0 \quad \text{versus} \quad H_A\colon p \neq p_0,$$

we compute $P(X \geq x)$ if $x \geq np_0$ or $P(X \leq x)$ if $x \leq np_0$.

For instance,

$$P(X \geq x) = \sum_{i=x}^{n} \binom{n}{i} p_0^i (1 - p_0)^{n-i}.$$

For the two-sided test, we double this probability to obtain the P-value.

8.1.2 Comparing Two Populations

In the previous section, we discussed significance testing when there was one population of interest. However, the same ideas apply when comparing two populations.

Example 8.4 We return to the North Carolina babies case study. The mean and standard deviation of the weights of the $n_1 = 898$ babies born to nonsmoking mothers are $\bar{x}_1 = 3472$ and $s_1 = 479$ g, whereas the mean and standard deviation of the weights of the $n_2 = 111$ babies born to smoking mothers are $\bar{x}_2 = 3257$ and $s_2 = 520$ g. Is the observed mean difference in weights of $\bar{x}_1 - \bar{x}_2 = 215$ g easily explained by chance, or is there a real difference in the mean weights of North Carolina babies born to nonsmoking and smoking mothers in 2004 (see Figure 8.1)?

Let μ_1 and μ_2 denote the true mean weight of babies born to nonsmoking and smoking mothers, respectively. We consider the hypotheses

$$H_0\colon \mu_1 = \mu_2 \quad \text{versus} \quad H_A\colon \mu_1 > \mu_2.$$

As in Section 7.1.3, if we assume that the distribution of weights is normal for babies born to both non-smoking and smoking mothers, then the statistic

$$t = \frac{(\bar{x}_1 - \bar{x}_2) - (\mu_1 - \mu_2)}{\sqrt{s_1^2/n_1 + s_2^2/n_2}}$$

has approximately a t distribution with degrees of freedom given by Equation 7.11.

For these data, the statistic is

$$t = \frac{3471.912 - 3256.910}{\sqrt{478.5524^2/898 + 520.4788^2/111}} = 4.14 \quad \text{with}$$

$$\nu = \frac{(479^2/898 + 520^2/111)^2}{\dfrac{(479^2/898)^2}{(898-1)} + \dfrac{(520^2/111)^2}{(111-1)}} = 134.011$$

degrees of freedom. If the null hypothesis is true ($\mu_1 - \mu_2 = 0$), then the chance of obtaining a statistic as extreme as 4.14 is $P(t \geq 4.1411) = 0.00003$. Thus, if there really is no difference in mean weights, then the samples we obtained are rare—random chance alone would give a test statistic that large, less than 3 out of a 100, 000

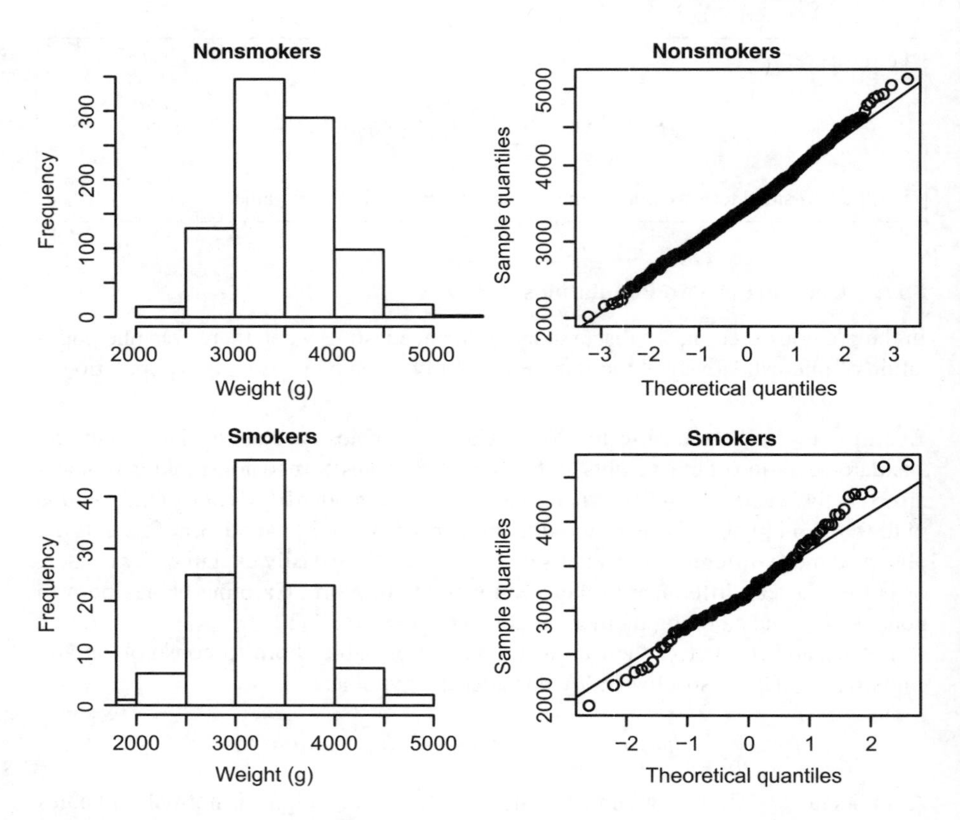

FIGURE 8.1 Distribution of weights of babies born to nonsmoking and smoking mothers.

times. Thus, we conclude that babies born to nonsmoking mothers do weigh, on average, more than babies born to smoking mothers.

R Note:

The `t.test` command calculates a two-sample t test.

```
> WeightNS <- subset(NCBirths2004, select=Weight,
                     subset=Smoker=="No", drop=T)
> WeightS  <- subset(NCBirths2004, select=Weight,
                     subset=Smoker=="Yes", drop=T)
> t.test(WeightNS, WeightS, alt="greater")

...
t = 4.1411, df = 134.011, p-value = 3.04e-05
alternative hypothesis: true difference in means is greater than 0
95 percent confidence interval:
 129.0090      Inf
```

```
sample estimates:
mean of x mean of y
 3471.912  3256.910
```

To test that the difference is a value different from 0, for example, H_0: $\mu_1 - \mu_2 = 50$ versus H_A: $\mu_1 - \mu_2 > 50$, add the argument `mu=50` to the `t.test` command.

From the R output, we can also see that with 95% confidence, babies born to nonsmoking mothers weigh at least 129 g more than babies born to smoking mothers. □

TWO-SAMPLE t-TEST FOR MEANS

Let $X_1, X_2, \ldots, X_{n_1} \sim N(\mu_1, \sigma_1^2)$ and $Y_1, Y_2, \ldots, Y_{n_2} \sim N(\mu_2, \sigma_2^2)$ be two independent random samples with sample means and standard deviations $\bar{X}$, S_1, $\bar{Y}$, S_2, respectively. To test

$$H_0\colon \mu_1 = \mu_2 \quad \text{versus} \quad H_A\colon \mu_1 \neq \mu_2,$$

we form the test statistic

$$T = \frac{\bar{X} - \bar{Y}}{\sqrt{S_1^2/n_1 + S_2^2/n_2}}.$$

If the null hypothesis is true, then T has approximately a t distribution with degrees of freedom given by Equation 7.11. The P-value is the probability that chance alone would produce a test statistic as extreme as or more extreme than the observed value if the null hypothesis is true.

Remark

- In the remark on page 181, we mentioned a version of the confidence interval for the difference between two-sample means when we could assume that the population variances were equal. Similarly, for the two sample t-test for means, if we can assume the population variances are equal, then we can use the test statistic $T = (\bar{X} - \bar{Y})/(S_p\sqrt{1/n_1^2 + 1/n_2^2})$, where S_p is the pooled sample variance. If the null hypothesis is true, then T follows a t distribution with $(n_1 + n_2 - 2)$ degrees of freedom. The same cautions that were voiced previously in making the assumption that the population variables are equal still hold.
- As with confidence intervals, you should check that your data follow the assumptions of the test. In the case of the one- or two-sample means test, plot your data to verify that they come from a normal distribution. In general, the advice we gave on the assumptions for t confidence intervals holds in the case of t tests for means.

In particular, the tests are sensitive to outliers, so you should always be on the lookout for them in your exploratory plots. In addition to comparing a numeric characteristic between two populations, we can test to see if the proportions of some binary characteristics of two populations are the same. ||

Example 8.5 Do men and women differ in their beliefs about an afterlife? In the 2002 General Social Survey Case Study in Section 1.6, participants were asked this question and of the 684 women who responded, 550 said yes (80.40%), compared to 425 of the 563 men (75.49%).

Let p_1 and p_2 denote the proportions of women and men, respectively, who believe that there is life after death. We wish to test

$$H_0\colon p_1 = p_2 \quad \text{versus} \quad H_A\colon p_1 \neq p_2.$$

Thus, we need to find a test statistic based on $\hat{p}_1$ and $\hat{p}_2$, the sample proportions, and an appropriate reference distribution. □

Before continuing with this example, let us review what we know about proportions, in particular, the sampling distribution of $\hat{p}_1 - \hat{p}_2$. We know that for large samples, $\hat{p}_1$ and $\hat{p}_2$ have approximate normal distributions, so their difference is also approximately normally distributed with $\mathrm{E}\left[\hat{p}_1 - \hat{p}_2\right] = p_1 - p_2$ and (assuming independence)

$$\mathrm{Var}[\hat{p}_1 - \hat{p}_2] = \mathrm{Var}[p_1] + \mathrm{Var}[p_2] = \frac{p_1(1-p_1)}{n_1} + \frac{p_2(1-p_2)}{n_2}$$

(Corollary 4.2 and Theorem A.10).

Standardizing gives

$$z = \frac{(\hat{p}_1 - \hat{p}_2) - (p_1 - p_2)}{\sqrt{p_1(1-p_1)/n_1 + p_2(1-p_2)/n_2}},$$

and this has an approximate standard normal distribution.

Now, let us return to the setting of a significance test. Under the null hypothesis, $p_1 = p_2 = p$. Now the test statistic becomes

$$z = \frac{(\hat{p}_1 - \hat{p}_2) - 0}{\sqrt{p(1-p)/n_1 + p(1-p)/n_2}}.$$

The problem is that this expression still involves the unknown parameter p. Thus, we need to find an estimate for p. Since we are assuming that the proportion of "successes" is the same in both populations, we pool the samples to get a better estimate of p. Let X_1 and X_2 denote the number of "successes" in each sample. Then a pooled estimate of p is given by

$$\hat{p}_{\mathrm{p}} = \frac{X_1 + X_2}{n_1 + n_2},$$

so the estimate of the standard error of the sampling distribution of $\hat{p}_1 - \hat{p}_2$ is

$$\mathrm{SE}[\hat{p}_1 - \hat{p}_2] = \sqrt{\frac{\hat{p}_p(1-\hat{p}_p)}{n_1} + \frac{\hat{p}_p(1-\hat{p}_p)}{n_2}}. \tag{8.1}$$

Thus, if the null hypothesis $p_1 = p_2$ is true, then

$$\begin{aligned} z &= \frac{(\hat{p}_1 - \hat{p}_2) - \mathrm{E}\left[\hat{p}_1 - \hat{p}_2\right]}{\mathrm{SE}[\hat{p}_1 - \hat{p}_2]} \\ &= \frac{\hat{p}_1 - \hat{p}_2}{\sqrt{\hat{p}_p(1-\hat{p}_p)/n_1 + \hat{p}_p(1-\hat{p}_p)/n_2}} \\ &= \frac{\hat{p}_1 - \hat{p}_2}{\sqrt{\hat{p}_p(1-\hat{p}_p)(1/n_1 + 1/n_2)}} \end{aligned}$$

follows approximately a standard normal distribution.

Example (GSS Continued) Going back to the afterlife question, we have $X_1 = 550$, $n_1 = 684$ and $X_2 = 425$, $n_2 = 563$, so the pooled estimate and standard error are

$$\hat{p} = \frac{550 + 425}{684 + 563} = 0.7819 \quad \text{and}$$

$$\mathrm{SE}[\hat{p}_1 - \hat{p}_2] = \sqrt{0.7819(1 - 0.7819)\left(\frac{1}{684} + \frac{1}{563}\right)} = 0.0235.$$

Thus, the test statistic is $z = (0.8040 - 0.7549)/0.0235 = 2.0936$ and the probability $P(z \geq 2.0936) = 0.0182$. For a two-sided test, we double this probability to get the P-value of 0.036. If the proportions of men and women who believe in an afterlife are the same, then about 3.6% of samples of the given sizes would lead to a random test statistic as extreme as that observed. Thus, the evidence suggests that men and women are different in how they view the afterlife.

TWO-SAMPLE-z TEST FOR PROPORTIONS

Let $X \sim \mathrm{Binom}(n_1, p_1)$ and $Y \sim \mathrm{Binom}(n_2, p_2)$ be independent. To test

$$H_0\colon p_1 = p_2 \quad \text{versus} \quad H_A\colon p_1 \neq p_2,$$

let $\hat{p}_p = \frac{X_1+X_2}{n_1+n_2}$ and form the test statistic

$$Z = \frac{\hat{p}_1 - \hat{p}_2}{\sqrt{\hat{p}_p(1-\hat{p}_p)\left(1/n_1 + 1/n_2\right)}}.$$

Then, under the null hypothesis, Z has an approximate standard normal distribution. The P-value is the probability of obtaining a random test statistic as extreme as or more extreme than the observed Z if the null hypothesis is true.

Remark It is not common to use a continuity correction when using the normal approximation to the two-sample proportions test, even though sometimes it would be helpful. This is tricky; the numerator $(\hat{p}_1 - \hat{p}_2)$ is of the form $X_1/n_1 - X_2/n_2$. If $n_1 = n_2$, the test statistic is strongly discrete, and we might do a continuity correction. Otherwise, the distribution is closer to continuous, depending in part on the greatest common denominator of n_1 and n_2 (see Exercise 13). ||

Example 8.6 Are young people more in tune with environmental issues than older people? The Pew Research Center for the People and the Press conducted a survey in 2009 asking a random sample of adults whether or not there is solid evidence of global warming. Of the 197 people aged 18–29 years old who responded, 126 said yes, compared to 223 out of 406 people aged 30–39 years old. Does this indicate that a higher proportion of younger people believe there is evidence of global warming? (`http://pewresearch.org/pubs/1386/cap-and-trade-global-warming-opinion`).

Solution Let p_1 and p_2 denote the proportion of 18–29 years old and 30–39 years old, respectively, who believe there is solid evidence of global warming. We wish to test H_0: $p_1 = p_2$ versus H_A: $p_1 > p_2$.

The pooled estimate of p and the standard error are

$$\hat{p}_p = \frac{126 + 223}{197 + 406} = 0.5788 \quad \text{and}$$

$$\text{SE}[\hat{p}_1 - \hat{p}_2] = \sqrt{0.5788(1 - 0.5788)\left(\frac{1}{197} + \frac{1}{406}\right)} = 0.0429.$$

Thus, the test statistic is $z = (0.64 - 0.55)/0.0429 = 2.097$. If the proportions are the same, then the one-sided P-value is $P(z \geq 2.097) = 0.0179$; that is, under the null hypothesis, only 1.8% of samples with these sizes would yield outcomes this extreme. Thus, we conclude that a higher proportion of 18–29 years believe there is solid evidence of global warming compared to the 30–39 years old. □

Remark Two-sided two-sample proportions test can be analyzed using contingency tables. For instance, the GSS example on the afterlife question can be summarized as

	Afterlife?	
Gender	Yes	No
Female	550	134
Male	425	138

The chi-square test statistic is 4.384, so comparing this to a chi-square distribution with 1 degree of freedom gives a P-value of 0.036, which is identical to the one computed in the GSS example. The reason for this is Theorem B.14, which states that the square of a standard normal random variable is a chi-square random variable. The z test statistic computed for the GSS example is $z = 2.0936$, so $z^2 = 4.383$. ||

R Note:

The `prop.test` command computes the chi-square test statistic rather than the z test statistic.

```
>  prop.test(c(550, 425), c(684, 563), correct=F)
  ...
X-squared = 4.3848, df = 1, p-value = 0.03626
alternative hypothesis: two.sided
95 percent confidence interval:
 0.002870906 0.095547134
sample estimates:
   prop 1    prop 2
0.8040936 0.7548845
```

We are 95% confident that between 0.2 and 9.6% more females than males believe in an afterlife.

8.2 TYPE I AND TYPE II ERRORS

With any hypothesis test, there are two errors we can make—to reject H_0 when it holds, or to fail to reject it when it does not hold. Let us look at these in more detail.

Definition 8.1 A *Type I error* occurs if we reject the null hypothesis when it is true. A *Type II error* occurs if we do not reject the null hypothesis when the alternative is true. ||

8.2.1 Type I Errors

To get an idea of the types of errors in a hypothesis test, we look at this in a courtroom setting. Suppose John Doe is on trial for murder. In the United States, accused are considered "innocent until proven guilty," and the proof must be "beyond a reasonable doubt." This corresponds to H_0: John Doe is innocent versus H_A: John Doe is guilty. Unless the evidence strongly shows otherwise, we accept the null hypothesis (Table 8.1).

Which error is more serious, convicting an innocent person (Type I) or freeing a guilty person (Type II)? Our justice system sidesteps this question; it holds that convicting an innocent person is bad, and the probability of a wrongful conviction must be small. The severity of a Type II error does not really enter the picture.

Similarly, in the classical approach to hypothesis testing, we do not adjust critical values to balance the two kinds of errors, taking into account their relative severity.

TABLE 8.1 Decisions by a Jury in a Murder Trial

Jury Decision	Truth	
	Innocent	Guilty
Guilty	Type I error	Correct
Not guilty	Correct	Type II error

Instead, we set thresholds to limit the probability of a Type I error to a pre-specified value, say 5%.

Example 8.7 A pharmaceutical company is testing a new drug that it hopes will lower cholesterol levels in patients. They conduct a study in which they test H_0: the drug is not effective in lowering cholesterol levels versus H_A: the drug is effective in lowering cholesterol levels. Describe the Type I and Type II errors in this case and the practical consequences of making these errors.

Solution A Type I error occurs if the researchers conclude that the drug is effective when in fact it is not. The practical consequences of this include endangering the health of patients who take the drug in the mistaken belief that it will lower their cholesterol, wasted costs, and potential liability claims filed against the company. A Type II error occurs if the researchers conclude the drug is not effective when in fact it is. A practical consequences of this are the failure to save lives or otherwise improve patients' health and lost revenue for the company. □

Example 8.8 We return to math SAT scores for 2009 college-bound seniors (scores are distributed $N(515, 116^2)$). Suppose local educators in a certain city wants to know how their students fare compare to the national average. The educators have also decided that if their students average much lower than the national average of 515 points, then they will request more money from the city council for new teaching reforms. The educators will obtain a random sample of scores and test H_0: $\mu = 515$ versus H_A: $\mu < 515$, where μ denotes the mean math SAT score in their city. If their sample size is 100 and they decide that a sample average of 505 or less is their criterion for making the funding request, what is the probability that they make a Type I error? Assume that the standard deviation of scores in the city is $\sigma = 116$.

Solution

$$\begin{aligned}
P(\text{Type I error}) &= P(\text{Reject } H_0 \mid H_0 \text{ true}) \\
&= P(\bar{X} < 505 \mid \mu = 515) \\
&= P\left(\frac{\bar{X} - 515}{116/\sqrt{100}} < \frac{505 - 515}{116/\sqrt{100}}\right) \\
&= P(Z < -0.8621) = 0.1943.
\end{aligned}$$

Thus, there is a 19.4% chance of the educators unnecessarily requesting funds from the city when, in fact, their students performance is in line with the national student body. □

Example 8.9 A company claims that only 3% of women who use their facial lotion develops an allergic reaction (rash). You are a bit suspicious of their claims since you think a higher proportion of women at your college are allergic to this lotion. You query a random sample of 50 women and ask them to try the lotion. If more than three people develop the rash, you will send a nasty email to the company CEO. What is the probability that you make a Type I error?

Solution If p denotes the true proportion of women at your college who are susceptible to the rash, you want to test H_0: $p = 0.03$ versus H_A: $p > 0.03$. Let X denote the number of women who develop the rash. If H_0 is true, then $X \sim \text{Binom}(50, 0.03)$, and

$$\begin{aligned} P(\text{Type I error}) &= P(X > 3 \mid H_0 \text{ true}) \\ &= P(X > 3 \mid X \sim \text{Binom}(50, 0.03)) \\ &= \sum_{i=4}^{50} \binom{50}{i} (0.03)^i (1 - 0.03)^{50-i} \\ &= 0.0627. \end{aligned}$$

Thus, there is a 6.3% chance of a Type I error. □

So what is an acceptable probability for making a Type I error? Obviously, it depends on the stakes: in the murder trial analogy, we would want the probability of sending an innocent person to prison (or worse) to be very small (preferably close to 0!) On the other hand, in the funding request example, perhaps a Type I error probability of 5–10% would be acceptable.

In practice, there are many situations where we do consider the severity of Type II errors. For example, in the early days of the AIDS/HIV epidemic, the Food and Drug Administration relaxed its criteria for accepting the first lifesaving drug, AZT, because a Type II error meant continuing to let people die.

Definition 8.2 The *significance level* of a hypothesis test is the largest value α that we find acceptable for the probability of a Type I error. ||

Recall that in Definition 3.4, we stated that 0.05 and 0.10 and 0.01 are common thresholds for declaring a result statistically significant.

Example 8.10 In the funding example, suppose the educators decide that they want the probability of a Type I error to be 10%. For what value of C would sample means $\bar{X} \leq C$ result in rejecting the null hypothesis? (Keep $n = 100$, $\sigma = 116$).

Solution

$$\begin{aligned} 0.10 &= P(\text{Reject } H_0 \mid H_0 \text{ true}) \\ &= P(\bar{X} \leq C \mid \mu = 515) \\ &= P\left(\frac{\bar{X} - 515}{116/\sqrt{100}} \leq \frac{C - 515}{116/\sqrt{100}}\right) \\ &= P\left(Z \leq \frac{C - 515}{11.6}\right). \end{aligned}$$

For the normal distribution, the 0.1 quantile is -1.2816. Thus, setting

$$-1.2816 = \frac{C - 515}{11.6}$$

and solving for C yields $C = 500.13$. Thus, sample means of 500.13 or less would result in the educators concluding their students score, on average, less than the national average in math SAT scores. □

Definition 8.3 Suppose we conduct a hypothesis test of H_0 versus H_A at the α significance level. Let $\mathcal{R}$ denote the set of all values of the test statistic for which we reject the null hypothesis. Then $\mathcal{R}$ is called the *critical region* and values at the endpoints (or boundary) of $\mathcal{R}$ are called *critical values* (see Figure 8.2). ||

In the previous example, $\mathcal{R} = (-\infty, 500.13]$ is the critical region and $C = 500.13$ is the critical value.

Example 8.11 Suppose you are losing badly at craps, so you begin to suspect that one die is loaded. You decide to test this theory by throwing the die 60 times and

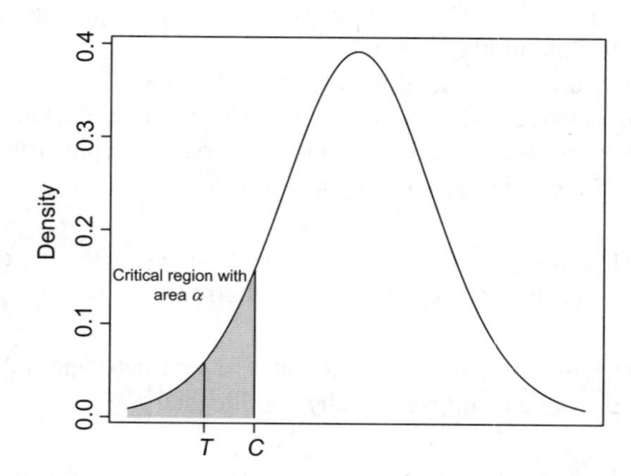

FIGURE 8.2 Critical region, critical value C, and a test statistic T for which $P(T \leq C) < \alpha$ for a test where the alternative is $\mu < \mu_0$.

TABLE 8.2 Cumulative Probabilities for Binom(60, 1/6)

C_1	0	1	2	3	4	5
$P(X \leq C_1)$	1.77×10^{-5}	2.307×10^{-4}	0.0015	0.0063	0.0202	0.0512
C_2	15	16	17	18	19	20
$P(X \geq C_2)$	0.0338	0.0164	0.0074	0.0031	0.0012	0.0005

recording the number of times 1 appears. If you decide to use a 5% significance level, for what number of occurrences of 1 would you conclude that the die is not fair?

Solution Let p denote the proportion of times the 1 appears. You test H_0: $p = 1/6$ versus H_A: $p \neq 1/6$. Let X denote the number of times 1 appears in 60 tosses of the die. If the die is fair, then $X \sim \text{Binom}(60, 1/6)$ and you would not reject H_0 if the number of times 1 appears is near 10. So the probability of a Type I error is

$$\begin{aligned} 0.05 &= P(\text{Reject } H_0 \mid H_0 \text{ true}) \\ &= P(X \text{ not close to } 10 \mid X \sim \text{Binom}(60, 1/6)). \end{aligned}$$

Since the die is probably biased if X is too low or too high, let us set these tail probabilities to be $0.05/2 = 0.025$ (Table 8.2):

$$0.025 = P(X \leq C_1 \mid X \sim \text{Binom}(60, 1/6)) = \sum_{i=0}^{C_1} \binom{60}{i} (1/6)^i (5/6)^{60-i}$$

and

$$0.025 = P(X \geq C_2 \mid X \sim \text{Binom}(60, 1/6)) = \sum_{i=C_2}^{60} \binom{60}{i} (1/6)^i (5/6)^{60-i}.$$

The critical region is $\mathcal{R} = \{0, 1, 2, 3, 4\} \cup \{16, 17, \ldots, 60\}$, and $P(X \leq 4) + P(X \geq 16) = 0.0366$. □

In this example, by the discrete nature of X, we cannot get an exact Type I error probability, so we are conservative, choosing critical values so that the actual Type I error rate is at most equal to the nominal significance level. Furthermore, for a two-sided test, the presumption is that the one-sided Type I error rates are each at most $\alpha/2$.

Some authors would allow the two critical values to be adjusted up or down to get the sum of the one-sided Type I error rates as close to α as possible. We feel this is misleading—if one of the one-sided Type I error rates is greater than $\alpha/2$, then it is easier to reject on that side than is implied by the nominal significance level. It opens the door for outright abuse, where one looks at the data before deciding which way to adjust the critical values.

Example 8.12 An analyst draws a random sample of size 8, $X_1, X_2, \ldots, X_8$, from a distribution with pdf $f(x;\theta) = (\theta + 1)x^{\theta}$, $0 \leq x \leq 1$, $\theta > 0$. She wants to test H_0:

$\theta = 2$ versus H_A: $\theta > 2$. As a decision rule, she records Y, the number of observations greater than or equal to 0.88. She rejects H_0 if $Y \geq 5$. What is the probability of a Type I error?

Solution She wants

$$P(\text{Reject } H_0 \mid H_0 \text{ true}) = P(Y \geq 5 \mid \theta = 2).$$

If $\theta = 2$ is true, then the pdf is $f(x; 2) = 3x^2$. First, we compute the probability of any observation drawn from this distribution to be greater than 0.88:

$$p = P(X_i \geq 0.88) \mid \theta = 2) = \int_{0.88}^{1} 3x^2\, dx = 0.3185.$$

Thus, the number of observations greater than or equal to 0.88 is $Y \sim \text{Binom}(8, 0.3185)$. We then compute

$$P(Y \geq 5 \mid \theta = 2) = \sum_{k=5}^{8} \binom{8}{k} (0.3185)^k (1 - 0.3185)^{8-k} = 0.0835. \qquad \square$$

8.2.2 Type II Errors and Power

We have seen that we can control the probability of Type I errors by setting the significance level α. We now consider Type II errors. Let β denote the probability of a Type II error: $\beta = P(\text{Do no reject} H_0 \mid H_A \text{ true})$. Now, in many settings, we are interested in the complement, $1 - \beta = P(\text{Reject } H_0 \mid H_A \text{ true})$. In other words, we want the probability of correctly rejecting a false null hypothesis.

Definition 8.4 *Power* is the probability of correctly rejecting a false null hypothesis.

$$1 - \beta = P(\text{Reject } H_0 \mid H_A \text{ true}). \qquad ||$$

Example 8.13 Suppose the heights of 2-year old girls are normally distributed with a mean of 30 in. and a standard deviation of 6 in. Let μ denote the true mean heights of these girls. A researcher wishes to test

$$H_0: \mu = 30 \quad \text{versus} \quad H_A: \mu < 30.$$

She plans to obtain a random sample of 30 girls and measure their heights, using $\alpha = 0.05$ for a significance level. What is the probability of correctly rejecting the null hypothesis if, in fact, the mean height of 2-year old girls in this community is 27 in.?

Solution If the null hypothesis holds, then the sampling distribution of $\bar{X}$ is normal with mean 30 and standard error $6/\sqrt{30} = 1.095$. Using a one-sided test at $\alpha = 0.05$, she rejects the null hypothesis if the z-score of her test statistic is $z \leq -1.645$ (0.05 quantile of the standard normal). This corresponds to $Z = (\bar{X} - 30)/1.095 \leq -1.645$

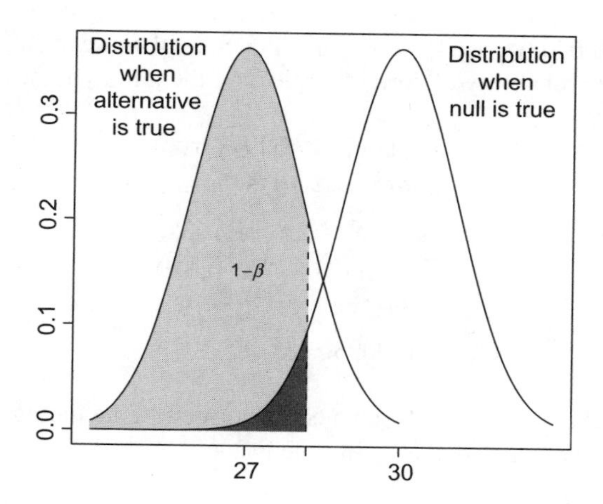

FIGURE 8.3 Distributions of the test statistic under the null and alternative hypotheses.

or $\bar{X} \leq 28.1987$. In other words, the critical region is $\mathcal{R} = (-\infty, 28.1987]$ and the critical value is $C = 28.1987$ (see Figure 8.3).

So, if the true mean height is 27 in., what is the probability of correctly rejecting the null hypothesis of $\mu = 30$?

$$\begin{aligned}1 - \beta &= P(\text{Reject } H_0 \mid H_A \text{ true})\\ &= P(\bar{X} \leq 28.1987 \mid \mu = 27)\\ &= P\left(\frac{\bar{X} - 27}{1.095} \leq \frac{28.1987 - 27}{1.095}\right)\\ &= P(Z \leq 1.0947)\\ &= 0.8632.\end{aligned}$$

Thus, she has about a 86% chance of correctly rejecting the null hypothesis if indeed the true mean height of 2-year old girls is 27 in. □

Example 8.14 A team of researchers plans a study to see if a certain drug can increase the speed at which mice move through a maze. An average decrease of 2 s through the maze would be considered effective, so the researchers would like to have a good chance of detecting a change this large or larger. Would 20 mice be a large enough sample? Assume the standard deviation is $\sigma = 3$ s and that the researchers will use a significance level of $\alpha = 0.05$.

Solution Let μ denote the true mean decrease in time through the maze. Then the researchers are testing H_0: $\mu = 0$ versus H_A: $\mu > 0$. If the null hypothesis holds, then the sampling distribution of $\bar{X}$ is normal with mean 0 and standard error $3/\sqrt{20} = 0.6708$. Using a one-sided test at $\alpha = 0.05$, the researchers will reject the null hypothesis if the z-score of the test statistic satisfies $Z \geq 1.645$. This corresponds to $Z = (\bar{X} - 0)/0.6708 \geq 1.645$ or $\bar{X} \geq 1.1035$. So, if the true mean decrease in time is 2 s, what is the probability of correctly rejecting the null hypothesis of $\mu = 0$?

$$\begin{aligned} 1 - \beta &= P(\text{Reject } H_0 \mid H_\text{A} \text{ true}) \\ &= P(\bar{X} \geq 1.1035 \mid \mu = 2) \\ &= P\left(\frac{\bar{X} - 2}{0.6708} \geq \frac{1.1035 - 2}{0.6708}\right) \\ &= P(Z \geq -1.3365) \\ &= 0.9093. \end{aligned}$$

Thus, the researchers have a 91% chance of correctly concluding that the drug is effective if the true average decrease in time is 2 s. □

In many instances, researchers decide on the power they want in a test and then determine the sample size needed to obtain that power.

Example 8.15 Suppose the researchers in the previous example want a 95% chance of rejecting H_0: $\mu = 0$ at $\alpha = 0.01$ if the true change is a 1.5 s decrease in time. What is the smallest number of mice that should be included in the study?

Solution On the standard normal curve, $q = 2.3264$ is the cutoff value for the upper 0.01 tail (i.e., the 0.99 quantile). Thus, we need $(\bar{X} - 0)/(3/\sqrt{n}) \geq 2.3264$ or $\bar{X} \geq 6.9792/\sqrt{n}$.

Thus,

$$\begin{aligned} 0.95 &= P\left(\bar{X} \geq \frac{6.9792}{\sqrt{n}} \mid \mu = 1.5\right) \\ &= P\left(\frac{\bar{X} - 1.5}{3/\sqrt{n}} \geq \frac{6.9792/\sqrt{n} - 1.5}{3/\sqrt{n}}\right) \\ &= P\left(Z \geq 2.3264 - \frac{1.5}{3/\sqrt{n}}\right). \end{aligned}$$

Using the 0.05 quantile for the standard normal,

$$-1.645 = 2.3264 - \frac{1.5}{3/\sqrt{n}}.$$

Thus, $n = 64$ is the smallest number of mice that the researchers should use. □

More generally, suppose we test H_0: $\mu = \mu_0$ versus H_A: $\mu > \mu_0$, where σ is known, at the α significance level. What is the power if the alternative is $\mu = \mu_1$ (Figure 8.4)? If H_0 is true, then the sampling distribution of $\bar{X}$ is normal with mean μ_0 and standard error $\sigma/\sqrt{n}$. Let q denote the $1 - \alpha$ quantile for the standard normal.

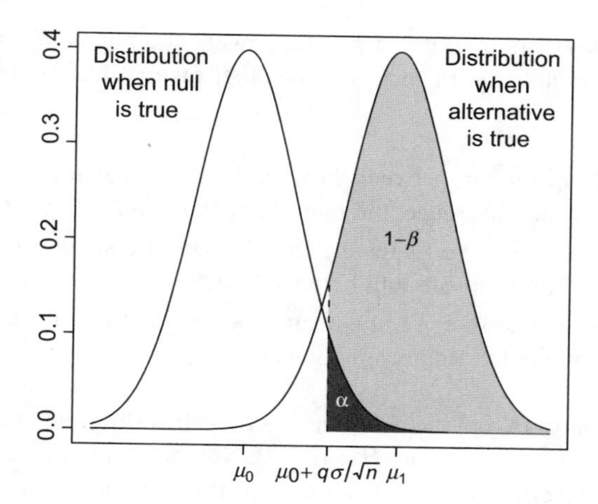

FIGURE 8.4 Distributions under the null and alternative hypotheses. Shaded regions represent power $(1-\beta)$ and significance level (α). Moving the critical value $\left(\mu_0 + q\sigma/\sqrt{n}\right)$ to the left increases power.

The corresponding critical value C for the sampling distribution of $\bar{X}$ is found by

$$\begin{aligned}\frac{C-\mu_0}{\sigma/\sqrt{n}} &= q,\\ C-\mu_0 &= q\frac{\sigma}{\sqrt{n}},\\ C &= \mu_0 + q\frac{\sigma}{\sqrt{n}}.\end{aligned}$$

Thus,

$$\begin{aligned}1-\beta &= P(\text{Reject } H_0 \mid H_A \text{ true})\\ &= P\left(\bar{X} > \mu_0 + q\frac{\sigma}{\sqrt{n}} \mid \mu = \mu_1\right)\\ &= P\left(\frac{\bar{x}-\mu_1}{\sigma/\sqrt{n}} > \frac{\mu_0-\mu_1+q(\sigma/\sqrt{n})}{\sigma/\sqrt{n}}\right)\\ &= P\left(Z > \frac{\mu_0-\mu_1}{\sigma/\sqrt{n}} + q\right)\\ &= P\left(Z > q - \frac{(\mu_1-\mu_0)}{\sigma/\sqrt{n}}\right).\end{aligned} \tag{8.2}$$

Thus, we see that the power of a test, $1-\beta$, is determined by the size of the lower bound of Z in Equation 8.2:

$$Z > q - \frac{(\mu_1-\mu_0)}{\sigma/\sqrt{n}}.$$

The smaller that lower bound, the larger the power. The population standard deviation σ is not controllable by the analyst. The other factors that determine the power are the

- *Effect size*: the difference between the hypothesized mean μ_0 and the actual mean μ_1. The larger the difference, the more likely we would detect the difference.
- *Denominator* $\sigma/\sqrt{n}$: The larger the sample size, the smaller the denominator, and hence the larger the amount being subtracted. This too increases power.
- *Significance Level* α: Increasing α decreases the quantile q and again pushes that lower bound to the left, increasing power.

Figure 8.5 displays two plots of power $1-\beta$ against different alternatives μ_1 for a hypothesis test H_0: $\mu = 0$ versus H_A: $\mu > 0$. Note that in both the $n = 50$ and the $n = 25$ cases, as the effect size increases (i.e., as the alternative value increases), the power increases. Also, for a fixed alternative value, the power is higher for the $n = 50$ sample.

In conducting a hypothesis test, a researcher will want to minimize all errors. We can set the Type I error by declaring the significance level before collecting the data. Minimizing the probability of a Type II error (β) means maximizing power. But as we can see from Figure 8.4, for a fixed sample size and standard deviation, increasing power results in increasing α, the probability of a Type I error. Thus, the only way to simultaneously decrease the probabilities of a Type I and a Type II error is to increase the sample size. Similar calculations as above can be done for the alternatives H_A: $\mu < \mu_0$ or H_A: $\mu \neq \mu_0$ (see Exercises 28 and 29).

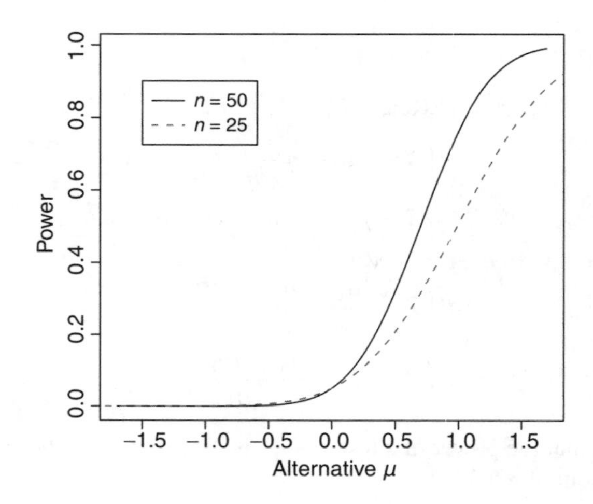

FIGURE 8.5 Plot of power against different alternatives μ_1 for $n = 50$ and $n = 25$ when hypothesized $\mu = 0$.

Remark

- For many studies, power is set to at least 80%.
- For hypothesis tests of a mean where the population standard deviation is not known, the sampling distribution of the test statistic $T = (\bar{X} - \mu_0)/(S/\sqrt{n})$ is a t distribution. If the alternative hypothesis is true, T no longer has a t distribution, but rather it has what is called a *noncentral T* distribution.
  In general, the calculations for power are more difficult for hypothesis tests that involve parameters other than the mean, and researchers often use specialized statistical software to do these calculations. ||

8.3 MORE ON TESTING

8.3.1 On Significance

Statistical Significance Versus Practical Importance Researchers designing a study typically have a theory that they are trying to confirm: does this new drug reduce heart attacks? Will this new teaching method improve test scores? Did the percentage of adults who have graduated from college change from 10 years ago? Thus, a statistically significant result—a small P-value— that supports the theory of researchers is a desired outcome: the efforts they put into their study have paid off. However, a small P-value that indicates that their result is not likely to be due to chance does not necessarily indicate that their result is of practical importance.

Example 8.16 A drug company is researching a new drug to help severely obese patients lose weight. They conduct a clinical trial to test H_0: $\mu = 0$ versus H_A: $\mu > 0$, where μ is the mean weight lost. Suppose after 1 year, the researchers measure an average weight lost of 4 pounds with standard deviation of 3 pounds for the 50 patients in the study who took the drug. The test statistic is $t = 4/(3/\sqrt{50}) = 9.4281$, so comparing this to a t distribution with 49 degrees of freedom, we find a P-value of 6.85×10^{-13}, essentially 0! Thus, there is strong evidence that obese patients will lose weight on this drug. On the other hand, a 95% confidence interval shows that, on average, the weight loss will be between 3.1 and 4.9 pounds.

The results are statistically significant—the positive weight lost by the patients is unlikely to be due to chance, so the drug does work —but it seems unlikely that the company will be able to successfully market a drug that promises severely obese patients that they can expect to lose, on average, at most 5 pounds! □

In practice, the null hypothesis is almost never exactly true, and with enough data, even very small, practically unimportant differences are statistically significant. A confidence interval for the parameter of interest is often more informative that the P-value and should be reported along with the P-value of the hypothesis test.

Lack of Significance In a criminal trial, the jury may decide on a "Not Guilty" verdict. Does this verdict mean the defendant is innocent? Not necessarily! In many instances, the prosecutor failed to present a case strong enough to convict the defendant, so a "Not Guilty" decision is not the same as a declaration of innocence!

The same is true in hypothesis testing. If you conduct a hypothesis test at a specified significance level and do not reject the null hypothesis, this does not necessarily mean that the null hypothesis is true. It could be true, or it could be that you just failed to gather enough evidence to support the alternative hypothesis. For instance, the power of your test may have been too low: perhaps your sample size was too small for the effect size you were trying to detect.

Searching for Significance As we noted earlier, researchers often have a vested interest in finding a statistically significant result in their studies. However, a finding of "no significance" does not mean that the researcher should go searching for significance.

For example, consider a pharmaceutical company testing a new drug for efficacy. They may find that the benefit of the drug is not statistically significant in their test as a whole. But if they then look at enough subpopulations, men, women, children, elderly, young adult, young women, Hispanic children, and so on, they have a high probability of finding some subpopulation in which the drug tests significant, purely by chance.

Even if the null hypothesis is true, with a 0.05 significance level, there is a 5% chance of incorrectly rejecting H_0. Suppose all null hypotheses are true. If we run two independent tests, each at the 5% level, there is nearly a 10% chance of getting at least one positive result ("reject the null hypothesis"): the probability of no positive result is $0.95^2 = 0.905$, so the probability of at least one positive result is $1 - 0.905$. Then for 20 independent tests, the probability of not rejecting any null hypotheses is $0.95^{20} = 0.3585$, so the chance of finding at least one "significant" result is $1 - 0.3585 = 0.6415$—in other words, there is a 64% chance that we will declare something significant even though there is nothing going on.

8.3.2 Adjustments for Multiple Testing

One way to adjust for such *multiple testing* is by using a *Sidak correction*. Suppose we wish to run k tests on a data set. Let α^* denote the nominal Type I error for each of these tests. Let α denote the desired overall Type I rate (the error in conducting all k tests). If each null hypothesis is true, and the tests are independent, then the probability of no rejections of the null hypothesis is $(1 - \alpha^*)^k$, so that the probability of at least one rejection of the null hypothesis is $1 - (1 - \alpha^*)^k$. Thus, to obtain an overall Type I error rate of α, we set

$$\alpha = 1 - (1 - \alpha^*)^k$$

and solve for α^*, yielding

$$\alpha^* = 1 - (1 - \alpha)^{1/k}.$$

For example, if we conduct five significance tests and we wish to keep the overall Type I error at 0.05, then we should set the significance level for each individual test at $\alpha^* = 1 - (1 - .05)^{1/5} = 0.01021$. Note that $\alpha^* \approx 0.01 = \alpha/5$. A quick approximation to the Sidak correction is the *Bonferroni correction*, which stipulates that if we want the overall Type I error to be α and we run k tests, then the Type I error for each of these tests should be set to α/k.

The Bonferroni correction is a bit conservative—that is, a smaller P-value (stronger evidence) is required to reject a null hypothesis. It can also be used when the tests are not independent, as the overall Type I error rate does not exceed α, even if the rejections are mutually exclusive. In practice, mutually exclusive tests are rare, while positively correlated tests are common; for example, a test for the effectiveness of a drug for the population as a whole, and for any subpopulation, are positively correlated because they share some of the same data. In this case, Bonferroni is even more conservative than necessary. However, in practice, that may not matter.

Example 8.17 *Google Web Optimizer^{TM}* allows web site owners to experiment with different versions of their web site. The web site owner wants site visitors to do something, for example, buy something, download software, donate, or send a letter to Congress. She may create different versions of the home page, with different text and images. She adds some JavaScript provided by Google to the default page; this JavaScript randomly leaves visitors on the default page or sends them to one of the alternate versions. Then more Google JavaScript records whether visitors *convert* (take the desired action). Google reports the results to the web site owner.

Google tested GWO itself and got 30% more downloads of *Picasa^{TM}* using a page that (1) did not include a screen shot, (2) had a "Try Picasa Now" button rather than "Free Download," and other minor changes.

Let $n_0, n_1, \ldots, n_k$ be the number of visitors allocated to the default page and k alternate pages, and $X_0, X_1, \ldots, X_k$ be the number of conversions on each page. Let p_j be the true conversion rate for the jth arm of this trial and $\hat{p}_j = X_j/n_j$ the estimated proportion. Google reports when an alternate page is significantly better than the default page. This corresponds to running k significance tests, with H_0: $p_j = p_0$ versus H_A: $p_j > p_0$.

The tests are positively correlated because they share some of the same data; if $\hat{p}_0$ happens to be low, then all tests are more likely to test significant. Hence, it may appear that Bonferroni would be too conservative. Some simulations showed that while it was conservative, using a more accurate correction made little difference. We omit details here because they have not been approved for public release, but you may wish to try your own simulations. One hint—when $k > 1$, it is best to let n_0 be larger than the other n_j values; since the default page is used in every comparison, it is worth using extra data to make $\hat{p}_0$ more accurate. This also makes the tests less correlated. □

8.3.3 *P*-values Versus Critical Regions

In this chapter and in Chapter 3, we conducted tests by computing a test statistic from the data and then finding the probability of obtaining a test statistic as extreme as the one observed, given that the null hypothesis is true. We used this probability, the P-value, to decide how likely it was that chance variation could explain our results. Small P-values supported the alternative hypothesis.

Another approach to hypothesis testing is to use the critical region introduced in Section 8.2.1 (Definition 3). For instance, consider the coffee example on page 213. For a one-sided test with $\alpha = 0.05$, to find the critical region, we set

$$\begin{aligned}
0.05 &= P(\text{Reject} H_0 \mid H_0 \text{ true}) \\
&= P(\bar{X} \leq C \mid \mu = 7) \\
&= P\Big(\frac{\bar{X} - 7}{0.8/\sqrt{15}} \leq \frac{C - 7}{0.8/\sqrt{15}}\Big) \\
&= P\Big(T \leq \frac{C - 7}{0.8/\sqrt{15}}\Big).
\end{aligned}$$

For the t distribution with 14 degrees of freedom, the 0.05 quantile is -1.76131. Thus, setting

$$-1.7613 = \frac{C - 7}{0.8/\sqrt{15}}$$

yields $C = 6.6362$. Thus, a sample mean less than 6.6362 would result in rejecting the null hypothesis.

The sample mean for the coffee example was 6.6, which is (barely) in the critical region of $(-\infty, 6.6362]$, so we would reject the null hypothesis.

A critical region approach to significance testing specifies a range of values of the test statistic for which we would reject H_0 at a specified significance level. We can compute the critical region without any data. Computing a P-value requires having the data; in addition, it does not require us to specify the significance level. Thus, in reporting a P-value, we allow the reader to determine whether the outcome is statistically significant.

8.4 LIKELIHOOD RATIO TESTS

In this section, we discuss a class of tests, based on likelihood ratios. Recall that in Chapter 6 we introduced the likelihood function $L(\theta \mid x_1, x_2, \ldots, x_n)$ that measures the consistency of the parameter θ and the observed data. In a likelihood ratio test (LRT), we decide whether data are more consistent with the alternative or null hypothesis by comparing L for values of θ from the two hypotheses. In certain situations, that of simple hypotheses, likelihood ratio tests are the best possible tests, obtaining the smallest possible Type II error rate for a given Type I error rate.

8.4.1 Simple Hypotheses and the Neyman–Pearson Lemma

Definition 8.5 A hypothesis is *simple* if it completely specifies the distribution of the population. Otherwise, it is *composite*. ||

For example, in the case of a sample drawn from a normal population with unknown mean but known standard deviation σ_0, the null hypothesis of H_0: $\mu = 4$ is a simple hypothesis. If the null hypothesis is true, then this specifies that the density is normal with $\mu = 4$ and $\sigma = \sigma_0$. On the other hand, if the sample is drawn from a normal population with unknown mean and unknown standard deviation, then H_0: $\mu = 4$ is a composite hypothesis because it includes multiple values of σ. Similarly, the alternative hypothesis H_A: $\mu \neq 4$ is composite.

Now, in the case that both hypotheses are simple,

$$H_0\colon \theta = \theta_0 \quad \text{versus} \quad H_A\colon \theta = \theta_A, \tag{8.3}$$

we create the *likelihood ratio test statistic*

$$T = \frac{L(\theta_0)}{L(\theta_A)} = \frac{L(\theta_0 \mid X_1, X_2, \ldots, X_n)}{L(\theta_A \mid X_1, X_2, \ldots, X_n)} \tag{8.4}$$

as the ratio of the two likelihood values. The *likelihood ratio test* rejects the null hypothesis for small values of T since small values of this ratio indicate that the alternative hypothesis is more likely. In particular, the test rejects H_0 if $T \leq c$ for some critical value c, which is practice is chosen based on the desired Type I error rate.

Example 8.18 Suppose $X_1, X_2, \ldots, X_9$ is drawn from the exponential distribution with pdf $f(X; \lambda) = \lambda e^{-\lambda X}$. To test

$$H_0\colon \lambda = 8 \quad \text{versus} \quad H_A\colon \lambda = 10,$$

we compute the likelihood ratio test statistic

$$\begin{aligned} T = \frac{L(\lambda = 8)}{L(\lambda = 10)} &= \frac{\prod_{i=1}^{9} f(X; 8)}{\prod_{i=1}^{9} f(X; 10)} \\ &= \frac{8^9 e^{-8 \sum_{i=1}^{8} X_i}}{10^9 e^{-10 \sum_{i=1}^{9} X_i}} \\ &= (0.8)^9 e^{2 \sum_{i=1}^{9} X_i}. \end{aligned}$$

We reject if

$$(0.8)^9 e^{2 \sum_{i=1}^{9} X_i} < c$$

or, equivalently,

$$\sum_{i=1}^{9} X_i < \frac{1}{2} \ln(1.25^9 \times c).$$

Call the right-hand side c_2; thus, the critical region is of the form

$$\mathcal{R} = \{(X_1, X_2, \ldots, X_9) \in \mathbf{R}^9 \mid \sum_{i=1}^{9} X_i < c_2, X_i \geq 0\}.$$

Suppose we wish to specify a significance level of $\alpha = 0.05$. Since $X_i \sim \text{Exp}(8) = \text{Gamma}(1, 8)$, by Theorem B.11, we have $\sum_{i=1}^{9} X_i \sim \text{Gamma}(9, 8)$. We set

$$0.05 = P((X_1, X_2, \ldots, X_9) \in \mathcal{R}; H_0) = P\left(\sum_{i=1}^{9} X_i < c_2 \mid H_0\right),$$

and then find the 0.05 quantile of Gamma(9, 8), which is $q = 0.5869 = c_2$. □

The Neyman–Pearson lemma indicates that this test is the best possible test.

Theorem 8.1 (**The Neyman–Pearson Lemma**) *Let $X_1, X_2, \ldots, X_n$ be a random sample from a distribution with parameter θ. Suppose we wish to test two simple hypotheses*

$$H_0\colon \theta = \theta_0 \quad \textit{versus} \quad H_A\colon \theta = \theta_A.$$

The likelihood ratio test rejects the null hypothesis if the test statistic $T = L(\theta_0)/L(\theta_A)$ satisfies $T < c$ at a significance level α. Then any test with significance level $\leq \alpha$ has power less than or equal to the power for this likelihood ratio test. That is, for a fixed α, the likelihood ratio test minimizes the probability of a Type II error.

See, for instance, Casella and Berger (2001) for a proof.

Example 8.19 Let $X_1, X_2, \ldots, X_n$ be n independent Bernoulli trials with parameter p. Suppose we wish to test H_0: $p = 0.4$ versus H_A: $p = 0.5$. Let $X = \sum_{i=1}^{n} X_i$. Then the likelihood ratio test statistic is

$$T = \frac{L(p = 0.4)}{L(p = 0.5)} = \frac{(0.4)^X (0.6)^{n-X}}{(0.5)^X (0.5)^{n-X}} = (\frac{2}{3})^X (1.2)^n.$$

We reject H_0 if

$$\left(\frac{2}{3}\right)^X (1.2)^n < c$$

or, equivalently,

$$X > \frac{\ln(c) - n \ln(1.2)}{\ln(2/3)}.$$

Let $c_2 = (\ln(c) - n\ln(1.2))/\ln(2/3)$. Then the critical region is

$$\mathcal{R} = \left\{ (X_1, X_2, \cdots, X_n) \in \{0, 1\}^n \mid \sum_{i=1}^{n} X_i > c_2 \right\}.$$

For a test with $\alpha = 0.05$, we want

$$0.05 = P((X_1, X_2, \ldots, X_n) \in \mathcal{R} \mid H_0) = P\left(\sum_{i=1}^{n} X_i > c_2 \right).$$

Then c_2 will be the 0.05 quantile of the binomial distribution with parameters $p = 0.4$ and n.

If $n = 10$, then $P(\sum_{i=1}^{10} X_i > 7) = 0.012$ while $P(\sum_{i=1}^{10} X_i > 6) = 0.055$, so in this case, we cannot achieve a test with significance level $\alpha = 0.05$ exactly. On the other hand, if we specify $\alpha = 0.012$, then any other significance test with $\alpha < 0.012$ would have less power than this likelihood ratio test (i.e., the test that rejects H_0 when $\sum_{i=1}^{10} X_i > 7$.) □

8.4.2 Generalized Likelihood Ratio Tests

Another version of the likelihood ratio test is useful when one or both hypotheses are composite. We cannot compute the simple ratio $T = L(\theta_0)/L(\theta_A)$ because there is no single θ_0 or single θ_A. Instead, for the numerator, we use the θ that maximizes L while satisfying the null hypothesis. For the denominator, we use the θ that maximizes L, regardless of whether that θ satisfies the null or alternative hypotheses (this tends to be simpler than limiting to the alternative hypothesis and ultimately makes little difference in the tests).

Let Ω denote the set of possible values for θ, and Ω_0 the subset that satisfies the null hypothesis. We test

$$H_0\colon \theta \in \Omega_0 \quad \text{versus} \quad H_A\colon \theta \in \Omega \setminus \Omega_0. \tag{8.5}$$

Definition 8.6 The *likelihood ratio test statistic* for testing the hypothesis given by Equation 8.5 is

$$T(x) = \frac{\max_{\Omega_0} L(\hat{\theta} \mid x)}{\max_{\Omega} L(\hat{\theta} \mid x)}. \tag{8.6}$$

A *likelihood ratio test* is any test that rejects the null hypothesis when $T \leq c$ for some number c, $0 \leq c \leq 1$. ||

Example 8.20 Let $X_1, X_2, \ldots, X_n$ be a random sample from $N(\mu, 1)$. Suppose we wish to test

$$H_0\colon \mu = 8 \quad \text{versus} \quad H_A\colon \mu < 8.$$

The likelihood of μ is (see Equation 6.5)

$$L(\mu) = \frac{1}{(\sqrt{2\pi})^n} e^{-(1/2)\sum_{i=1}^{n}(X_i-\mu)^2}.$$

In particular, suppose we have $X_1 = 7$, $X_2 = 7$, $X_3 = 6.6$, and $X_4 = 6$. Under the null hypothesis, the likelihood is

$$\begin{aligned} L(\mu = 8) &= \frac{1}{(\sqrt{2\pi})^4} e^{-(1/2)\left((7-8)^2+(7-8)^2+(6.6-8)^2+(6-8)^2\right)} \\ &= 0.000473. \end{aligned}$$

In this case the null hypothesis is simple, so we do not need to maximize across a set of possible θ values.

The maximum likelihood estimate of μ is the sample mean, $\hat{\mu} = \bar{X} = 6.65$ (see Exercise 5), so the likelihood is

$$\begin{aligned} L(\hat{\mu}) &= \frac{1}{(\sqrt{2\pi})^4} e^{-(1/2)\left((7-6.65)^2+(7-6.65)^2+(6.6-6.65)^2+(6-6.65)^2\right)} \\ &= 0.0181. \end{aligned}$$

The likelihood ratio test statistic is $T = 0.00473/0.181 = 0.259$. In rough terms, this means that the null hypothesis is only one-fourth as consistent with the data as is the alternative hypothesis. But is that beyond normal chance variation—enough to reject the null hypothesis? We need to know more about the distribution of the test statistic when the null hypothesis is true. To do this, we will do more general calculations and avoid plugging in our specific numbers too early. The hypotheses may be written as

$$H_0\colon \mu = \mu_0 \quad \text{versus} \quad H_A\colon \mu < \mu_0.$$

There is a simple null hypothesis, so the numerator of the test statistic is

$$L(\mu = \mu_0) = \frac{1}{(\sqrt{2\pi})^n} e^{-(1/2)\sum_i (X_i-\mu_0)^2}.$$

For the denominator, we assume that $\bar{X} < \mu_0$ in the following calculations; otherwise $T = 1$ and we would just accept H_0. The likelihood evaluated at the maximum likelihood estimate of $\hat{\mu} = \bar{X}$ is

$$L(\hat{\mu} = \bar{X}) = \frac{1}{(\sqrt{2\pi})^n} e^{-(1/2)\sum_i (X_i-\bar{X})^2}.$$

Thus, the ratio of the two likelihoods is

$$
\begin{aligned}
T(X) &= \frac{\exp\left(-(1/2)\sum_{i=1}^n (X_i - \mu_0)^2\right)}{\exp\left(-(1/2)\sum_{i=1}^n (X_i - \bar{X})^2\right)} \\
&= \exp\left[\frac{1}{2}\left(-\sum_{i=1}^n (X_i - \mu_0)^2 + \sum_{i=1}^n (X_i - \bar{X})^2\right)\right] \\
&= \exp\left[-n(\bar{X} - \mu_0)^2/2\right],
\end{aligned}
$$

where the last equality comes from the calculation

$$\sum_{i=1}^n (X_i - \mu_0)^2 = \sum_{i=1}^n (X_i - \bar{X})^2 + n(\bar{X} - \mu_0)^2.$$

A likelihood ratio test rejects H_0 when $T(X)$ is sufficiently small. Thus, we can specify a critical region $\mathcal{R}$ to be sample means $\bar{X}$ such that $\bar{X} < \mu_0$ and

$$T(x) = \exp\left[-\frac{n}{2}(\bar{X} - \mu_0)^2\right] \leq c,$$

or $\bar{X} < \mu_0$ and

$$|\bar{X} - \mu_0| \geq \sqrt{-\frac{2}{n}\ln(c)}.$$

In other words, the critical region is when

$$\bar{X} < \mu_0 - \sqrt{-\frac{2}{n}\ln(c)}$$

for some $0 < c < 1$ or, equivalently, when

$$\bar{X} < \mu_0 - d$$

for some $d > 0$. In this example, the likelihood ratio test reduces to a one-sided test based on $\bar{X}$. Since $\bar{X} \sim N(\mu_0, 1/n)$ when H_0 is true, a suitable critical value for a size α test would be $d = z_\alpha/\sqrt{n}$. For this example, we would reject at the 5% level when $\bar{X} < 8 - 1.644/2 = 7.18$. Since $\bar{x} = 6.65$, we conclude that the mean μ is less than 8. □

8.5 EXERCISES

For all hypothesis tests, state your answer in the context of the problem (i.e., do not just state "reject null" or "accept null").

1. Calcium levels in healthy adults is normally distributed with mean 9.5 mg/dL per deciliter and unknown standard deviation. A physician in a rural community

suspects that the mean calcium levels are different for women in this community. He collects measurements from 20 healthy women and finds $\bar{x} = 9.2$ with sample standard deviation $s = 1.1$. Does this support his hypothesis?

2. The mean cholesterol level for vegetarians in the United States is 161 mg/dL, and the distribution is normal. You suspect that vegetarians from Sodor are different from vegetarians in the United States You conduct a study. For a random sample of 24 vegetarians from Sodor, you find a sample mean cholesterol level of 164 mg/dL with a standard deviation of 5 mg/dL. Do these data support the hypothesis that cholesterol levels for vegetarians in Sodor are different from the United States?

3. What is normal body temperature? The standard has been 98.6°F. Suppose a medical worker suspects that body temperatures in children in Sodor are higher than the norm. She obtains measurements from a random sample of 18 children and finds the following:

98.0	98.9	99.0	98.9	98.8	98.6	99.1	98.9	98.5
98.9	98.9	98.4	99.0	99.2	98.6	98.8	98.9	98.7

   (a) What are the hypotheses to test?
   (b) Carry out the test and state a conclusion in a complete sentence.

4. Suppose the number of births per month in Sodor can be modeled by a Poisson random variable with parameter $\lambda > 0$. The town elder claims that, on average, there are 15 births per month, but you suspect it is more. If in 1 month, there are 20 births, does this support your hypothesis?

5. Geologists wonder if the surface soil pH levels at two different locations of a certain desert site are similar. The scientists obtained the following pH levels at randomly selected points within each of the two locations.

Location A	8.53	8.52	8.01	7.99	7.93	7.89	7.85	7.82	7.80
Location B	7.85	7.73	7.58	7.40	7.35	7.30	7.27	7.27	7.23

   Do the data suggest that the true mean soil pH values differ for the two locations? Set up and carry out a significance test and state your conclusion in a complete sentence.

6. One of the factors of interest in the Spruce Case Study in Section 1.9 was whether competition would affect the growth rate of the seedlings. The biologist removed weeds and other growth from some of the plots ("No Competition") but not others ("Competition").
   (a) Conduct a test to see if the average change in diameter of the seedlings over the 5-year period of the study is different between the seedlings in the two groups (data set `Spruce`).
   (b) Recall that in the study design, the seedlings were randomly assigned to the "No Competition" or "Competition" plots. What does this imply for the conclusion reached in part (a)?

7. The data set for the `Alelager` contains calories and alcohol content (by volume) for a sample of domestic and international ales and lager beers (per 12 ounces). Investigate the hypothesis that ales have more calories than lagers.
8. You suspect that men in town A weigh, on average, less than men in town B. You decide to carry out a study. Suppose the following are the weights (in pounds) of a random sample of men in these two towns:

A	192.0	183.0	194.5	185.0	214.0	203.5	194.0	199.0	109.0
B	195.0	200.0	209.0	198.0	200.5	207.5	199.0	205.0	220.0

   (a) First, create exploratory plots of the data in the two towns. What do you observe?
   (b) Carry out the hypothesis test. Is this evidence that men in town A weigh less, on average, than men in town B?
   (c) There is an outlier in one of the towns. Remove this observation and repeat the test. Is this outlier influential?
9. In 1987, 39.1% of sex workers in Bamako, Mali, were HIV positive. A sociologist suspects that this proportion is higher today. To check this, she surveys a random sample of $n = 130$ sex workers from Bamako and finds $\hat{p} = 0.48$. Is this sufficient evidence to conclude that the proportion of sex workers in Bamako who are HIV positive is greater than 0.391?
10. According to the 2004 American Community Survey, 29.1% of residents of Illinois have completed a bachelor's degree. A college professor suspects that this percentage is lower in a certain county in the state. He surveys a random sample of adults in this county and finds that of the 350 in his sample, 87 have bachelor's degree. Does this support the professor's hypothesis?
11. Are there regional differences in support for same-sex marriage? A 2009 survey conducted by the Pew Center for the People & the Press and the Pew Forum on Religion & Public Life found that 38% of 505 people from the Midwest supported same-sex marriage compared to 31% of 773 people from the South (`http://pewresearch.org/pubs/1375/gay-marriage-civil-unions-opinion`). Conduct a test to see if this difference is statistically significant.
12. Infections following surgery is a serious concern that can have a major impact on a patient's road to recovery. One approach to counter infection is to kill surgical pathogens by oxidation. Greif et al. (2000) randomly assigned 250 patients to receive 30% inspired oxygen and 250 patients to receive 80% inspired oxygen. All patients were undergoing surgery for colorectal resection. Of the patients receiving 30% inspired oxygen, 28 had a surgical wound infection compared to 13 patients who received the 80% inspired oxygen treatment.
   (a) Conduct a test to see if this difference is statistically significant. If it is, find a 95% confidence interval for the difference and give a sentence interpreting this interval.

(b) There was no control group— a group that received no inspired oxygen—in this study. What is the implication of this?

13. In the remark on page 220, we indicated that is not common to use a continuity correction when using the normal approximation to the two-sample proportions test. Use simulation to explore the sampling distribution of the Z statistic, for
    (a) $p = 0.5, n_1 = 10, n_2 = 10$,
    (b) $p = 0.5, n_1 = 10, n_2 = 11$,
    (c) $p = 0.5, n_1 = 10, n_2 = 12$,
    (d) $p = 0.5, n_1 = 10, n_2 = 15$,
    (e) your choice of four other combinations of p, n_1, and n_2.

    In all cases, use at least 10^4 replications and create normal quantile plots with the option `pch="."` (for better resolution). Describe the distributions.

14. A pharmaceutical company is conducting a clinical trial to determine the effectiveness of a new drug to lower cholesterol levels. If μ denotes the mean change (before–after) in cholesterol levels, they will test H_0: $\mu = 0$ versus H_A: $\mu > 0$. Describe the Type I and Type II errors that could be made and the practical consequences of making these errors.

15. The Environmental Protection Agency (EPA) sets a standard for arsenic levels in water to be no more than 10 ppb (parts per billion). You suspect that, on average, arsenic levels in your community's water are much higher than this standard. You will study this by measuring arsenic levels in water samples selected from 15 households and testing H_0: $\mu = 10$ ppb versus H_A: $\mu > 10$ ppb, where μ denotes the mean arsenic level (ppb) in your community.
    (a) Describe the Type I and Type II errors that could be made in this study and the practical consequences of making these types of errors.
    (b) You gather your sample, compute the sample mean and standard deviation, then have your intern carry out the one-sample test for a mean. He uses the standard normal distribution instead of the t distribution and claims to reject the null hypothesis. If you had used the t distribution, would you have necessarily come to the same conclusion? Explain.

16. Suppose you are interested in the lengths of a certain species of snake in Lyman Lake. Assume the lengths (in centimeters) are normally distributed with unknown mean μ but known standard deviation $\sigma = 4$. You decide to test H_0: $\mu = 25$ versus H_A: $\mu > 25$ at $\alpha = 0.05$ with a sample size of 30. What is the power of the test if, in fact, $\mu = 27$?

17. The Food and Drug Administration sets an action level for mercury in fish at 1 ppm—that is, if mercury levels are higher than this value in commercial fish, the FDA will take action to remove the fish from stores. Suppose a public health official is worried about mercury levels in the Fox River. She will obtain a random sample of n fish and find the average amount of mercury. If μ denotes the mean amount of mercury in fish in the Fox river, she will test H_0: $\mu \leq 1$ versus H_A: $\mu > 1$ at $\alpha = 0.01$. Suppose the mercury levels

in fish in this river is normally distributed with $\sigma = 0.3$. How large should the sample be if she wants a 90% chance of detecting a mean of $\mu = 1.4$ (or higher)?

18. In the town of Sodor, the average weight of mice is 340 g with standard deviation 30 g. The mayor in Shetland suspects that mice in his town weigh more. If the true average weight of mice in Shetland is 354 g (or more), he will contact the public health officials. He plans to catch a sample of n mice to test this theory. If he conducts a test at the 0.1 significance level, what must n be if he wants an 85% chance of detecting this weight?

19. Your favorite brand of cereal comes in boxes of 570 g. You suspect that the company is under-filling the boxes. If the true average weight is 561 g (or less), you will contact the Better Business Bureau. If you decide to test your hypothesis at a 0.05 significance level, how many boxes of cereal would you need to sample if you want an 80% chance of detecting a mean of 561 g? Assume for the null hypothesis that cereal weights are distributed normally with mean $\mu = 570$ and standard deviation $\sigma = 14$ g.

20. Let $X_1, X_2, \ldots, X_{12}$ be a random sample from a Bernoulli distribution with unknown success probability p. We will test H_0: $p = 0.3$ versus H_A: $p < 0.3$, rejecting the null if the number of successes, $Y = \sum_{i=1}^{12} X_i$, is 0 or 1.
    (a) Find the probability of a Type I error.
    (b) If the alternative is true, find an expression for the power, $1 - \beta$, as a function of p.
    (c) Plot the power against p.

21. Let $X_1, X_2, \ldots, X_{50}$ be a random sample from a Bernoulli distribution with unknown success probability p. We will test H_0: $p = 0.6$ versus H_A: $p \neq 0.6$, rejecting the null hypothesis at a $\alpha = 0.05$ significance level. Find the critical region for this test.

22. Let $X_1, X_2, \ldots, X_5$ be a random sample from the Poisson distribution with $\lambda > 0$. A researcher wishes to test H_0: $\lambda = 2$ versus H_A: $\lambda > 2$. She will reject H_0 if $\sum_{i=1}^{5} X_i \geq 16$.
    (a) Compute the probability of a Type I error. *Hints*: (a) Theorem B.6; (b) the `R` command `ppois` computes probabilities for the Poisson distribution.
    (b) If, in fact, $\lambda = 4$, what is the probability of a Type II error?

23. Suppose a single measurement is taken from a distribution with pdf $f(x) = \lambda e^{-\lambda x}$, $x > 0$. The hypotheses are H_0: $\lambda = 1$ versus H_A: $\lambda < 1$ and the null hypothesis is rejected if $x \geq 3.2$.
    (a) Calculate the probability of committing a Type I error.
    (b) Calculate the probability of committing a Type II error if, in fact, $\lambda = 1/5$.

24. A researcher draws a random sample $X_1, X_2, \cdots, X_{10}$ from an exponential distribution with $\lambda > 0$. He wishes to test H_0: $\lambda = 0.25$ versus H_A: $\lambda < 0.25$. He decides that he will reject the null hypothesis if at least three of the values are greater than 9.

(a) Compute the probability of a Type I error.

(b) If the true λ is 0.15, what is the power of his test?

25. Let $X_1, X_2, \ldots, X_{15}$ be a random sample from the exponential distribution with $\lambda > 0$. To test H_0: $\lambda = 1/5$ versus H_A: $\lambda < 1/5$, you will use $X_{\min}$ as a test statistic. If $X_{\min} \geq 1$, you will reject the null hypothesis.

    (a) Compute the probability of a Type I error.

    (b) Find the power of the test if, in fact, $\lambda = 1/25$.

    See Exercise 24 of Chapter 4.

26. Suppose that a random sample of size 5 is drawn from a uniform distribution on $[0, \theta]$. To test H_0: $\theta = 2$ versus H_A: $\theta > 2$, you will use $X_{\max}$ as a test statistic. If you reject the null hypothesis when $X_{\max} \geq k$, what value must k be to make the probability of a Type I error equal to 0.05?

27. A sample of size 1 from a distribution with pdf $f(x) = (1+\theta)x^\theta$, $0 \leq x \leq 1$, $\theta > 0$, is drawn to test H_0: $\theta = 2$ versus H_A: $\theta > 2$. The null hypothesis is rejected if $x \geq 3/4$.

    (a) Find an expression for power, $1 - \beta$, as a function of θ.

    (b) Graph the power against θ in the interval (2, 10).

28. Derive an expression for power like Equation 8.2 for the alternative hypothesis H_A: $\mu < \mu_0$. Include a graph similar to Figure 8.4.

29. Derive an expression for power like Equation 8.2 for the alternative hypothesis H_A: $\mu = \mu_0$. Include a graph similar to Figure 8.4.

30. Suppose you have a large data set and you plan to conduct 12 significance tests. If you want the overall Type I error to be 0.1, what should the Type I error rate be for each individual test?

31. The data set for the GSS Case Study in Section 1.6 has 19 variables. Suppose an ambitious student wants to perform a chi-square test of independence for every possible pair of variables. If she wants the overall Type I error rate to be 0.05, what significance level should she use for each test?

32. Let X be a random variable with pdf $f(x;\theta) = \theta x^{\theta-1}$, $0 < x < 1$ and $\theta > 0$. Consider H_0: $\theta = 1/2$ versus H_A: $\theta = 1/4$.

    (a) Derive the most powerful test, using $\alpha = 0.05$.

    (b) Compute the power of this test.

33. Suppose $X_1, X_2, \ldots, X_n$ are a random sample from a population with an exponential distribution λ. Derive the most powerful test for H_0: $\lambda = 7$ versus H_a: $\lambda = 5$.

34. Consider a sequence of n independent Bernoulli random variables $X_1, X_2, \ldots, X_n$. Derive the most powerful test for H_0: $p = 0.4$ versus H_A: $p = 0.3$.

35. Consider a sequence of n independent Bernoulli random variables $X_1, X_2, \ldots, X_n$. Derive the most powerful test for H_0: $p = p_0$ versus H_A: $p = p_a$ for $p_0 < p_a$.

36. In genetics, certain characteristics of a plant (among other living things) are governed by three possible genotypes, denoted by AA, Aa, and aa. Suppose the probabilities of each type in a particular population is θ^2, $2\theta(1-\theta)$, $(1-\theta)^2$, respectively, where $0 < \theta < 1$. In a random sample of n plants, let X_1, X_2, X_3 denote the numbers of each genotype. Derive the most powerful test to test H_0: $\theta = 0.4$ versus H_A: $\theta = 0.7$. The critical region can be specified by an expression involving X_1 and X_2.
37. Let $X_1, X_2, \ldots, X_n$ be a random sample from a normal distribution with unknown mean μ and unknown variance σ^2. Suppose we wish to test H_0: $\mu = \mu_0$ versus H_A: $\mu < \mu_0$ at significance level α. Show that the likelihood ratio test reduces to a one-sided t test.
38. Let $X_1, X_2, \cdots, X_n$ be a random sample from the uniform distribution over $[0, \theta]$. Suppose we wish to test H_0: $\theta = 5$ versus H_A: $\theta < 5$ at significance level $\alpha = 0.05$. Use the likelihood ratio test to find the critical region.
39. Let $X_1, X_2, \ldots, X_n \sim N(\mu_1, \sigma^2)$ and $Y_1, Y_2, \ldots, Y_m \sim N(\mu_2, \sigma^2)$ are random samples, where μ_1, μ_2, and σ^2 are unknown. Suppose we wish to test H_0: $\mu_1 = \mu_2$ versus H_A: $\mu_1 < \mu_2$ at significance level α. Show that the likelihood ratio test reduces to a one-sided, two-sample t test.

9

REGRESSION

In the Black Spruce Seedings Case Study in Section 1.9, the biologist was interested in how much the seedlings grew over the course of the study. Let $(x_1, y_1), (x_2, y_2), \ldots, (x_{72}, y_{72})$ denote the height and diameter change, respectively, for each of the 72 seedlings. In Figure 9.1, we see that there is a strong, positive, and linear relationship between height and diameter changes.

In this chapter, we will describe a method to model this relationship, that is, we will find a mathematical equation that best explains the linear relationship between the change in height and the change in diameter.

9.1 COVARIANCE

In Chapter 2, we introduced the scatter plot as a graphical tool to explore the relationship between two numeric variables. For example, referring to Figure 9.2a, we might describe the relationship here between the two variables as positive, linear, and moderate to moderately strong.

Now, consider the graph in Figure 9.2b. How would you describe the relationship here? This relationship would be described as linear, positive, and strong.

In fact, these two graphs are of the same two variables! The difference in the two impressions are due to the y-axis scaling. In the first graph, the range of the y-axis is roughly -2.5 to 2.5; in the second graph, the y-axis range is roughly -4.5 to 4.5. Graphs are excellent tools for exploring data, but issues such as scaling can distort

Mathematical Statistics with Resampling and R, First Edition. By Laura Chihara and Tim Hesterberg.

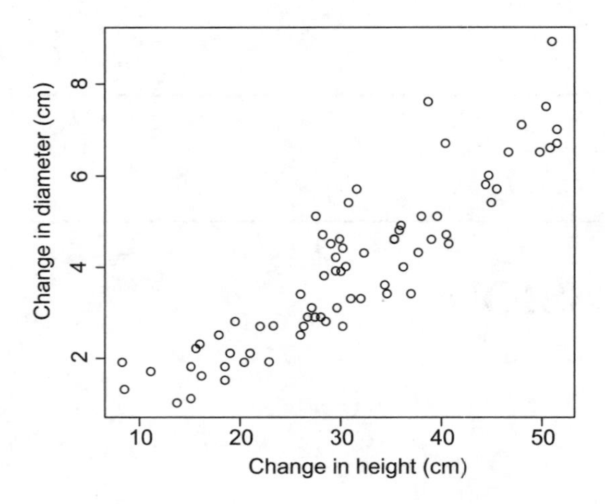

FIGURE 9.1 Change in diameter against change in height of seedlings over a 5-year period.

our perception of underlying properties and relationships. Thus, we will consider a numeric measure that indicates the strength of a linear relationship between the two variables.

We return to the Black Spruce data and recreate the scatter plot of diameter change against height change, adding a vertical line to mark the mean of the height changes ($\bar{x}$) and a horizontal line to mark the mean of the diameter changes ($\bar{y}$) (Figure 9.3).

For points in quadrant I, both $x_i - \bar{x}$ and $y_i - \bar{y}$ are positive, so $(x_i - \bar{x})(y_i - \bar{y})$ is positive. Similarly, in quadrant III, $(x_i - \bar{x})(y_i - \bar{y})$ is positive since each of the factors is negative. In quadrants II and IV, $(x_i - \bar{x})(y_i - \bar{y})$ is negative since the factors have opposite signs. For the Spruce data, on average, $(x_i - \bar{x})(y_i - \bar{y})$ is positive since most of the points are in quadrants I and III.

This motivates the following definition, a measure of how X and Y are related.

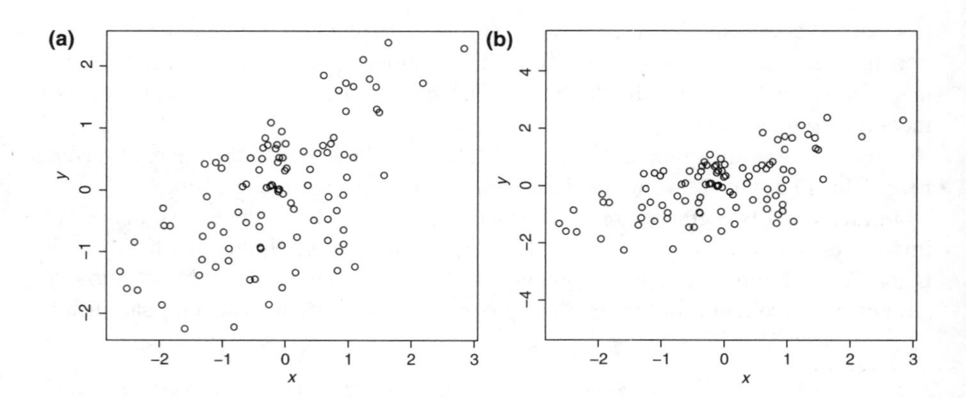

FIGURE 9.2 Two scatter plots of two numeric variables.

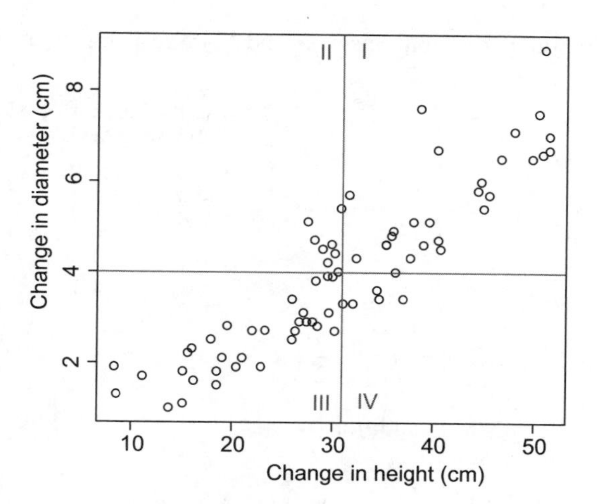

FIGURE 9.3 Change in diameter against change in height of seedlings over a 5-year period with vertical and horizontal lines at the mean values.

Definition 9.1 The *covariance of X and Y* is

$$\text{Cov}[X, Y] = \text{E}\left[(X - \mu_X)(Y - \mu_Y)\right]. \qquad ||$$

If $\text{Cov}[X, Y] > 0$, then we say that X and Y are *positively correlated*: on average, X and Y are both either greater or less than their respective means. If $\text{Cov}[X, Y] < 0$, then we say that X and Y are *negatively correlated*: on average, one of X or Y is greater than and one less than their mean.

Note that the definition implies that

$$\text{Cov}[X, X] = \text{E}\left[(X - \mu_X)^2\right] = \text{Var}[X].$$

Proposition 9.1 $\text{Cov}[X, Y] = \text{E}[XY] - \text{E}[X]\,\text{E}[Y]$.

Proof

$$\begin{aligned}
\text{Cov}[X, Y] &= \text{E}\left[(X - \mu_X)(Y - \mu_Y)\right] \\
&= \text{E}\left[(XY - \mu_Y X - \mu_X Y + \mu_X \mu_Y)\right] \\
&= \text{E}[XY] - \mu_Y \text{E}[X] - \mu_X \text{E}[Y] + \mu_X \mu_Y \\
&= \text{E}[XY] - \mu_Y \mu_X - \mu_X \mu_Y + \mu_X \mu_Y \\
&= \text{E}[XY] - \text{E}[X]\,\text{E}[Y].
\end{aligned}$$

□

Example 9.1 Let X and Y have the joint distribution:

$$f(x, y) = \begin{cases} \frac{3}{2}(x^2 + y^2), & 0 < x < 1,\ 0 < y < 1, \\ 0, & \text{otherwise.} \end{cases}$$

Find the covariance of X and Y.

Solution The marginal distributions of X and Y are respectively as follows:

$$f_X(x) = \int_0^1 \frac{3}{2}(x^2 + y^2)\,dy = \frac{3}{2}\left(x^2 + \frac{1}{3}\right),$$

$$f_Y(y) = \int_0^1 \frac{3}{2}(x^2 + y^2)\,dx = \frac{3}{2}\left(y^2 + \frac{1}{3}\right).$$

Thus,

$$\mathrm{E}[X] = \int_0^1 x \cdot \frac{3}{2}\left(x^2 + \frac{1}{3}\right)\,dx = \frac{5}{8}$$

and, similarly, $\mathrm{E}[Y] = 5/8$. In addition,

$$\mathrm{E}[XY] = \int_0^1 \int_0^1 xy \cdot \frac{3}{2}(x^2 + y^2)\,dxdy = \frac{3}{8}.$$

Thus,

$$\mathrm{Cov}[X, Y] = \frac{3}{8} - \frac{5}{8} \cdot \frac{5}{8} = -\frac{1}{64}.$$

□

Corollary 9.1 *If X and Y are independent, then* $\mathrm{Cov}[X, Y] = 0$.

Remark The converse is false: $\mathrm{Cov}[X, Y] = 0$ does not imply that X and Y are independent. For example, if X has a symmetric distribution and $Y = (X - \mu_X)^2$, then Y completely depends on X, but the covariance is zero. ||

Covariances of sums of random variables add up; this yields a useful expression for the variance of a sum of random variables.

Theorem 9.1

$$\mathrm{Cov}\left[\sum_{i=1}^{n} X_i, \sum_{j=1}^{m} Y_j\right] = \sum_{i=1}^{n}\sum_{j=1}^{m} \mathrm{Cov}[X_i, Y_j].$$

Proof

$$\mathrm{Cov}\left[\sum_{i=1}^{n} X_i, \sum_{j=1}^{m} Y_j\right] = \mathrm{E}\left[\left(\sum_{i=1}^{n} X_i - \mathrm{E}\left[\sum_{i=1}^{n} X_i\right]\right)\left(\sum_{j=1}^{m} Y_j - \mathrm{E}\left[\sum_{i=1}^{n} Y_j\right]\right)\right]$$
$$= \mathrm{E}\left[\sum_{i=1}^{n} (X_i - \mu_{X_i}) \sum_{j=1}^{m} (Y_j - \mu_{Y_j})\right]$$

$$= \sum_{i=1}^{n} \sum_{j=1}^{m} \mathrm{E}\left[(X_i - \mu_{X_i})(Y_j - \mu_{Y_j})\right]$$

$$= \sum_{i=1}^{n} \sum_{j=1}^{m} \mathrm{Cov}[X_i, Y_j].$$

□

Corollary 9.2 *If $X_1, X_2, \ldots, X_n$ are random variables, then*

$$\mathrm{Var}\left[\sum_{i=1}^{n} X_i\right] = \sum_{i=1}^{n} \sum_{j=1}^{n} \mathrm{Cov}[X_i, X_j] = \sum_{i=1}^{n} \mathrm{Var}[X_i] + 2 \sum_{1 \le i < j \le n} \mathrm{Cov}[X_i, X_j].$$

In particular, $\mathrm{Var}[X + Y] = \mathrm{Var}[X] + \mathrm{Var}[Y] + 2\mathrm{Cov}[X, Y]$.

Thus, from the Corollary 9.1, we have Corollary 9.3.

Corollary 9.3 *If X and Y are independent, then*

$$\mathrm{Var}[X + Y] = \mathrm{Var}[X] + \mathrm{Var}[Y].$$

9.2 CORRELATION

In the Black Spruce Case Study, the biologist made his measurements using centimeters. Now, suppose he decides to convert his measurements to inches.

Let X, Y denote the heights and diameters in centimeters, while X', Y' denote the measurements in inches. There are 0.3937 in. to the centimeter, so

$$\begin{aligned} \mathrm{Cov}[X', Y'] &= \mathrm{Cov}[0.3937X, 0.3937Y] \\ &= \mathrm{E}[(0.3937X)(0.3937Y)] - \mathrm{E}[0.3937X]\,\mathrm{E}[0.3937Y] \\ &= 0.3937^2\, \mathrm{Cov}[X, Y]. \end{aligned}$$

Thus, the covariance decreases by a factor of 0.155.

But changing the measurement units really does not affect how strongly the variables are related. We could even do a scatter plot that would look exactly the same, except for axis labels. We need a measure of the relationship that is unitless.

Definition 9.2 The *correlation coefficient* of random variables X and Y is

$$\rho(X, Y) = \frac{\mathrm{Cov}[X, Y]}{\sigma_X \sigma_Y}.$$

||

Correlation is not affected by adding constants or multiplying by positive constants.

Proposition 9.2 *Let $X' = a + bX$ and $Y' = c + dY$ for constants a, $b \geq 0$, and c, $d \geq 0$. Then*

$$\rho(X', Y') = \rho(a + bX, c + dY) = \rho(X, Y).$$

Proof Exercise. □

Since correlation is not affected by linear transformations, the correlation may be expressed in terms of the correlation of standardized variables, which in turn equals the covariance of the standardized variables.

Corollary 9.4

$$\rho(X, Y) = \rho\left(\frac{X - \mu_X}{\sigma_X}, \frac{Y - \mu_Y}{\sigma_Y}\right) = \mathrm{Cov}\left[\frac{X - \mu_X}{\sigma_X}, \frac{Y - \mu_Y}{\sigma_Y}\right].$$

The correlation is bounded by -1 and 1.

Proposition 9.3 $|\rho(X, Y)| \leq 1$.

Proof Let $Z_X = (X - \mu_X)/\sigma_X$ and $Z_Y = (Y - \mu_Y)/\sigma_Y$, so $\rho(X, Y) = \mathrm{Cov}[Z_X, Z_Y]$.

$$\begin{aligned} \mathrm{Var}[Z_X \pm Z_Y] &= \mathrm{Var}[Z_X] + \mathrm{Var}[Z_Y] \pm 2\mathrm{Cov}[Z_X, Z_Y] \\ &= 2 \pm 2\rho(X, Y). \end{aligned}$$

But, variances are always nonnegative, so

$$\begin{aligned} 2 \pm 2\rho(X, Y) &\geq 0 \\ \pm\rho(X, Y) &\geq -1 \\ |\rho(X, Y)| &\leq 1. \end{aligned}$$ □

Proposition 9.4 $|\rho(X, Y)| = 1$ *if and only if* $Y = a + bX$ *for some real numbers* a *and* b.

Proof From the proof of the previous proposition, if $\rho(X, Y) = 1$, then $\mathrm{Var}[Z_X - Z_Y] = 0$.

Thus, $Z_X - Z_Y = C$ for some constant C.

$$\begin{aligned} Z_Y &= Z_X - C, \\ \frac{Y - \mu_Y}{\sigma_Y} &= \frac{X - \mu_X}{\sigma_X} - C, \\ Y &= \frac{\sigma_Y}{\sigma_X}X - \sigma_Y C + \mu_Y - \mu_X\frac{\sigma_Y}{\sigma_X}, \\ Y &= aX + B. \end{aligned}$$

Similarly, for $\rho(X, Y) = -1$.

We leave the converse as an exercise. □

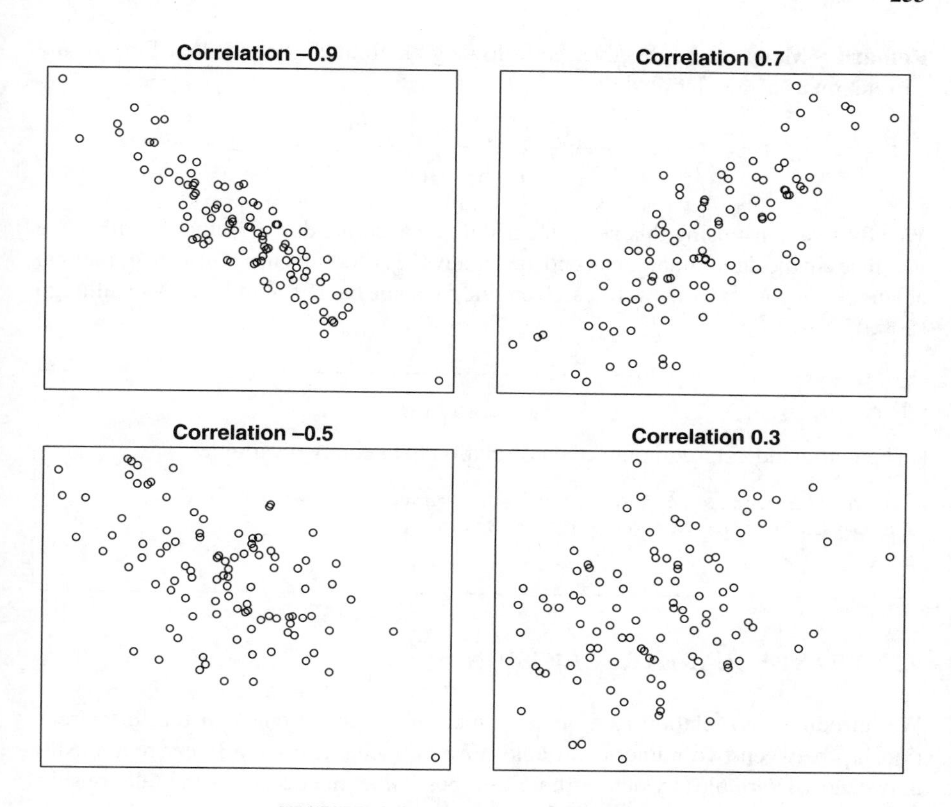

FIGURE 9.4 Examples of correlation.

The *sample correlation* for data $(X_1, Y_1), (X_2, Y_2), \ldots, (X_n, Y_n)$ is obtained by plugging in sample moments for population moments (Figure 9.4). The population correlation is

$$\rho(X, Y) = \frac{\text{Cov}[X, Y]}{\sigma_X \sigma_Y}.$$

The sample correlation is

$$\begin{aligned} r &= \frac{(1/n)\sum_{i=1}^n (x_i - \bar{x})(y_i - \bar{y})}{\sqrt{(1/n)\sum_{i=1}^n (x_i - \bar{x})^2}\sqrt{(1/n)\sum_{i=1}^n (y_i - \bar{y})^2}} \\ &= \frac{(1/n)\sum_{i=1}^n (x_i - \bar{x})(y_i - \bar{y})}{\sqrt{(1/(n-1))\sum_{i=1}^n (x_i - \bar{x})^2}\sqrt{(1/(n-1))\sum_{i=1}^n (y_i - \bar{y})^2}} \\ &= \frac{\sum_{i=1}^n (x_i - \bar{x})(y_i - \bar{y})}{\sqrt{\sum_{i=1}^n (x_i - \bar{x})^2 \sum_{i=1}^n (y_i - \bar{y})^2}}. \qquad (9.1) \end{aligned}$$

It does not matter whether you use a divisor of n or $n - 1$, provided you are consistent.

Remark Many textbooks give the following algebraically equivalent form of the correlation as a calculation aid:

$$r = \frac{(1/n)\sum_{i=1}^{n} x_i y_i) - \bar{x}\bar{y}}{\sqrt{((1/n)\sum_{i=1}^{n} x_i^2) - \bar{x}^2}\sqrt{((1/n)\sum_{i=1}^{n} y_i^2) - \bar{y}^2}}.$$

We advise against using this version since it is inaccurate due to roundoff error—the variances in the denominator can end up negative! Try looking up "software numerical accuracy" in your favorite web search engine for some background or see McCullough (2000). ||

R Note:

The command `cor` computes the correlation between two variables.

```
> plot(Spruce$Di.change, Spruce$Ht.change)  # x first, then y
> cor(Spruce$Di.change, Spruce$Ht.change)
[1]  0.9021
```

9.3 LEAST-SQUARES REGRESSION

We introduced correlation as a numeric measure of the strength of the linear relationship between two numeric variables. We now characterize a linear relationship between two variables by determining the "best" line that describes the relationship.

What do we mean by "best"? In most statistical applications, we pick the line $y = a + bx$ to make the vertical distances from observations to the line small, as shown in Figure 9.5. The reason using vertical distances is that we typically use one

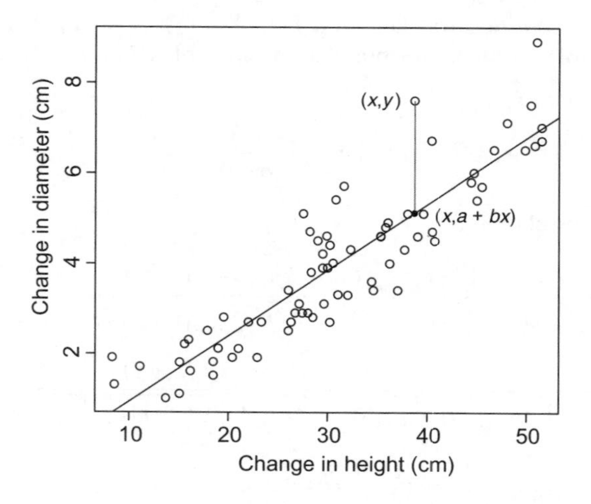

FIGURE 9.5 "Best fit" line.

variable, the x variable, to predict or explain the other (y) and we try to make the prediction errors as small as possible.

Next, we need some way to measure the overall error, taking into account all vertical distances. The most natural choice would be to add the distances,

$$\sum_i |y_i - (a + bx_i)|, \tag{9.2}$$

and in some applications this is a good choice. But more common is to choose a and b to minimize the *sum of squared distances*

$$g(a, b) = \sum_{i=1}^{n} (y_i - (a + bx_i))^2. \tag{9.3}$$

To minimize, we set the partial derivatives equal to zero:

$$\frac{\partial g}{\partial a} = 2\sum_{i=1}^{n} (y_i - a - bx_i)(-1) = 0,$$
$$\frac{\partial g}{\partial b} = 2\sum_{i=1}^{n} (y_i - a - bx_i)(-x_i) = 0,$$

and solve for a and b; this simplifies to

$$b = \frac{\sum_{i=1}^{n} (x_i - \bar{x})(y_i - \bar{y})}{\sum_{i=1}^{n} (x_i - \bar{x})^2}, \tag{9.4}$$

$$a = \bar{y} - b\bar{x}. \tag{9.5}$$

The line $\hat{y} = a + bx$ is called the *least-squares regression* line. In practice, we use statistical software to calculate the coefficients.

Example 9.2 For the Spruce data, let x denote height change and y denote the diameter change. Then, the least-squares line is

$$\hat{y} = -0.519 + 0.146x.$$

For every centimeter increase in the change in height, there is an associated increase of 0.146 cm in the change in diameter.

None of the seedlings in the study grew 25 cm over the course of 5 years, so we compute $\hat{y} = -0.519 + 0.149 \times 25 = 3.206$; that is, we predict that a seedling growing 25 cm in height would grow 3.206 cm in diameter. □

Definition 9.3 For any x, let $\hat{y} = a + bx$, then $\hat{y}$ is called a *predicted value* or a *fitted value*. ||

Note that Equation 9.5 can be written as $\bar{y} = a + b\bar{x}$, which implies that $(\bar{x}, \bar{y})$ lies on the least-squares line $y = a + bx$.

Proposition 9.5 $(1/n)\sum_{i=1}^{n} \hat{y}_i = \bar{y}$. *That is,* $\bar{\hat{y}} = \bar{y}$. *The mean of the predicted y's is the mean of the observed y's.*

Proof Exercise. □

Next, we see that there is a relationship between correlation and least-squares regression: in particular, the slope of the least-squares line is proportional to the correlation.

Let

$$\text{ss}_x = \sum_{i=1}^{n} (x_i - \bar{x})^2, \tag{9.6}$$

$$\text{ss}_y = \sum_{i=1}^{n} (y_i - \bar{y})^2, \tag{9.7}$$

$$\text{ss}_{xy} = \sum_{i=1}^{n} (x_i - \bar{x})(y_i - \bar{y}). \tag{9.8}$$

Then, from Equation 9.4, we re-express the estimated slope as

$$b = \frac{\sum_{i=1}^{n}(x_i - \bar{x})(y_i - \bar{y})}{\sum_{i=1}^{n}(x_i - \bar{x})^2} = \frac{\text{ss}_{xy}}{\text{ss}_x}.$$

We can also re-express the correlation (page 253) as

$$r = \frac{\sum_{i=1}^{n}(x_i - \bar{x})(y_i - \bar{y})}{\sqrt{\sum_{i=1}^{n}(x_i - \bar{x})^2}\sqrt{\sum_{i=1}^{n}(y_i - \bar{y})^2}} = \frac{ss_{xy}}{\sqrt{ss_x ss_y}}$$

Therefore, $\text{ss}_{xy} = r\sqrt{\text{ss}_x}\sqrt{\text{ss}_y}$, so we have

$$\begin{aligned} b &= r\frac{\sqrt{ss_x}\sqrt{ss_y}}{ss_x} \\ &= r\frac{\sqrt{ss_y}}{\sqrt{ss_x}} \\ &= r\frac{(1/(n-1))\sum_{i=1}^{n}(y_i - \bar{y})^2}{\sqrt{(1/(n-1))\sum_{i=1}^{n}(x_i - \bar{x})^2}} \\ &= r\frac{s_y}{s_x} \end{aligned}$$

where s_x and s_y denote the sample standard deviations for x and y.

Example 9.3 The correlation between the diameter change and height change is 0.9021. The standard deviation of the diameter changes and height changes are 1.7877 and 11.0495, respectively. Thus, $b = 0.9021 \times 1.7877/11.0495 = 0.146$, which agrees with what we obtained on page 255. □

LEAST-SQUARES REGRESSION

Let $(x_1, y_1), (x_2, y_2), \ldots, (x_n, y_n)$ be n observations. The least-squares regression line is $\hat{y} = a + bx$, where

$$b = \frac{\sum_{i=1}^{n}(x_i - \bar{x})(y_i - \bar{y})}{\sum_{i=1}^{n}(x_i - \bar{x})^2}, \tag{9.9}$$

$$a = \bar{y} - b\bar{x}. \tag{9.10}$$

In addition,

$$b = r\frac{s_y}{s_x}, \tag{9.11}$$

where r is the correlation and s_x and s_y are the standard deviations of the x_i's and y_i's, respectively. The variable being predicted, y, is called the "outcome", "response", or fitted variable. The variable used for predicting is called the "predictor" or "explanatory" variable.

Some authors refer to the y and x variables as "dependent" and "independent," respectively. However, this can be confused with independence and dependence of random variables.

R Note:

```
> spruce.lm <- lm(Di.change~Ht.change, data=Spruce)
> spruce.lm
...
Coefficients:
(Intercept)     Ht.change
    -0.5189        0.1459
> plot(Spruce$Ht.change, Spruce$Di.change)
> abline(spruce.lm)
```

To obtain the predicted values,

```
> fitted(spruce.lm) # or predict(spruce.lm)

        1         2         3         4         5         6
6.0488081 4.7644555 3.8595707 4.7060758 4.6331013 5.9612386
...
```

To calculate sums of squares in R, a shortcut is to use the relationship $ss_x = (n-1)s_x^2$, where the sample variance s_X^2 is calculated in R using var. For example, to find $\sum_{i=1}^{72}(x_i - \bar{x})^2$ for the height change variance in the Spruce data set,

```
> (nrow(Spruce) -1 ) * var(Spruce$Ht.change)
[1] 8668.38
```

9.3.1 Regression Toward the Mean

If x and y are perfectly correlated ($r = 1$), then the slope is s_y/s_x; so every change of one standard deviation in x results in a change of one standard deviation of y. But if $r \neq 1$, then for a change of one standard deviation in x, the vertical change is less than one standard deviation of y; so $\hat{y}$ is less responsive to a change in x. If $\rho = 0$, then the regression line is flat. See Figure 9.6.

This phenomenon is the origin of the name "regression." Sir Francis Galton studied the heights of parents and children. He found that although the children of tall parents were tall, on average, they were less so than their parents. Similarly, children of short parents averaged less short than their parents. The data, from Verzani (2010), are shown in Figure 9.7. He termed this "regression toward mediocrity," with the implication that in the long run, everyone would become of average height. This is of course not true (just look at the coauthors of this book!)[1] The regression line does

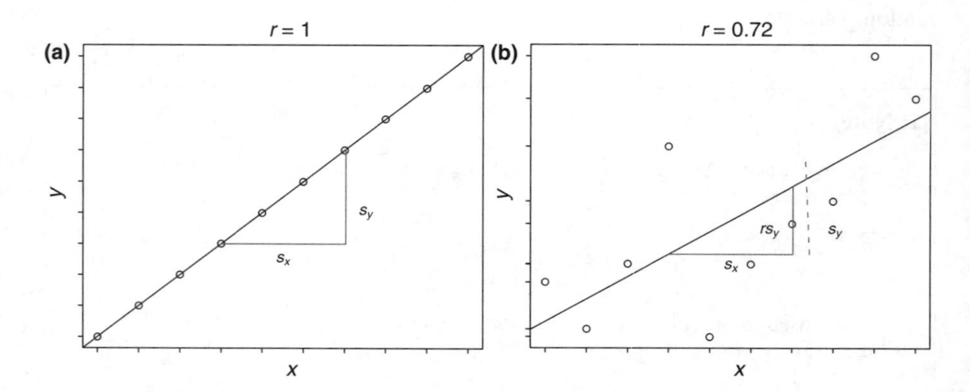

FIGURE 9.6 (a) The relationship between the regression slope, s_y and s_x with perfectly correlated data (b) The relationship without perfect correlation. For every change in one standard deviation of x, $\hat{y}$ changes less than one standard deviation in y.

[1]One of us is 5 ft 2 in., the other 6 ft 3 in.!

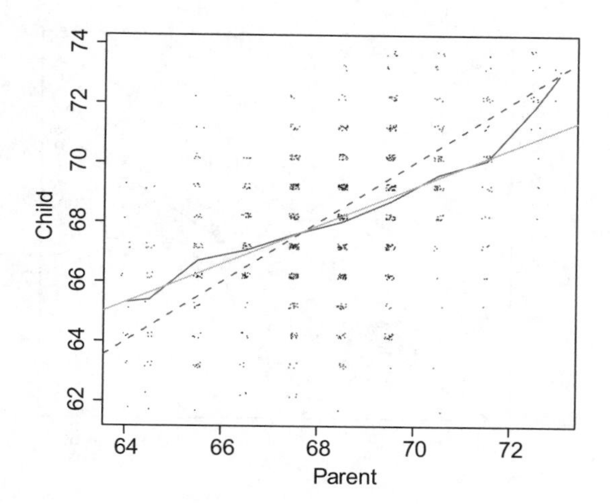

FIGURE 9.7 Heights of parents and children. The data are *jittered*—a small amount of random noise is added so that multiple points with the same x and y are visible. The x-axis contains the "midparent" height—the average of the father's height and 1.08 times the mother's height. The y-axis contains the average adult child's height, with female heights multiplied by 1.08. The dashed line is the 45° line. The solid line is the least-squares regression line and the zigzag line connects the mean child height with each midparent height.

not give the whole picture. There is also substantial variability above and below the regression line. Some offspring of average-height parents end up tall or short, so over time the variability above and below the mean remains roughly constant.

9.3.2 Variation

Let us look at the different components of variation in regression in more detail to see where Galton went wrong. We will use the Spruce data.

We partition the difference between an observed y value and the mean of the observed y values into two parts (Figure 9.8),

$$y_i - \bar{y} = (y_i - \hat{y}_i) + (\hat{y}_i - \bar{y}).$$

By algebraic manipulation, we can show that

$$\sum_{i=1}^{n}(y_i - \bar{y})^2 = \sum_{i=1}^{n}(y_i - \hat{y}_i)^2 + \sum_{i=1}^{n}(\hat{y}_i - \bar{y})^2. \tag{9.12}$$

This is related to the Pythagorean theorem for the sides of triangles; the vectors $(y_1 - \hat{y}_i, \ldots, y_n - \hat{y}_i)$ and $(\hat{y}_1 - \bar{y}, \ldots, \hat{y}_n - \bar{y})$ are orthogonal. We say that the *total variation* (of the y's) equals the *variation of the residuals* plus the *variation of the predicted values*.

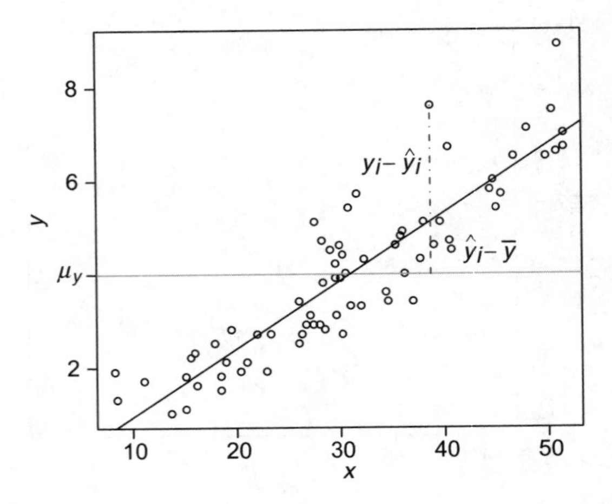

FIGURE 9.8 The partitioning of the variability. The horizontal line is the mean of the observed y's.

Also,

$$\begin{aligned}
\sum_{i=1}^{n} (\hat{y}_i - \bar{y})^2 &= \sum_{i=1}^{n} (a + bx_i - (a + b\bar{x}))^2 \\
&= \sum_{i=1}^{n} \big(b(x_i - \bar{x})\big)^2 \\
&= b^2 \sum_{i=1}^{n} (x_i - \bar{x})^2 \\
&= b^2 \mathrm{ss}_x \\
&= r^2 \mathrm{ss}_y. \qquad (9.13)
\end{aligned}$$

Thus,

$$\begin{aligned}
r^2 &= \frac{\sum_{i=1}^{n} (\hat{y}_i - \bar{y})^2}{\sum_{i=1}^{n} (y_i - \bar{y})^2} \\
&= \frac{1/(n-1) \sum_{i=1}^{n} (\hat{y}_i - \bar{y})^2}{1/(n-1) \sum_{i=1}^{n} (y_i - \bar{y})^2} \\
&= \frac{\text{variance of predicted } y\text{'s}}{\text{variance of observed } y\text{'s}},
\end{aligned}$$

r^2 is the proportion of the variation of the observed $y's$ that is explained by the regression line.

R-SQUARED = PROPORTION OF VARIANCE EXPLAINED

The R-squared coefficient is the square of the correlation r^2. In other words, $r^2 \times 100\%$ of the variance, or variation, of y is explained by the linear regression. We say that r^2 is the proportion of the variance explained by the regression model.

Example 9.4 For the Spruce Case Study, $r^2 = 0.9021^2 = 0.8138$, so about 81% of the variability in the diameter changes is explained by this model. □

9.3.3 Diagnostics

We can fit a linear regression line to any two variables, whether or not there is a linear relationship. But for predictions from the line to be accurate, the relationship should be approximately linear. A linear relationship is also required (together with some additional assumptions) for the standard errors and confidence intervals in Section 9.4 to be correct. So our next step is to check if it is appropriate to model the relationship between these two variables with a straight line.

Definition 9.4 Let (x_i, y_i) be one of the data points. The number $y_i - \hat{y}_i$ is called a *residual*.

A *residuals plot* is a plot of $y_i - \hat{y}_i$ against x_i for $i = 1, 2, \ldots, n$. ||

The residual is the difference between an observed y value and the corresponding fitted value; it provides information on how far off the least-squares line is in predicting the y_i value at a particular data point x_i. If the residual is positive, then the predicted value is an underestimate, whereas if the residual is negative, then the predicted value is an overestimate.

Example 9.5 In the example on page 255, we computed the least-squares line $\hat{y} = -0.519 + 0.146x$. Thus, for the first tree in the data set, the predicted diameter change is $\hat{y} = -0.519 + 0.146 \times 5.416 = 6.049$; so the corresponding residual is $5.416 - 6.049 = -0.633$ (Table 9.1). The least-squares line overestimates the diameter change for this tree. □

The plot of residuals against the predictor variable $(x_i, y_i - \hat{y}_i)$ provides visual information on the appropriateness of a straight line model (Figure 9.9). Ideally, points should be scattered randomly about the reference line $y = 0$.

TABLE 9.1 Partial View of `Spruce` Data

Tree	Height Change	Diameter Change	Predicted $\hat{y}$	Residual
1	45.0	5.416	6.049	−0.633
2	36.2	4.009	4.764	−0.755
3	30.0	3.914	3.859	0.054
4	35.8	4.813	4.706	0.106
5	35.3	4.6125	4.633	−0.021

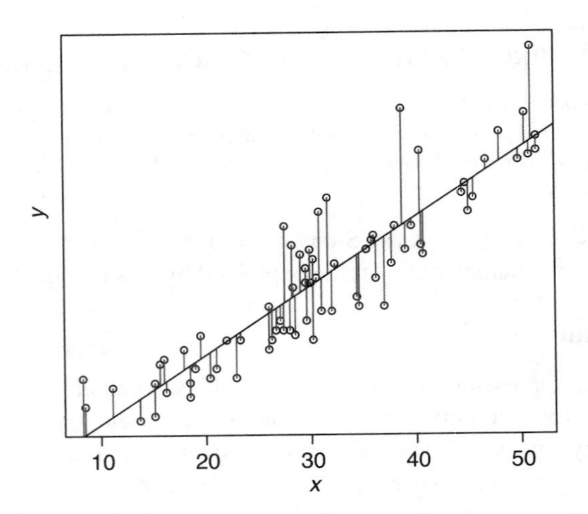

FIGURE 9.9 Residuals are the (signed) lengths of the line segments drawn from each observed y to the corresponding predicted $\hat{y}$.

Residuals plots are useful for the following:

- Revealing curvature—that is, for indicating that the relationship between the two variables is not linear.
- Spotting outliers.

If outliers are noticed, then you should check if they are influential: Does their removal dramatically change the model?

See Figure 9.10 for examples illustrating these points.

For the Spruce data, the residuals plot reveals that the distribution of the residuals is positively skewed—most residuals are negative but small, and there are a smaller number of positive residuals, but they are larger (Figure 9.11). This does not mean that a linear relationship is inappropriate, but it does cause problems for some methods that assume that residuals are normally distributed.

A bigger issue is whether there may be curvature. Here, some caution is needed—the human eye can be good at creating patterns out of nothing (the ancient star constellations are one example). Here, ignoring the most negative residual, the residuals appear to have a curved bottom. This impression is reinforced by a second set of points slightly higher, also curved upward. But these may be purely random artifacts. There do appear to be a large number of negative residuals in the middle—but there are also a number of even bigger positive residuals in the middle. There do appear to be a lack of large negative residuals on both sides—but this may be simply because there are fewer observations on each side.

A more effective way to judge curvature is to add a *scatter plot smooth* to the plot, a statistical procedure that tries to find a possibly curved relationship in the data. There are many such procedures, for example, the "connect-the-dot" procedure shown in

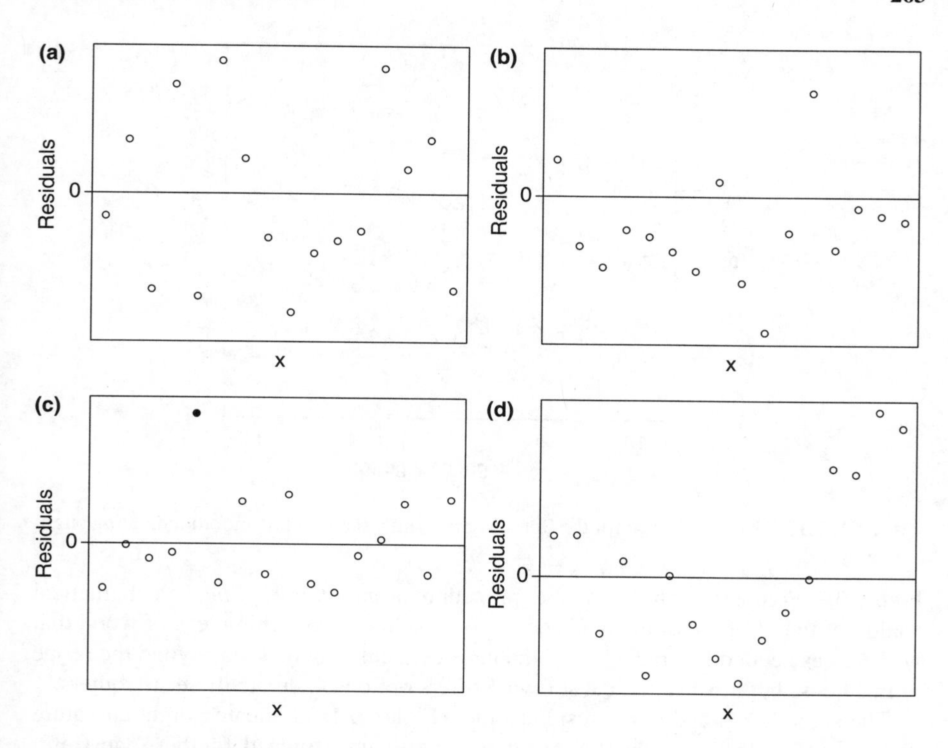

FIGURE 9.10 Examples of residual plots. (a) A good straight line fit. (b) The regression line is consistently overestimating the y values. (c) An outlier. (d) Curvature—a straight line is not an appropriate model for the relationship between the two variables.

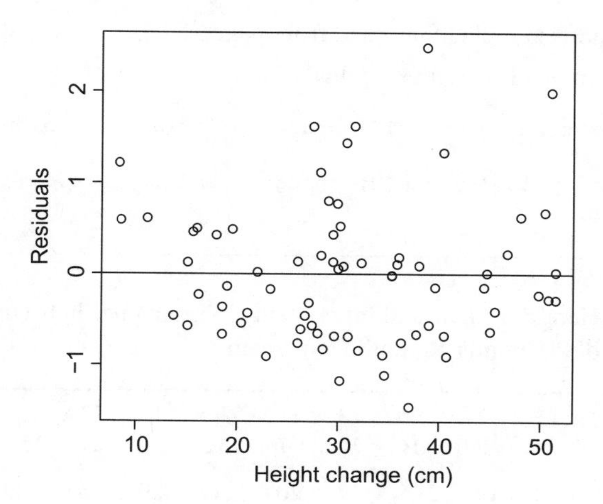

FIGURE 9.11 Residuals plot for the Spruce data.

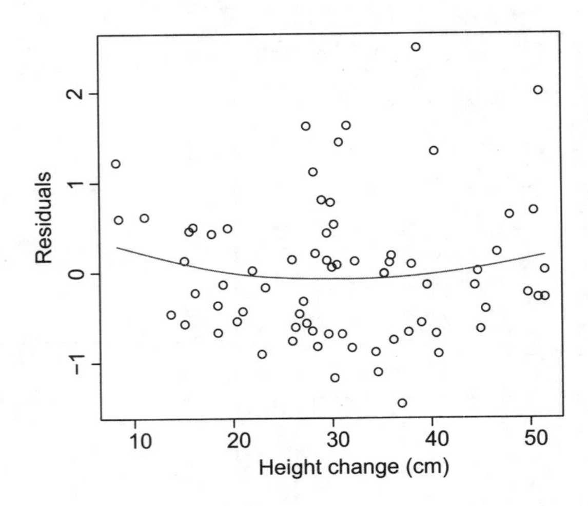

FIGURE 9.12 Residuals plot for the Spruce data, with a scatter plot smooth indicating slight curvature.

Figure 9.7. Figure 9.12 shows another procedure, a *smoothing spline*, a mathematical analog of the old-time draftsman's device—a "spline" was a thin piece of wood that was bent as needed for tracing smooth curves. Calculating these is beyond the scope of this book, but most statistical software offers options to create these smoothers.

The smoother added to the residuals plot (Figure 9.12) indicates slight curvature that should lead the researcher to some more investigation. Indeed, for these data (refer to the description on page 8), the observations do not come from one population since the seedlings were planted under different conditions.

R Note

Spruce regression example, continued from page 257.

The `resid` command gives the residuals for a regression.

```
plot(Spruce$Ht.change, resid(spruce.lm), ylab = "Residuals")
abline(h=0)
lines(smooth.spline(Spruce$Ht.change, resid(spruce.lm), df = 3),
      col="blue")
```

Example 9.6 Here are sugar and fat content (in grams per half cup serving) for a random sample of 20 brands of vanilla ice cream.

Sugar	15	13	20	23	11	21.5	12	23	23.0	19
Fat	8	8	16	14	7	15.5	8	21	15.5	16
Sugar	19.0	19	21.8	17	20	17	20	16	11	12
Fat	4.5	13	13.5	12	16	8	15	8	7	6

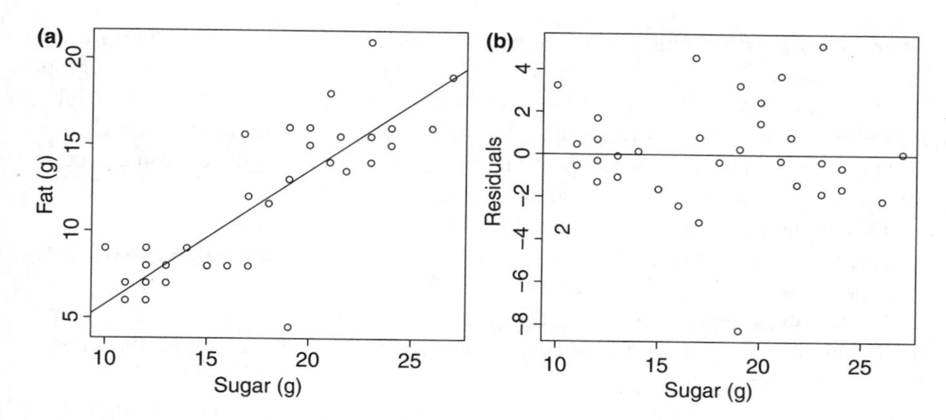

FIGURE 9.13 (a) Scatter plot of fat content against sugar content in ice cream. (b) Residuals plot for the least-squares regression.

The mean and standard deviation of the fat values are 11.6 and 4.5236 g, respectively, while those for sugar are 17.665 and 4.1323 g. The correlation between fat and sugar is 0.792.

For a least-squares line of fat on sugar, the slope of the least-squares line is $b = 0.792 \times (4.5236/4.1323) = 0.867$, while the intercept is $a = 11.6 - 0.867 \times 17.655 = -3.707$. Thus, the equation is $\hat{\texttt{fat}} = -3.707 + 0.867 \cdot \texttt{sugar}$.

For every gram increase in sugar, there is an associated increase of 0.867 g in fat content.

About 62.7% of the variability in fat content can be explained by this least-squares line.

The residuals plot (Figure 9.13), reveals a large negative outlier at about 19 g of sugar.

Removing this observation results in a least-squares line of $\hat{\texttt{Fat}} = -3.912 + 0.903 \cdot \texttt{Sugar}$. The slope of the regression line does not change very much, although the proportion of the variability in fat that is explained by the model does change (78.3%). □

9.3.4 Multiple Regression

The ideas of linear regression can be applied in the case where there are multiple predictors; then instead of a prediction equation $\hat{y} = a + bx$, the typical equation is of the form $\hat{y} = a + b_1x_1 + \cdots, b_px_p$, where p is the number of predictors. In some cases, such as when Google uses regression to improve web search answers, the number of predictors can be in millions.

One special case is when there are multiple groups in the data. In the Spruce example, there are two additional predictors—whether the tree was fertilized or not and whether the tree faced competition. One relatively simple model in this case is

$$\hat{y} = 0.51 + 0.104 \cdot \texttt{Ht.change} + 1.03 \cdot \texttt{Fertilizer} - 0.49 \cdot \texttt{Competition},$$

where we convert the categorical predictors to *dummy variables*— 1 if Fertilizer = "F" and 0 for "NF"; 1 if Competition = "C" and 0 for "NC." This equation suggests that trees that grew taller tended to grow thicker; that for a given change in height, these trees that were fertilized tended to grow thicker; and that for a given change in Height, trees that were in competition did not grow as thick—it seems they spend more energy growing taller rather than thicker.

This model has a single slope and different intercepts for the four groups defined by Fertilizer and Competition. Other models can be fit, for example, we may allow different slopes in different groups.

The formulas for calculating multiple regression coefficients are beyond the scope of this course, but can be performed using statistical software; for example, the R command for the model above is

```
lm(Di.change ~ Ht.change + Fertilizer + Competition, data = Spruce).
```

For more about multiple regression, see Kutner et al. (2005), Weisberg (2005), and Draper and Smith (1998).

9.4 THE SIMPLE LINEAR MODEL

The least-squares regression line is the "best fit" line for a set of n ordered pairs, $(x_1, y_1), (x_2, y_2), \ldots, (x_n, y_n)$ irrespective of whether this set represents a population or is a sample from a larger population. If this set is just a sample from a larger population, then the least-squares line is an estimate of a "true" least-squares line fit to the entire population.

Thus, in the case of a sample, after we calculate sample estimates such as the sample correlation or regression slope, we often want to quantify how accurate these estimates are using, for example, standard errors, confidence intervals, and significance tests. We will do so here using our usual bag of tricks—permutation tests, bootstrapping, and formulas based on certain assumptions.

Example 9.7 In figure skating competitions, skaters perform twice: a 2 min short program and a 4 min free skate program. The scores from the two segments are combined to determine the winner. What is the relationship, if any, between the score on the short program and the score on the free skate portion? We will investigate this by looking at the scores of 24 male skaters who competed in the 2010 Olympics in Vancouver. We will consider these observations as a sample taken from a larger population of all Olympic-level male figure skaters.

Figure 9.14a displays the scores, together with the least-squares regression line. The scores are highly correlated, with a correlation of 0.84. The regression line for predicting the score on the free skate program, based on the short program scores, is $\widehat{\texttt{Free}} = 7.97 + 1.735 \cdot \texttt{Short}$.

But how accurate are these numerical results? Are the correlation and regression slope significantly different from zero? What are standard errors or confidence intervals for the correlation, slope, or the prediction for $\hat{Y}$ at a particular value of x?

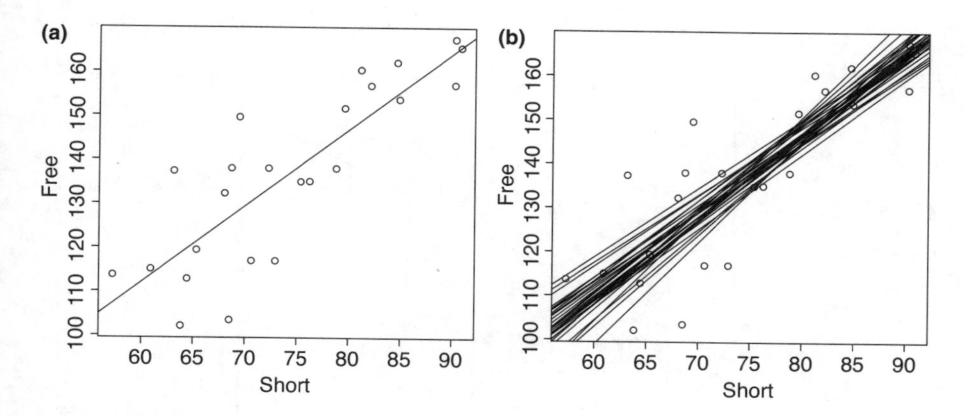

FIGURE 9.14 Scores of the 24 finalists in the 2010 Olympics men's figure skating contest for the short program and free program. (a) The least-squares regression line. (b) The regression lines from 30 bootstrap samples.

Figure 9.14b shows regression lines from 30 bootstrap samples. This gives a useful impression of the variability of the regression predictions. The predictions are most accurate for values of x near the center, and become less accurate as we extrapolate more in either direction. □

In the following sections, we will obtain standard errors and confidence intervals in two different ways, first using formulas and then using resampling. These are complementary—the bootstrap is better at visual impressions, while the formula approach gives mathematical expressions that quantify the visual impressions.

The least-squares regression line is derived without making any assumption about the data. That is, we do not require one or both variables to be an independent random sample drawn from any particular distribution or even for the relationship to be linear. However, in order to draw inferences or calculate confidence intervals, we need to make some assumptions.

ASSUMPTIONS FOR THE SIMPLE LINEAR MODEL

Let $(x_1, Y_1), (x_2, Y_2), \ldots, (x_n, Y_n)$ be points with fixed x values and random Y values, independent of other Y's, where the distribution of Y given x is normal, $Y_i \sim N(\mu_i, \sigma^2)$, with

$$\mu_i = \mathrm{E}[Y_i] = \alpha + \beta x_i,$$

for constants α and β.

In other words, we assume the following:

- x values are fixed, not random.
- Relationship between the x values and the means μ_i is linear, $\mathrm{E}[Y_i|x_i] = \alpha + \beta x_i$.
- Residuals $\epsilon_i = Y_i - \mu_i$ are independent.

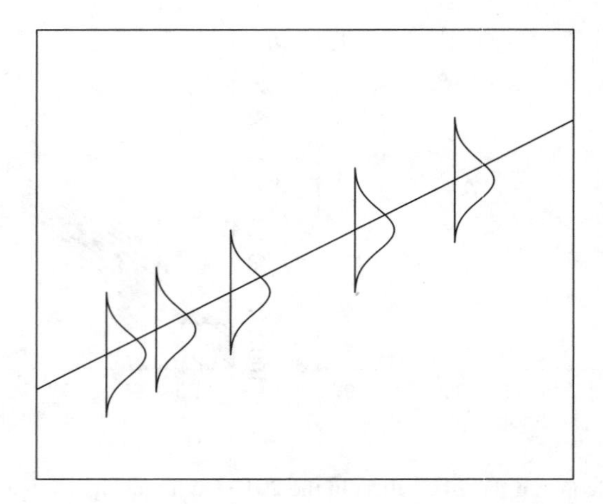

FIGURE 9.15 Linear regression assumptions: each Y_i is normal with mean $\alpha + \beta x_i$ and constant variance. Assumptions not shown are that the x values are fixed and observations are independent.

- Residuals have constant variance.
- Residuals are normally distributed.

See Figure 9.15.

In practice, the linear and independence assumptions are very important, the others less so—the reasons will be explained in Section 9.4.3.

Theorem 9.2 *Let* $(x_1, Y_1), (x_2, Y_2), \ldots, (x_n, Y_n)$ *satisfy the assumptions for a linear model. Then, the maximum likelihood estimates are*

$$\hat{\beta} = \frac{\sum(x_i - \bar{x})(Y_i - \bar{Y})}{\sum(x_i - \bar{x})^2}, \tag{9.14}$$

$$\hat{\alpha} = \bar{Y} - \hat{\beta}\bar{x}, \tag{9.15}$$

$$\hat{\sigma}^2 = \frac{1}{n}\sum(Y_i - \hat{Y})^2. \tag{9.16}$$

Proof Since the Y_i's are normally distributed, we can form the likelihood function:

$$\begin{aligned} L(\alpha, \beta, \sigma) &= \prod_1^n \frac{1}{\sqrt{2\pi}\sigma} e^{-(Y_i-\mu_i)^2/(2\sigma^2)} \\ &= \prod_1^n \frac{1}{\sqrt{2\pi}\sigma} e^{-(Y_i-\alpha-\beta x_i)^2/(2\sigma^2)} \\ &= \frac{1}{(\sqrt{2\pi}\sigma)^n} e^{-1/(2\sigma^2)\sum_1^n(Y_i-\alpha-\beta x_i)^2}. \end{aligned}$$

Thus, the log-likelihood is

$$\ln(L) = -n\ln(\sqrt{2\pi}\sigma) - \frac{1}{2\sigma^2}\sum_1^n (Y_i - \alpha - \beta x_i)^2.$$

We take the partial derivatives with respect to α, β, and σ, respectively:

$$\frac{\partial \ln(L)}{\partial \alpha} = \frac{-1}{\sigma^2}\sum_1^n (Y_i - \alpha - \beta x_i)(-1),$$

$$\frac{\partial \ln(L)}{\partial \beta} = \frac{-1}{\sigma^2}\sum_1^n (Y_i - \alpha - \beta x_i)(-x_i),$$

$$\frac{\partial \ln(L)}{\partial \sigma} = \frac{-n}{\sigma} + \frac{1}{\sigma^2}\sum_1^n (Y_i - \alpha - \beta x_i)^2.$$

Equating each partial derivative to 0 and doing some algebra yield the maximum likelihood estimates:

$$\hat{\beta} = \frac{\sum_{i=1}^n (x_i - \bar{x})(Y_i - \bar{Y})}{\sum (x_i - \bar{x})^2},$$

$$\hat{\alpha} = \bar{Y} - \hat{\beta}\bar{x},$$

$$\hat{\sigma}^2 = \frac{1}{n}\sum_{i=1}^n (Y_i - \hat{Y}_i)^2.$$

□

Note that these maximum likelihood estimates for β and α are exactly the same as the least-squares estimates (page 257).

We state without proof:

Theorem 9.3 *Suppose $(x_1, Y_1), (x_2, Y_2), \ldots, (x_n, Y_n)$ satisfy the assumptions for a linear model. Then,*

1. *$\hat{\sigma}^2$, $\hat{\beta}$, and $\bar{Y}$ are mutually independent.*
2. *$n\,\hat{\sigma}^2/\sigma^2$ has a chi-square distribution with $n - 2$ degrees of freedom.*

We would not actually use $\hat{\sigma}^2$ much; instead, we will use an unbiased version:

Corollary 9.5

$$S^2 = \frac{n}{n-2}\hat{\sigma}^2 = \frac{1}{n-2}\sum (Y_i - \hat{Y}_i)^2$$

is an unbiased estimator of σ^2.

We call S the *residual standard deviation* or *residual standard error*. Note that it is computed with a divisor of $n-2$, corresponding to the degrees of freedom (this is because the means are affected by two estimated parameters $\hat{\alpha}$ and $\hat{\beta}$).

Proof From Theorem B.12, the expected value of a chi-square distribution with $n-2$ degrees of freedom is $n-2$. Thus, from Theorem 9.3 (2), $\mathrm{E}\left[n\,\hat{\sigma}^2/\sigma^2\right] = n-2$, or upon rearranging,

$$E[S^2] = \mathrm{E}\left[\frac{n}{n-2}\hat{\sigma}^2\right] = \sigma^2.$$

□

9.4.1 Inference for α and β

We now consider some properties of the maximum likelihood estimators $\hat{\alpha}$ and $\hat{\beta}$ of the intercept and slope for the linear model $\mathrm{E}[Y] = \alpha + \beta x$.

Theorem 9.4 *Let $(x_1, Y_1), (x_2, Y_2), \ldots, (x_n, Y_n)$ satisfy the assumptions for a simple linear model, and let $\hat{\alpha}$ and $\hat{\beta}$ denote the estimators of α and β, respectively. Then,*

1. *$\hat{\alpha}$ and $\hat{\beta}$ are normal random variables,*
2. $\mathrm{E}[\hat{\alpha}] = \alpha$ *and* $\mathrm{E}\left[\hat{\beta}\right] = \beta$,
3. $\mathrm{Var}[\hat{\beta}] = \sigma^2/\mathrm{ss}_x$,
4. $\mathrm{Var}[\hat{\alpha}] = \sigma^2\left[1/n + (x-\bar{x})^2/\mathrm{ss}_x\right]$,

where ss_x is given in Equation 9.6.

Proof

$$\begin{aligned}\hat{\beta} &= \frac{\sum_{i=1}^n (x_i-\bar{x})(Y_i-\bar{Y})}{\sum_{i=1}^n (x_i-\bar{x})^2} \\ &= \frac{1}{\mathrm{ss}_x}\sum_{i=1}^n ((x_i-\bar{x})Y_i - (x_i-\bar{x})\bar{Y}) \\ &= \frac{1}{\mathrm{ss}_x}\left(\sum_{i=1}^n (x_i-\bar{x})Y_i - \sum_{i=1}^n (x_i-\bar{x})\bar{Y}\right) \\ &= \frac{1}{\mathrm{ss}_x}\sum_{i=1}^n (x_i-\bar{x})Y_i.\end{aligned}$$

Note that $\hat{\beta}$ is a linear combination of independent normal random variables, so is also a normal random variable (Theorem A.11).

$$\begin{aligned}
\mathrm{E}\left[\hat{\beta}\right] &= \mathrm{E}\left[\frac{1}{\mathrm{ss}_x}\sum_{i=1}^{n}(x_i-\bar{x})Y_i\right] \\
&= \frac{1}{\mathrm{ss}_x}\sum_{i=1}^{n}(x_i-\bar{x})\mathrm{E}\left[Y_i\right] \\
&= \frac{1}{\mathrm{ss}_x}\sum_{i=1}^{n}(x_i-\bar{x})(\alpha+\beta x_i) \\
&= \alpha\frac{\sum_{i=1}^{n}(x_i-\bar{x})}{\mathrm{ss}_x} + \beta\frac{\sum_{i=1}^{n}(x_i-\bar{x})x_i}{\mathrm{ss}_x} \\
&= \beta,
\end{aligned}$$

where the last equality follows from $\sum_{i=1}^{n}(x_i-\bar{x}) = 0$ and $\mathrm{ss}_x = \sum_{i=1}^{n}(x_i-\bar{x})^2 = \sum_{i=1}^{n}(x_i-\bar{x})x_i$.

Thus, $\hat{\beta}$ is an unbiased estimator of β.

The variance is

$$\begin{aligned}
\mathrm{Var}[\hat{\beta}] &= \mathrm{Var}\left[\frac{1}{\mathrm{ss}_x}\sum_{i=1}^{n}(x_i-\bar{x})Y_i\right] \\
&= \frac{1}{(\mathrm{ss}_x)^2}\sum_{i=1}^{n}\mathrm{Var}[(x_i-\bar{x})Y_i] \\
&= \frac{1}{(\mathrm{ss}_x)^2}\sum_{i=1}^{n}(x_i-\bar{x})^2\mathrm{Var}[Y_i] \\
&= \frac{1}{(\mathrm{ss}_x)^2}\sum_{i=1}^{n}(x_i-\bar{x})^2\sigma^2 \\
&= \frac{\sigma^2}{\mathrm{ss}_x}.
\end{aligned}$$

Thus, the sampling distribution of $\hat{\beta}$ is normal with mean β and variance σ^2/ss_x.

The proof for $\hat{\alpha}$ is similar. □

Now, since $\hat{\beta}$ follows a normal distribution, we can form the z statistic,

$$Z = \frac{\hat{\beta}-\beta}{\sqrt{\sigma^2/\mathrm{ss}_x}} = \frac{\hat{\beta}-\beta}{\sigma/\sqrt{\mathrm{ss}_x}},$$

which follows a standard normal distribution.

In practice, σ is unknown, so we plug in the estimate S to obtain

$$\frac{\hat{\beta}-\beta}{S/\sqrt{\mathrm{ss}_x}}.$$

As in earlier chapters, replacing the population standard deviation with an estimate results in a t rather than standard normal distribution.

Let $\hat{\text{SE}}[\hat{\beta}] = S/\sqrt{\text{ss}_x}$, the estimate of the standard error of $\hat{\beta}$; then, we have the following theorem:

Theorem 9.5 *Let $(x_1, Y_1), (x_2, Y_2), \ldots, (x_n, Y_n)$ satisfy the assumptions for the simple linear model. Then,*

$$T = \frac{\hat{\beta} - \beta}{\hat{\text{SE}}[\hat{\beta}]}$$

follows a t distribution with $n - 2$ degrees of freedom.

Proof From Theorem 9.4

$$Z = \frac{\hat{\beta} - \beta}{\sigma/\sqrt{\text{ss}_x}}$$

follows a standard normal distribution. Also, from Theorem 9.3,

$$\frac{n\hat{\sigma}^2}{\sigma^2} = \frac{(n-2)S^2}{\sigma^2}$$

has a χ^2 distribution with $n - 2$ degrees of freedom, and Z and $(n-2)S^2/\sigma^2$ are independent. Thus, from Theorem B.17, the ratio

$$\frac{Z}{\sqrt{((n-2)S^2/\sigma^2)/(n-2)}} = \frac{\hat{\beta} - \beta}{S/\sqrt{\text{ss}_x}}$$

has a t distribution with $n - 2$ degrees of freedom. □

In practice, we are often interested in testing if the slope β is zero or we will want a confidence interval for β.

INFERENCE FOR β

To test the hypothesis H_0: $\beta = 0$ versus H_a: $\beta \neq 0$, form the test statistic

$$T = \frac{\hat{\beta}}{\hat{\text{SE}}[\hat{\beta}]}.$$

Under the null hypothesis, T has a t distribution with $n - 2$ degrees of freedom.

A $(1 - \alpha) \times 100\%$ confidence interval for β is given by

$$\hat{\beta} \pm q\,\hat{\text{SE}}[\hat{\beta}],$$

where q is the $1 - \alpha/2$ quantile of the t distribution with $n - 2$ degrees of freedom and $\hat{\text{SE}}[\hat{\beta}] = S/\sqrt{\text{ss}_x}$.

Example 9.8 The data set `Skating2010` contains the scores from the short program and free skate for men's figure skating in the 2010 Olympics.

R Note:

```
> skate.lm <- lm(Free ~ Short, data=Skating2010)
> summary(skate.lm)
...
Coefficients:
            Estimate Std. Error t value Pr(>|t|)
(Intercept)   7.9691    18.1175   0.440    0.664
Short         1.7347     0.2424   7.157 3.56e-07 ***
---
Signif. codes:  0 '***' 0.001 '**' 0.01 '*' 0.05 '.' 0.1 ' ' 1

Residual standard error: 11.36 on 22 degrees of freedom
Multiple R-squared: 0.6995,     Adjusted R-squared: 0.6859
F-statistic: 51.22 on 1 and 22 DF,  p-value: 3.562e-07
```

From the R output, we obtain $S = 11.36$, the estimate for σ.

To test H_0: $\beta = 0$ versus H_A: $\beta \neq 0$, we use $t = \hat{\beta}/\hat{\text{SE}}[\hat{\beta}] = 1.7347/0.2424 = 7.157$. We compare this to a t distribution with 22 degrees of freedom to obtain a P-value of $2 \times 1.780036 \times 10^{-7} = 3.56 \times 10^{-7}$. Thus, we conclude that $\beta \neq 0$.

To compute a 95% confidence interval for the true β, we first find the 0.975 quantile for the t distribution with 22 degrees of freedom, $q = 2.0738$. Then,

$$1.7347 \pm 2.0738 \times 0.2424 = (1.232, 2.2374).$$

Thus, we are 95% confident that the true slope β is between 1.23 and 2.34. □

Similarly, we could give a standard error for $\hat{\alpha}$ and calculate a t statistic for testing H_0: $\hat{\alpha} = 0$. Statistical software routinely provides these. We caution, however, that it is rarely appropriate to do that test. It may be tempting to test whether one can simplify a regression model by omitting the intercept. But unless you have a physical model that omits the intercept, you should include the intercept in describing a linear relationship. And even when there is such a physical model, in practice including the intercept provides a useful fudge factor for adjusting the discrepancy between theory and reality.

9.4.2 Inference for the Response

In many applications, we will be interested in estimating the mean response for a specific value of x, say $x = x_s$. If $\hat{Y}_s = \hat{\alpha} + \hat{\beta}x_s$ denotes the point estimate of $\text{E}[Y_s]$, we need the sampling distribution of $\hat{Y}_s$.

We state results for both $\bar{Y}$ and $\hat{Y}$.

Theorem 9.6 *Let $(x_1, Y_1), (x_2, Y_2), \ldots, (x_n, Y_n)$ satisfy the assumptions for a simple linear model. Then,*

1. *$\bar{Y}$ is a normal random variable.*
2. $\mathrm{E}\left[\bar{Y}\right] = \alpha + \beta\bar{x}$.
3. $\mathrm{Var}[\bar{Y}] = \sigma^2/n$.
4. *$\hat{Y}_{\mathrm{s}}$ is a normal random variable.*
5. $\mathrm{E}\left[\hat{Y}_{\mathrm{s}}\right] = \mathrm{E}\left[Y_{\mathrm{s}}\right] = \alpha + \beta x_{\mathrm{s}}$.
6. $\mathrm{Var}[\hat{Y}_{\mathrm{s}}] = \sigma^2\left[1/n + (x_{\mathrm{s}} - \bar{x})^2/\mathrm{ss}_x\right]$.

Proof We leave the proof for the normality, mean, and variance of $\bar{Y}$ as an exercise. From Theorem 9.4,

$$\mathrm{E}\left[\hat{Y}_{\mathrm{s}}\right] = \mathrm{E}\left[\hat{\alpha} + \hat{\beta}x_{\mathrm{s}}\right] = \mathrm{E}\left[\hat{\alpha}\right] + \mathrm{E}\left[\hat{\beta}\right]x_{\mathrm{s}} = \alpha + \beta x_{\mathrm{s}}.$$

Using Equation 9.15, we have $\hat{Y}_{\mathrm{s}} = \bar{Y} + (x_{\mathrm{s}} - \bar{x})\hat{\beta}$, which is a linear combination of two independent normal variables. Also, by Theorems 9.3 and 9.4,

$$\begin{aligned}\mathrm{Var}[\hat{Y}_{\mathrm{s}}] &= \mathrm{Var}[\bar{Y}_{\mathrm{s}} + (x_{\mathrm{s}} - \bar{x})\hat{\beta}] \\ &= \mathrm{Var}[\bar{Y}_{\mathrm{s}}] + (x_{\mathrm{s}} - \bar{x})^2\mathrm{Var}[\hat{\beta}] \\ &= \frac{\sigma^2}{n} + (x_{\mathrm{s}} - \bar{x})^2\frac{\sigma^2}{\mathrm{ss}_x}.\end{aligned}$$

□

Again, using the residual standard error S as an estimate of σ, we have that

$$T = \frac{\hat{Y}_{\mathrm{s}} - \mathrm{E}\left[\hat{Y}_{\mathrm{s}}\right]}{S\sqrt{1/n + (x_{\mathrm{s}} - \bar{x})^2/\mathrm{ss}_x}}$$

follows a t distribution.

Let $\widehat{\mathrm{SE}}[\hat{Y}_{\mathrm{s}}] = S\sqrt{1/n + (x_{\mathrm{s}} - \bar{x})^2/\mathrm{ss}_x}$, the estimate of the standard error of $\hat{Y}_{\mathrm{s}}$. We summarize without formal proof:

Theorem 9.7 *Let $(x_1, Y_1), (x_2, Y_2), \ldots, (x_n, Y_n)$ satisfy the assumptions for a simple linear model. Let $x = x_{\mathrm{s}}$ be a specific value of the independent variable and $\hat{Y}_{\mathrm{s}} = \hat{\alpha} + \hat{\beta}x_{\mathrm{s}}$. Then,*

$$T = \frac{\hat{Y}_{\mathrm{s}} - \mathrm{E}\left[\hat{Y}_{\mathrm{s}}\right]}{\widehat{\mathrm{SE}}[\hat{Y}_{\mathrm{s}}]}$$

follows a t distribution with $n - 2$ degrees of freedom.

CONFIDENCE INTERVAL FOR $E[Y_s]$

A $(1-\alpha)\times 100\%$ confidence interval for $E[Y_s]$ at $x = x_s$ is given by

$$\hat{Y}_s \pm q\,\hat{SE}[\hat{Y}_s] = \hat{Y}_s \pm q\,S\sqrt{\frac{1}{n} + \frac{(x_s-\bar{x})^2}{ss_x}},$$

where q is the $1-\alpha/2$ quantile of the t distribution with $n-2$ degrees of freedom and S is the residual standard error.

We see that the variance of $\hat{Y}_s$ is smallest at $x_s = \bar{x}$ and increases as $(x_s-\bar{x})^2$ increases. In other words, the farther the x_s from $\bar{x}$, the less accurate the predictions.

Example 9.9 In the Olympic skating in Example 9.8, suppose we consider a short program score of 60. Then the estimate of the mean free skate score is $\hat{E}[Y_s] = 7.969 + 1.735 \times 60 = 112.07$. From the data set, we find the mean and standard deviation of the short score to be $\bar{x} = 74.132$ and $s_x = 9.771$, respectively. Thus, with $n = 24$, $S = 11.36$, and $ss_x = (n-1)s_x^2 = 2195.691$, the standard error is

$$11.36\sqrt{\frac{1}{24} + \frac{(60-74.132)^2}{2195.61}} = 4.137;$$

the 0.975 quantile for the t distribution with 22 degrees of freedom $q = 2.074$. Thus, the 95% confidence interval for the mean free skate score when the short score is 60 is

$$112.07 \pm 2.074 \times 4.137 = (103.5, 120.7).$$

We conclude that with 95% confidence, the expected free skate score is between 103.5 and 120.7 when the short program score is 60 points. □

What if instead of the mean free skate score corresponding to a short score of 60, we want an estimate of an individual free skate score? In this case, we need to take into account the uncertainty in the expected value as well as the random variability of a single observation. Thus, the variance of the prediction error is

$$\text{Var}[Y-\hat{Y}] = \text{Var}[Y] + \text{Var}[\hat{Y}] = \sigma^2 + \sigma^2\left[\frac{1}{n} + \frac{(x-\bar{x})^2}{ss_x}\right].$$

Thus, the estimate of the prediction standard error is

$$\hat{SE}[\text{prediction}] = S\sqrt{1 + \frac{1}{n} + \frac{(x-\bar{x})^2}{ss_x}}. \tag{9.17}$$

PREDICTION INTERVAL FOR Y_s

A $(1 - \alpha) \times 100\%$ prediction interval for Y_s at $x = x_s$ is given by

$$\hat{Y}_s \pm q\,\widehat{\mathrm{SE}}[\text{prediction}] = \hat{Y}_s \pm q\,S\sqrt{1 + \frac{1}{n} + \frac{(x_s - \bar{x})^2}{\mathrm{ss}_x}}, \tag{9.18}$$

where q is the $1 - \alpha/2$ quantile of the t distribution with $n - 2$ degrees of freedom and S is the residual standard error.

This interval is very sensitive to normality—if the residual distribution is not normal, this should not be used even if n is huge.

Example 9.10 Suppose a male skater scores 60 on his short program. Find a 95% prediction interval for his score on the free skate.

Solution Referring to Example 9.9, we have $\hat{Y}_s = 112.07$. The standard error of prediction is

$$11.36\sqrt{1 + \frac{1}{24} + \frac{199 \times 7393}{2195.61}} \approx 12.09.$$

Thus,

$$112.07 \pm 2.074 \times 12.09 = (86.995, 137.146).$$

Note that the prediction interval is much wider than the confidence interval; also see Figure 9.16. □

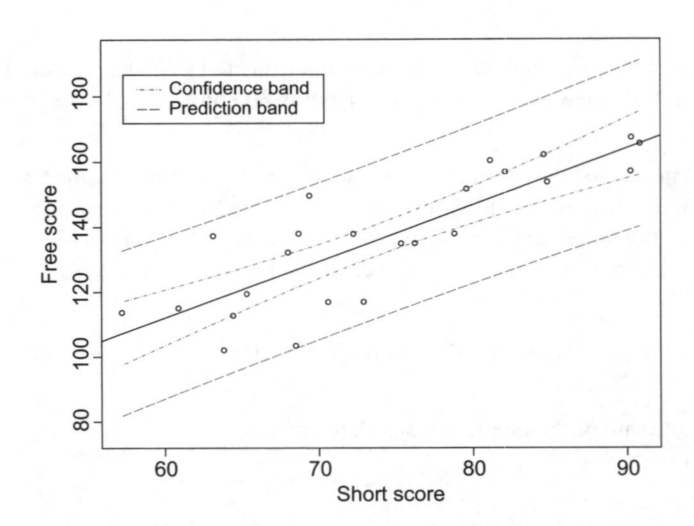

FIGURE 9.16 Plot of pointwise confidence and prediction intervals for the Olympic skating data.

Example 9.11 A wildlife biologist is interested in the relationship between the girth and length of grizzly bears. Suppose from a sample of 17 bears, she finds the following linear relationship: `Girth` $= 32.67 + 0.55$`Length`, where the measurements are in centimeters. Suppose the residual standard error and r-squared are 4.69 and 0.792, respectively. In addition, suppose the mean and standard deviation of the length measurements are 141.59 and 16.19 cm, respectively.

(a) Find the standard deviation of the girth measurements.

(b) Find a 95% confidence interval for the average girth of a bear whose length is 120 cm.

(c) Find a 95% prediction interval for the girth of an individual bear whose length is 120 cm.

Solution We are given $s = 4.69$, $r = \sqrt{0.792} = 0.890$, $\hat{\beta} = 0.55$, and $s_x = 16.19$. Thus,

(a) The standard deviation of the girth measurements is $s_y = \hat{\beta} s_x / r = 0.55 \times 16.19/0.890 = 10.01$.

(b) $Y_s = 32.67 + 0.55 \times 120 = 98.67$. Since $\text{ss}_x = (n-1)s_x^2$, we have

$$\hat{\text{SE}}[\bar{Y}_s] = 4.69\sqrt{\frac{1}{17} + \frac{(120 - 141.59)^2}{(16 \times 16.19^2)}} = 1.934.$$

The 0.975 quantile for a t distribution on 15 degrees of freedom is $q = 2.131$, so a 95% confidence interval for the mean girth of grizzlies that are 120 cm in length is $98.67 \pm 2.131 \times 1.934 = (94.5, 102.8)$ cm.

(c) The standard error of prediction is

$$\hat{\text{SE}}[\text{prediction}] = 4.69\sqrt{1 + \frac{1}{17} + \frac{(120 - 141.59)^2}{(16 \times 16.19^2)}} = 5.07.$$

Thus, a 95% prediction interval for the girth of an individual grizzly that is 120 cm in length is $98 \pm 2.131 \times 5.07 = (87.9, 108.5)$ cm.

□

9.4.3 Comments About Assumptions for the Linear Model

We began Section 9.4 with certain assumptions. We now discuss how important these assumptions are.

The x Values are Fixed In practice, this assumption holds in some designed experiments, say when estimating the relationship between crop yield and fertilizer, where the amount of fertilizer applied to each plot or plant is specified in advance. It does not hold in the more common case that both the X and Y values are random, and the pairs (X_i, Y_i) are drawn at random from a joint distribution.

This assumption played a key role in derivations of the properties of $\hat{\beta}$ and other estimates. But in practice this assumption is relatively unimportant. We routinely use the output from regression printouts, even when the X values are random.

In fact, it is typically better to do the analysis as if the X values are fixed, even if they are random. In doing so, we are *conditioning on the observed information.* Information is a concept that relates to how accurately we can make estimates. For example, a larger sample size corresponds to more information and more precise estimates. In simple linear regression, the information also depends on how spread out the x values are. Recall that $\text{Var}[\hat{\beta}] = \sigma^2/\text{ss}_x$—the more spread out the x values, the more accurate the estimate of slope.

Now, what does it mean to condition on the observed information? Suppose you are planning a survey and your roommate agrees to help. Each of you will poll 100 people to end up with a total sample size of 200. However, she gets sick and cannot help. When analyzing the results of your survey and computing standard errors, should you take into account that the eventual sample size was random, with a high probability of being 200? No, you should not. You should just analyze the survey based on the amount of information you have, not what might have been. Similarly, when computing standard errors for the regression slope, it is generally best to compute them based on how spread out the X values actually are, rather than adjusting for the fact that they could have been more or less spread out.

The Relationship Between the Variables Is Linear This is critical. Suppose, for example, that the real relationship is quadratic. Then the residuals standard error is inflated, because $\sum(Y_i - \hat{Y}_i)^2$ includes not only the random deviations but also the systematic error, the differences between the line and the curve.

In practice, this linearity assumption is often violated. If the violation is small, we may proceed anyway, but if it is larger, then standard errors, confidence intervals, and P-values are all incorrect.

The Residuals are Independent This assumption is also critical. This assumption is often violated when the observations are collected over time. Often data are collected over time, and successive residuals are positively correlated, in which case the actual variances are larger than indicated by the usual formulas.

Remark This issue of variances being larger when observations are correlated arises in other contexts too. I (Hesterberg) consulted for Verizon in a case for the Public Utilities Commission of New York. I was a young guy and on the other side was an eminent statistician, who used ordinary two-sample t tests but neglected to take the correlation of the observations into account. I showed that this completely invalidated the results, using simulation to help the PUC understand. An eye-witness reported that the other side was "furious, but of course they couldn't refute any of it," and Verizon won handily. ||

The Residuals Have Constant Variance This is less important when doing inferences for $\hat{\beta}$. The assumption is often violated: in particular, we often see the residual variance increase (or decrease) with x, with an average value in the middle. Then, the differences between reality and the assumptions tend to cancel out in computing $\text{Var}[\hat{\beta}]$.

However, when computing $\text{Var}[\hat{Y}]$ when $x \neq \bar{x}$ or $\text{Var}[\hat{\alpha}]$, this assumption does matter. We will see an example below.

The Residuals Are Normally Distributed Here we benefit from a version of the Central Limit Theorem—if the sample size is large and the information contained in ss_x is not concentrated in a small number of observations, then $\hat{\alpha}$, $\hat{\beta}$, and $\hat{Y}$ are approximately normally distributed even when the residuals are not normal and confidence intervals are approximately correct.

Prediction intervals are another story. They are a prediction for a single value, not an average, and a large sample size does not make these approximately correct if the residual distribution is nonnormal.

SUMMARY OF ASSUMPTIONS FOR LINEAR MODEL

The critical assumptions are that the relationship between the two variables is linear and that the observations are independent. The constant variance assumption can be important, but the most common violations have little effect on inferences for $\hat{\beta}$. Normality and fixed X values are relatively unimportant for confidence intervals, but normality matters when computing a prediction interval.

9.5 RESAMPLING CORRELATION AND REGRESSION

Another approach to obtaining inferences is to resample. We begin with the bootstrap for standard errors and confidence intervals and then use permutation tests for hypothesis testing.

To bootstrap, we treat these skaters as a random sample of the population of all Olympic-quality male skaters. Then, to create a bootstrap sample, we resample the skaters. For each bootstrap sample, we calculate the statistic(s) of interest.

Here is the general bootstrap procedure for two variables:

BOOTSTRAP FOR TWO VARIABLES

Given a sample of size n from a population with two variables,

1. Draw a resample of size n with replacement from the sample; in particular, draw n bivariate observations (x_i, y_i). If the observations are rows and variables are columns, we resample whole rows.
2. Compute a statistic of interest, such as the correlation, slope, or for prediction, $\hat{\text{E}}[\hat{Y}] = \hat{\alpha} + \hat{\beta}x$ *at a specific value of* x.
3. Repeat this resampling process many times, say 10,000.
4. Construct the bootstrap distribution of the statistic. Inspect its spread, bias, and shape.

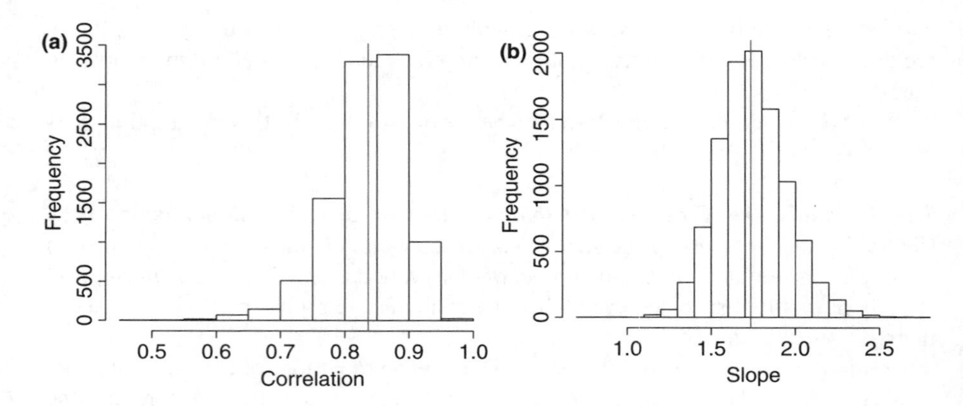

FIGURE 9.17 Bootstrap distributions of correlations (a) and slopes (b) for scores from the 2010 Olympics men's figure skating competition. The vertical lines are the corresponding statistics for the original data.

Figure 9.17a shows the bootstrap distribution for the correlation coefficient. This distribution is skewed, and indicates bias—the mean of the bootstrap distribution is smaller than the correlation of the original data. This is typical for correlations—they are biased toward zero. This is reasonable—if the original correlation is near 1 (or -1), the correlation for a sample cannot get much larger (or smaller), but can get much smaller (larger).

The bootstrap standard errors are 0.57 for the correlation and 0.20 for the slope. As before, these are the standard deviations of the bootstrap distributions.

We can use the range of the middle 95% of the bootstrap values as a rough confidence interval. For instance, in one simulation for the skating data, we found a 95% percentile confidence interval for the correlation to be $(0.70, 0.93)$ and for the slope β of the regression line to be $(1.38, 2.18)$. In Example 9.8, using the t distribution, we found a 95% confidence interval for the slope to be $(1.23, 2.34)$.

In this case, the classical interval is probably more accurate. In most of the regression problems you will encounter, the bootstrap does not offer much improvement in accuracy over classical intervals, except when the assumptions behind classical intervals are violated (one such example is the Bushmeat Case Study on page 283). Still, the bootstrap offers a way to check your work and provides graphics that may help you understand confidence intervals and standard errors in regression problems.

R Note:

The script for bootstrapping the correlation, slope, and mean response is given below. Since we want to resample the observations (x_i, y_i), we will resample the corresponding row numbers: that is, we will draw samples of size 24 (the number of skaters) with replacement from $1, 2, \ldots, 24$ and store these in the vector `index`. The command `Skating2010[index, ]` creates a new data frame from the original with rows corresponding to the rows in `index` and keeping all the columns.

```
N <- 10^4
cor.boot <- numeric(N)
beta.boot <- numeric(N)
alpha.boot <- numeric(N)
yPred.boot <- numeric(N)
n <- 24                                    # number of skaters
for (i in 1:N)
{
  index <- sample(n, replace = TRUE) # sample from 1,2,...,n
  Skate.boot <- Skating2010[index, ] # resampled data

  cor.boot[i] <- cor(Skate.boot$Short, Skate.boot$Free)

  #recalculate linear model estimates
  skateBoot.lm <- lm(Free ~ Short, data = Skate.boot)
  alpha.boot[i] <- coef(skateBoot.lm)[1] # new intercept
  beta.boot[i] <- coef(skateBoot.lm)[2]  # new slope
  yPred.boot[i] <- alpha.boot[i] + 60 * beta.boot[i] # recompute Y^
}

mean(cor.boot)
sd(cor.boot)
quantile(cor.boot, c(.025,.975))

hist(cor.boot, main="Bootstrap distribution of correlation",
  xlab = "Correlation")
observed <- cor(Skating2010$Short, Skating2010$Free)
abline(v = observed, col = "blue")    # add line at original cor.
```

The commands for the summaries and plot of the slope and response results are similar.

For a specific x, bootstrapping gives a percentile confidence interval for the expected value $E[Y]$.

Figure 9.18 shows the bootstrap distribution for the mean free skate program score corresponding to a short program score of 60. This distribution is centered at the original prediction and is roughly normally distributed, perhaps with a long left tail. The range of the middle 95% of the bootstrap lines (for a given x) gives the percentile confidence interval for that x. For instance, for the skating data, a 95% bootstrap percentile confidence interval for $E[Y]$ at $x = 60$ is (103.3, 120.2): we are 95% confident that at $x = 60$, the corresponding mean Y value is between 103.3 and 120.2.

On the other hand, a prediction interval gives a range for an individual—for a male who scores 60 on the short program, a 95% prediction interval should have a 95%

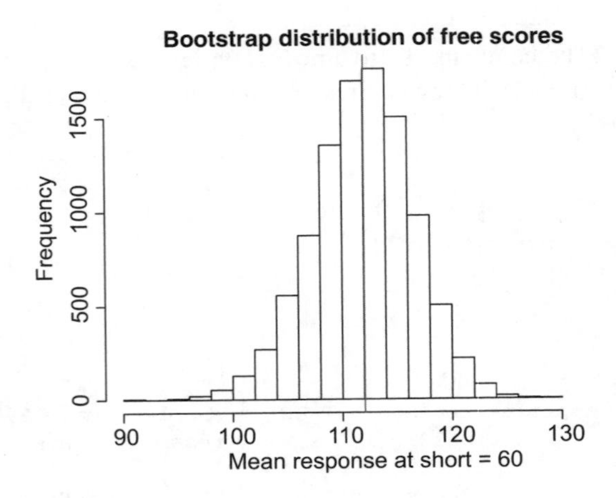

FIGURE 9.18 Distribution of bootstrapped free program scores when the short program score is 60.

chance of containing the free program score for that individual. The algorithm for a prediction interval for a response at a given x is more involved since we need to take into account the variability of an individual, and the Central Limit Theorem does not apply. See Davison and Hinkley (1997) for a way to compute prediction intervals.

9.5.1 Permutation Tests

To test whether there is a relationship between x and y or whether they are independent, we turn to permutation tests. The procedure here is to create a permutation sample by randomly permuting just one (not both) of the two variables and then computing a statistic such as correlation or slope.

PERMUTATION TEST FOR INDEPENDENCE OF TWO VARIABLES

Given a sample of size n from a population with two variables,

1. Draw a permutation resample of size n without replacement from one of the variables; keep the other variable in its original order.
2. Compute a statistic that measures the relationship, such as the correlation or slope.
3. Repeat this resampling process many times, say 9999.
4. Calculate the P-value.

For the Skating scores, the P-values are essentially zero; the probability of random chance alone producing a correlation as strong as 0.84 is minuscule, so we conclude that the two scores are not independent.

R Note:

Script for testing to see whether the short program score and the free skate score are independent. We permute just one of the variables (`Short`) while leaving `Free` fixed.

```
N <- 9999
n <- nrow(Skating2010)   # number of observations
result <- numeric(N)
observed <- cor(Skating2010$Short, Skating2010$Free)
for (i in 1:N)
{
  index <- sample(n, replace=FALSE)
  Short.permuted <- Skating2010$Short[index]
  result[i] <- cor(Short.permuted, Skating2010$Free)
}

(sum(observed <= result) + 1) / (N + 1)   # P-value
```

9.5.2 Bootstrap Case Study: Bushmeat

Many species of wildlife are going extinct due to habitat loss, climate change, and hunting. Brashares et al. (2004) found evidence of a direct link between fish supply (in kg) and subsequent demand for bushmeat[2] in Ghana. Table 9.2 and Figure 9.19 contain data of 30 years of local fish supply and biomass of 41 species in nature preserves.

TABLE 9.2 Bushmeat: Local Supply of Fish Per Capita and Biomass of 41 Species in Nature Preserves

Year	Fish	Biomass	Year	Fish	Biomass	Year	Fish	Biomass
1970	28.6	942.54	1980	21.8	862.85	1990	25.9	529.41
1971	34.7	969.77	1981	20.8	815.67	1991	23.0	497.37
1972	39.3	999.45	1982	19.7	756.58	1992	27.1	476.86
1973	32.4	987.13	1983	20.8	725.27	1993	23.4	453.80
1974	31.8	976.31	1984	21.1	662.65	1994	18.9	402.70
1975	32.8	944.07	1985	21.3	625.97	1995	19.6	365.25
1976	38.4	979.37	1986	24.3	621.69	1996	25.3	326.02
1977	33.2	997.86	1987	27.4	589.83	1997	22.0	320.12
1978	29.7	994.85	1988	24.5	548.05	1998	21.0	296.49
1979	25.0	936.36	1989	25.2	524.88	1999	23.0	228.72

[2]From Wikipedia: Bushmeat is the term commonly used for meat of terrestrial wild animals, killed for subsistence or commercial purposes throughout the humid tropics of the Americas, Asia, and Africa.

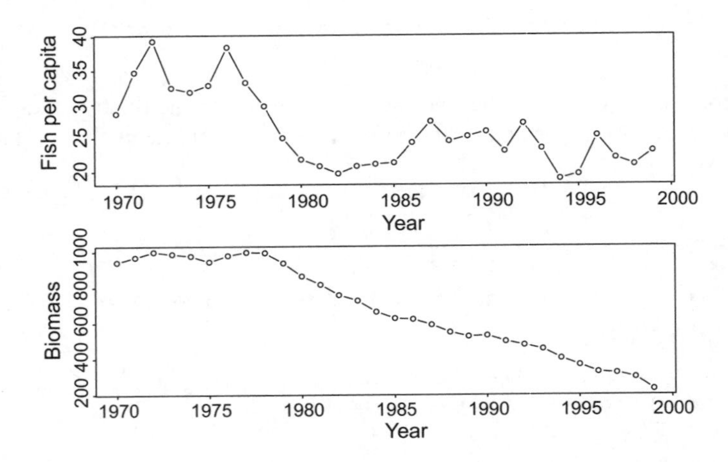

FIGURE 9.19 Bushmeat data; fish per capita and biomass of 41 species of wildlife in nature preserves for 30 years.

There is a general decline in biomass over the study period, but a closer look suggests that the decline is steeper in years with a small supply of fish. Rather than looking at the biomass for each year, we look at the percentage change. In Figure 9.20, we observe a positive relationship between fish supply and percentage change in biomass. The correlation is 0.67 and a regression of percent change in biomass against fish supply gives a slope of 0.64, suggesting that each increase of 1 kg fish per capita results in 0.64% loss of biomass and that with sufficiently large fish supplies, estimated at 33.3 (the x intercept of the least-squares line), there would be no loss in biomass.

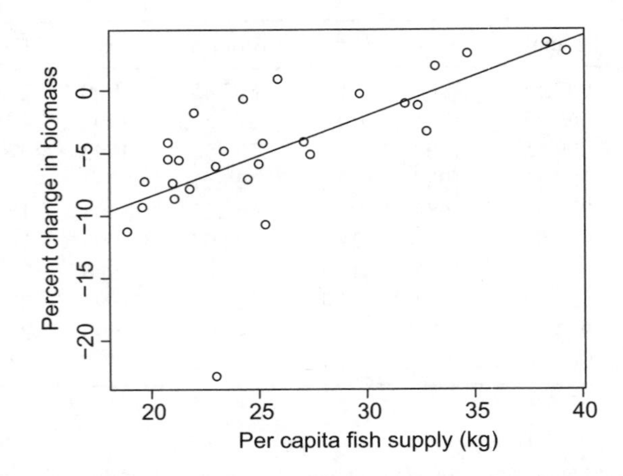

FIGURE 9.20 Scatter plot of percent change in biomass against fish supply with least-squares line imposed.

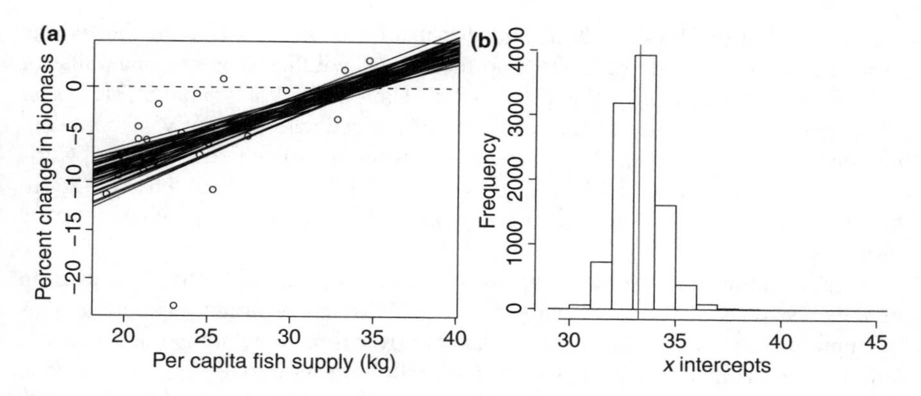

FIGURE 9.21 (a) Regression lines from 40 bootstrap samples of the bushmeat data. (b) Bootstrap distribution of the x intercept.

However, these are estimates based on a limited amount of data, so we turn to the bootstrap for more accuracy.

Figure 9.21 shows two views of the bootstrap output. Figure 9.21a is a graphical bootstrap—for 40 bootstrap samples, we calculate the slopes and intercepts and draw the corresponding lines over the original data and original line. We note that there appears to be a moderate amount of variability in the regression slopes, but not to the extent that any slopes are negative. There is also variability in the height of the regression lines, especially as we move to the right or to the left. If we extrapolate to the left, all the way to zero fish, it would show the variability in the intercept $\hat{\alpha}$.

Figure 9.21a shows that the regression lines have the smallest variability in the middle; the farther one goes to either side, the less accurate the answers are. However, the smallest variance occurs not at $\bar{x} = 26.1$, as implied by Theorem 9.6, but farther right. This is because the assumption of constant variance is violated. Looking back at the original scatter plot in Figure 9.20, we see that the residuals are smaller on the right. This is probably not just random variation, for two reasons, related to the numerator and denominator of the definition of relative change. When fish are more plentiful, there is less demand for bushmeat and correspondingly less variability in that demand; at the extreme, if there is zero demand, the variance is zero. Second, the observations on the left typically occur in the later years when the denominator is smaller, hence, even constant variability in the numerator results in greater variability in the ratio.

The results suggest that increasing the fish supply would reduce bushmeat harvest. An important question is what level of fish would stop the loss of wildlife? Based on the original regression line, the value would be 33.25 (the x intercept of the line). We can use the bootstrap to get an idea how accurate that number is.

We will use more bootstrap samples to do this. We use only 40 samples for plotting the regression lines because otherwise the figure becomes a mass of black ink. But now, for better accuracy in estimating the intercept, we will use 10^4 samples.

Figure 9.21b displays the bootstrap distribution of the x intercept, that is, the estimated supply of fish needed to stop the loss of wildlife. The original value of 33.25 falls in the middle of this distribution. The middle 95% range is 31.78 and 35.04, giving a rough idea of the reliability of the estimate. We are 95% confident that the supply of fish needed to forestall loss of biomass lies in that interval. Curiously, the interval $(31.5, 35.43) = (33.25 - 1.75, 33.25 + 2.18)$ stretches farther to the right, which gives a pessimistic story—it takes a lot of fish to gain confidence on the positive side.

We must admit that the bootstrap we just did, sampling with replacement from the data, assumes that the original data are i.i.d. from a bivariate population. This assumption is violated, because the data occur over time; they are neither independent nor identically distributed. The 22% drop in biomass in one year, for example, occurs in the final year when the denominator is small, making a large change in either direction relatively easy. There are procedures intended for time series data, both bootstrap and formula based; these are more complicated and are beyond the scope of this book.

9.6 LOGISTIC REGRESSION

According to the Centers for Disease Control and Prevention, the leading cause of death for people under the age of 34 is motor vehicle-related injuries.[3] Since many of these accidents are due to impaired driving—driving while under the influence of alcohol or drugs—there is a lot of emphasis on educating young drivers on the dangers of combining drinking and driving. But is there evidence that in fatal accidents, drinking and age are linked?

The Fatal Analysis Reporting System (FARS) (`http://www.nhtsa.gov/FARS`) database contains data on all fatal traffic accidents in the United States, the District of Columbia and Puerto Rico, since 1975. FARS is maintained by the National Center for Statistics and Analysis, part of the National Highway Traffic Safety Administration. We investigate the relationship between the involvement of alcohol and age of the driver in a random sample of 100 driver fatalities in 2009 in Pennsylvania; the drivers were driving a car, SUV, or light pickup truck (vehicles such as motor homes, convertibles, or commercial vehicles are excluded) (Table 9.3). One variable is a binary variable coded 1 if alcohol was involved and 0 otherwise; another variable is age of the driver, in years.

Let Y_i denote the alcohol involvement variable and x_i the independent variable age, $i = 1, 2, \ldots, 100$. For these individuals, we assume that the Y_i's are independent Bernoulli random variables with $P(Y_i = 1) = p_i$.

We want to understand the relationship between $p_i = \mathrm{E}[Y_i]$ and x_i. In Section 9.4, we considered linear regression of the form $\mathrm{E}[Y_i] = p_i = \alpha + \beta x_i$. But if β is nonzero, for sufficiently large and small x this would give probabilities less than zero or greater

[3] `http://www.cdc.gov/Motorvehiclesafety/`

TABLE 9.3 Part of the Data on Driver Fatalities in Pennsylvania

ID	Alcohol	Age
1	0	86
2	0	38
3	0	40
4	0	20
5	1	27

than one. Furthermore, linear regression assumes the same variances for every observation, but for Bernoulli data $\mathrm{Var}[Y_i] = p_i(1 - p_i)$.

So linear regression is not appropriate for these data. In this chapter, we discuss a type of regression suitable for zero–one data, *logistic regression*. We begin by introducing odds.

Definition 9.5 Let p denote the probability of some event. The *odds* of the event is defined by $p/(1 - p)$. ||

For instance, if $p = 0.8$ is the probability of a soccer team winning its next game, then $0.8/(1 - 0.8) = 4$ is its odds of winning the next game: the odds of the team winning (to not winning) the next game is 4 to 1. If $p = 0.25$ is the probability of dying from a certain disease, then $0.25/0.75 = 0.33$: the odds of dying (to not dying) is 1 to 3.

Let $(x_1, Y_1), (x_2, Y_2), \ldots, (x_n, Y_n)$ be a set of ordered pairs where $x_1, x_2, \ldots, x_n$ are fixed and $Y_1, Y_2, \ldots, Y_n$ are Bernoulli random variables with $P(Y_i = 1) = p_i$. In logistic regression, we model the logarithm of the odds as a linear function of x:

$$\ln\left(\frac{p_i}{1 - p_i}\right) = \alpha + \beta x_i, \quad i = 1, 2, \ldots, n. \tag{9.19}$$

Equivalently,

$$p_i = \frac{e^{\alpha+\beta x_i}}{1 + e^{\alpha+\beta x_i}}. \tag{9.20}$$

This gives an S-shaped relationship between x and $\mathrm{E}[Y]$ (see Figure 9.22).

The slope coefficient β describes how quickly the estimated probability increases; the maximum slope is $\beta/4$ (where $p = 0.5$). We can also interpret β by evaluating the odds at two different values, say x and $x + \Delta x$,

$$\frac{p_1}{1 - p_1} = e^{\alpha+\beta x},$$
$$\frac{p_2}{1 - p_2} = e^{\alpha+\beta(x+\Delta x)}.$$

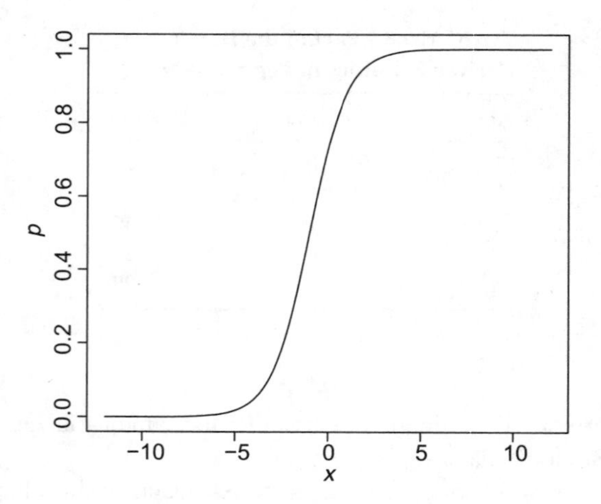

FIGURE 9.22 Plot of a typical logistic curve, Equation 9.20.

The *odds ratio* is

$$\frac{p_2/(1-p_2)}{p_1/(1-p_1)} = \frac{e^{\alpha+\beta(x+\Delta x)}}{e^{\alpha+\beta x}} = e^{\beta \Delta x}$$

and the *log odds ratio* is $\beta\Delta x$. So β measures how quickly the log odds ratio changes.

The parameters α and β are estimated using maximum likelihood. The likelihood function is given by

$$L(\alpha, \beta) = \prod_{i=1}^{n} p_i^{Y_i}(1-p_i)^{1-Y_i},$$

and then taking the logarithm, we find

$$\begin{aligned}\ln(L(\alpha, \beta)) &= \sum_{i=1}^{n} Y_i \ln(p_i) + (1-Y_i)\ln(1-p_i) \\ &= \sum_{i=1}^{n} Y_i \ln\left(\frac{p_i}{1-p_i}\right) + \ln(1-p_i).\end{aligned}$$

Setting the partial derivatives with respect to α and β (using the chain rule, because p_i is a function of α and β) equal to 0 and simplifying yields equations:

$$\begin{aligned}\frac{\partial \ln(L)}{\partial \alpha} &= \sum_{i=1}^{n} y_i - \frac{e^{\alpha+\beta x_i}}{1+e^{\alpha+\beta x_i}} = \sum_{i=1}^{n} y_i - p_i = 0, \\ \frac{\partial \ln(L)}{\partial \beta} &= \sum_{i=1}^{n} y_i x_i - \frac{e^{\alpha+\beta x_i} x_i}{1+e^{\alpha+\beta x_i}} = \sum_{i=1}^{n} (y_i - p_i) x_i = 0.\end{aligned}$$

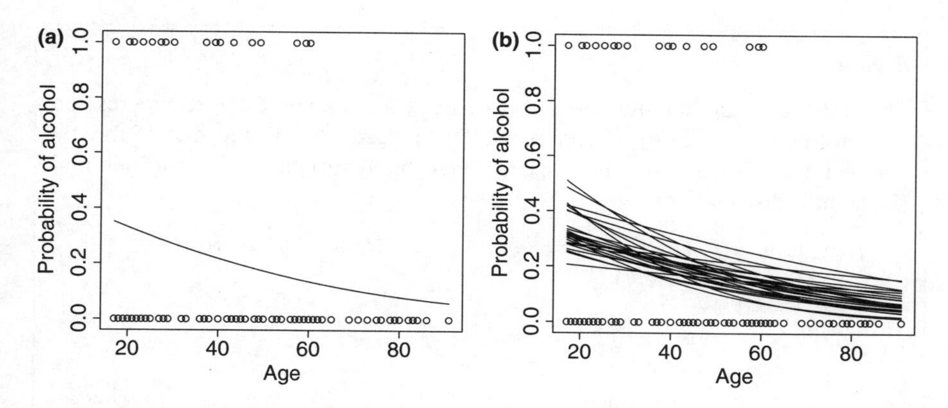

FIGURE 9.23 (a) Plot of estimated probability of alcohol being involved against age of driver. (b) Estimates from 25 bootstrap samples (see Section 9.6.1).

There is no closed form solution to these two equations, so a numerical algorithm must be used to find estimates $\hat{\alpha}$ and $\hat{\beta}$. For instance, `R` uses a procedure called iteratively reweighted least-squares, a multivariate version of Newton's method for finding the zero of a function; for those who have taken linear algebra, it uses the gradient and Hessian of $\ln L(\alpha, \beta)$.

Example 9.12 For the FARS fatalities data from Pennsylvania, the estimated logistic equation is

$$\ln\left(\frac{\hat{p}}{1-\hat{p}}\right) = -0.123 - 0.029x$$

(see Figure 9.23). For a 25-year-old driver, the estimated probability that alcohol was involved is

$$\hat{p} = \frac{e^{-0.123-0.029\times 25}}{1+e^{-0.123-0.029\times 25}} = 0.3$$

and the odds of alcohol being involved to not being involved is

$$\frac{\hat{p}}{1-\hat{p}} = e^{-0.123-0.029\times 25} = 0.428.$$

Similarly, we find the odds of alcohol being involved in the case of a 35-year-old driver, $\hat{p}_2/(1-\hat{p}_2) = 0.32$. The odds ratio is $0.428/0.32 = 1.34$. The odds of alcohol being involved in the fatal accident of a 25-year-old driver is 1.34 times greater than the odds of alcohol being involved in the fatal accident of a 35-year-old driver. Equivalently, we can say that the odds of alcohol being involved in the fatal accident of a 25-year-old driver is 34% higher than the odds of alcohol being involved in the fatal accident of a 35-year-old driver.

R Note:

The FARS data are in a file called `Fatalities.csv`. In `R`, logistic regression is performed using the `glm` command. The syntax is similar to that of the `lm` command, except that we must specify that the Y variable follows a binomial (Bernoulli) distribution:

```
> glm(Alcohol ~ Age, data = Fatalities, family = binomial)
...
Coefficients:
(Intercept)          Age
   -0.12262     -0.02898

x <- seq(17, 91, length=500)       # vector spanning the age range
y <- exp(-.123-.029*x) / (1+exp(-.123-.029*x))

plot(x, y, type = "l", ylim = c(0,1), xlab = "age",
 ylab = "Probability of alcohol")
points(Fatalities$Age, Fatalities$Alcohol)  # observations
```

□

Example 9.13 Suppose a hospital conducts a study to see if there is a link between patients getting an infection ($y = 1$ if yes) and their length of stay in the hospital (x, in days). A logistic regression performed on their data gives

$$\ln\left(\frac{\hat{p}}{1-\hat{p}}\right) = -1.942 + 0.023x.$$

How do the odds of getting an infection change for somebody who stays an additional week in this hospital?

Solution To compare the odds of infection for somebody who stays $\Delta x = 7$ days longer than another patient, compute $e^{0.023 \cdot 7} = 1.175$. Thus, staying an additional week increases the odds of infection by about 17.5%. Alternatively, we can express this result: the odds of infection are 1.175 times greater for every extra 7 days in the hospital. □

Logistic regression is a special case of a *generalized linear model*; another common version is *Poisson regression* in which Y has a Poisson distribution with mean given by $\exp(\alpha + \beta x)$; see Collett (2003), Dobson (2002), Kutner et al. (2005), McCullagh and Nelder (1989), for details. One of us (Hesterberg) consults at Google on a project to predict web traffic for billions of search phrases based on time of day and day of week, using Poisson regression. Searches that receive more traffic than predicted may be flagged for *Google Trends*. For example, a current hot search on November 21, 2010 was "ben roethlisberger punched."

9.6.1 Inference for Logistic Regression

Standard errors in logistic regression are based on assumptions similar to those in Section 9.4; we assume the following:

- x values are fixed, not random.
- Relationship between the x values and $\log(p/(1-p))$ is linear.
- Y values have Bernoulli distributions, with parameters p_i.
- Y values are independent.

We will rely on software to do the calculations for standard errors for coefficients and predictions ($\hat{p}_i$).

We can use the standard errors to produce confidence intervals, t statistics, P-values, and hypothesis tests, but be warned that sample sizes may need to be quite large for these to be accurate. Some software calculates t statistics, but then admirably declines to print P-values based on these t statistics because the P-values cannot be trusted.

Alternatively, we may bootstrap for standard errors and confidence intervals. We resample individuals, that is resample paired values (age, alcohol). For each bootstrap data set, we estimate the logistic regression parameters and calculate the desired predictions. Figure 9.23b shows the predictions from 25 bootstrap samples. This gives a rough idea of variability and suggests that some distributions are skewed—for example, predictions for the probability of alcohol involvement at age 80 are mostly near zero, but with a few larger values.

Figure 9.24 shows a histogram and normal quantile plot for b for 1000 bootstrap samples, and Figure 9.25 shows a histogram and normal quantile plot for the probabilities of alcohol involvement at age 20.

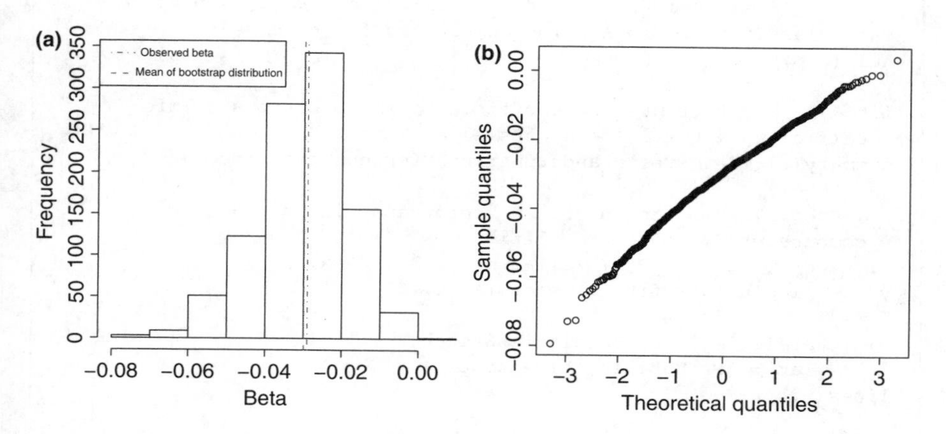

FIGURE 9.24 Histogram and normal quantile plot for b, the logistic regression slope coefficient for alcohol involvement versus age of driver.

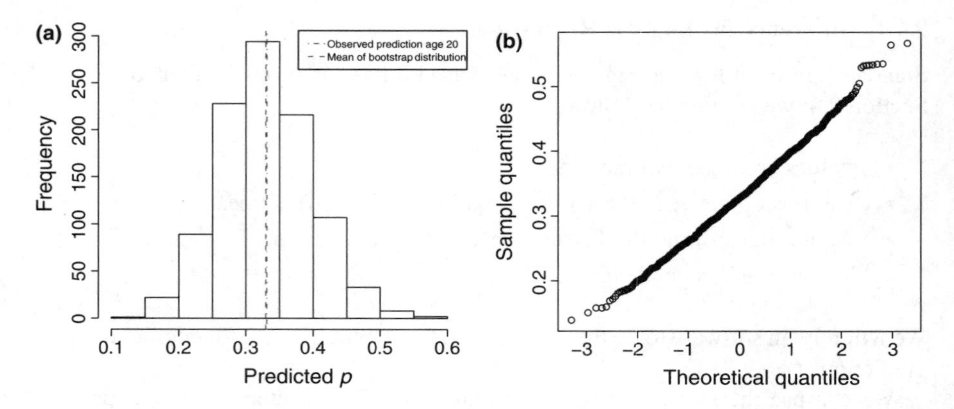

FIGURE 9.25 Histogram and normal quantile plot for probability of involvement at age 20.

Bootstrap percentile confidence intervals are $(-0.056, -0.008)$ for β and $(0.20, 0.47)$ for the prediction of alcohol involvement at age 20. These are quite wide, a range of 20–47% for the probability, and a ratio of 7:1 for the slope. For comparison, we note that intervals based on the formula standard errors are similar: $(-0.056, -0.002)$ for β and $(0.187, 0.475)$ for the probability.

R Note:

The command to extract coefficients from a `glm` object is `coef`. The command `plogis` is the cdf for a logistic random variable and is a handy way to compute $\exp(x)/(1+\exp(x))$.

```
fit <- glm(Alcohol ~ Age, data = Fatalities, family = binomial)
data.class(fit)  # is a "glm" object, so for help use:
help(glm)

fit            # prints the coefficients and other basic info
coef(fit)      # the coefficients as a vector
summary(fit) # gives standard errors for coefficients, etc.

x <- seq(17, 91, length=500) # vector spanning the age range
# compute predicted probabilities
y1 <- exp(-.123-.029*x) / (1+exp(-.123-.029*x))
y2 <- plogis(coef(fit)[1] + coef(fit)[2] * x)

plot(Fatalities$Age, Fatalities$Alcohol,
     ylab = "Probability of alcohol")
lines(x, y2)

# Full bootstrap - slope coefficient, and prediction at age 20
N <- 10^3
```

```
n <- nrow(Fatalities)                        # number of observations
alpha.boot <- numeric(N)
beta.boot <- numeric(N)
pPred.boot <- numeric(N)

for (i in 1:N)
{
  index <- sample(n, replace = TRUE)
  Fatal.boot <- Fatalities[index, ]          # resampled data

  fit.boot <- glm(Alcohol ~ Age, data = Fatal.boot,
                  family = binomial)
  alpha.boot[i] <- coef(fit.boot)[1]         # new intercept
  beta.boot[i] <- coef(fit.boot)[2]          # new slope
  pPred.boot[i] <- plogis(alpha.boot[i] + 20 * beta.boot[i])
}

quantile(beta.boot, c(.025, .975))          # 95% percentile intervals
quantile(pPred.boot, c(.025, .975))

par(mfrow=c(2,2))                            # set layout
hist(beta.boot, xlab = "beta", main = "")
qqnorm(beta.boot, main = "")

hist(pPred.boot, xlab = "p^", main = "")
qqnorm(pPred.boot, main = "")
# Figures in the text also use abline to add lines, and legend.
par(mfrow=c(1,1))                            # reset layout
```

The predict command can also be used to give predicted values. The required arguments of predict are the model object (the output from the glm command) and a data frame containing the values of the explanatory variable at which you wish to predict. By default, predict returns the predicted $a + bx$. To obtain the predicted probabilities, provide the argument type = "response". We illustrate the use of predict in reproducing Figure 9.23:

```
help(predict.glmm)                     # for more help on predict

n <- nrow(Fatalities)                  # number of observations
x <- seq(17, 91, length=500)           # vector spanning the age range
df.Age <- data.frame(Age = x)          # data frame to hold
    # explanatory variables, will use this for making predictions

plot(Fatalities$Age, Fatalities$Alcohol,
     ylab = "Probability of alcohol")
for (i in 1:25)
 {
  index <- sample(n, replace = TRUE)
  Fatal.boot <- Fatalities[index, ]    # resampled data
```

```
    fit.boot <- glm(Alcohol ~ Age, data = Fatal.boot,
                    family = binomial)
    pPred <- predict(fit.boot, newdata = df.Age, type = "response")
    lines(x, pPred)
  }
```

9.7 EXERCISES

1. Let X and Y be random variables with joint probability density function given by

$$f(x, y) = \begin{cases} \frac{6}{5}(x + y^2), & 0 \le x \le 1,\ 0 \le y \le 1, \\ 0, & \text{otherwise.} \end{cases}$$

   Find the covariance $\mathrm{Cov}[X, Y]$.
2. Let X be a random variable with mean μ and variance σ^2. Let $Y = C$, a constant. Find the covariance between X and Y.
3. Let X and Y be random variables with $\mathrm{Var}[X] = 4$, $\mathrm{Var}[Y] = 9$, and $\mathrm{Cov}[X, Y] = 3$. Find the variance of $2X + 3Y$.
4. Let X and Y be random variables with $\mathrm{Var}[X] = 5$, $\mathrm{Var}[Y] = 7$, and $\mathrm{Cov}[X, Y] = 2$. Find the variance of $2X - 5Y$.
5. Let X, Y, and Z be random variables with $\mathrm{Var}[X] = 3$, $\mathrm{Var}[Y] = 2$, and $\mathrm{Var}[Z] = 3$ and $\mathrm{Cov}[X, Y] = -2$, $\mathrm{Cov}[X, Z, =] - 4$, and $\mathrm{Cov}[Y, Z] = 7$. Find $\mathrm{Var}[5X - Y + 2Z]$.
6. Import the data from data set `corrExerciseA`.
   (a) Create a scatter plot of X and Y and then find the correlation between X and Y.
   (b) Note that there is another variable Z that puts each observation into group A or B. Create a scatter plot of A points and then of B points. Describe the relationship between X and Y for each group.
   (c) Find the correlation between X and Y for each group.
   (d) What is the lesson here?
7. Import the data from data set `corrExerciseB`.
   (a) Find the correlation between X and Y.
   (b) Note that observations are put into one of four groups: A, B, C, and D. Find the mean of X and the mean of Y for each group.
   (c) Create a scatter plot of the means $(\bar{X}_A, \bar{Y}_A)$, $(\bar{X}_B, \bar{Y}_B)$, ... and then find the correlation between the $\bar{X}'$s, and the $\bar{Y}'$s. Compare to (a).

   Ecological Correlation Correlations based on rates or groups are often higher than correlations based on individuals. This is a common problem in the social/behavior sciences where many data sets are based on summaries (e.g., census data for the 50 states: mean income levels, mean literacy rate, etc.).

8. Compare the roundoff error of two ways of computing sample variances. Write functions that compute `(mean(x^2)- mean(x)^2) * n/(n-1)` and `mean((x-mean(x))^2) * n/(n-1)` and calculate the variance of $x_1 = c$, $x_2 = c + 1$, $x_3 = c + 2$ for $c = 0$, $c = 10^5$, $c = 10^6$, ... using both. What do you find? Compare to the R `var` function.
9. Verify that $\sum_{i=1}^{n}(y_i - \hat{y}_i) = 0$ — that is, the average of the residuals sum to 0. *Hint*: Use the fact that a and b are solutions to $\partial g/\partial a = 0$, where $g(a, b) = \sum_{i=1}^{n}(y_i - (a + bx_i))^2$.
10. Let $s_{\hat{y}}$ denote the standard deviation of the predicted y's; that is, the standard deviation of $\hat{y}_1, \hat{y}_2, \ldots, \hat{y}_n$.
    (a) Verify that $s_{\hat{y}} = rs_y$ (Equation 9.13).
    (b) Let s_e denote the standard deviation of the residuals, $e_i = y_i - \hat{y}_i$, $i = 1, 2, \ldots, n$. Show that
    $$s_e = \sqrt{1 - r^2} s_y.$$
11. Suppose the height and weight of 30 girls in Sodor are measured. The mean and standard deviation of the heights are 46 and 7 in., respectively, and the mean and standard deviation of the weights are 94 and 15 pounds, respectively. Suppose the correlation between height and weight is 0.75.
    (a) Find the equation of the least-squares regression line of weight against height.
    (b) Find the predicted weight of a girl who is 5 ft. tall.
    (c) Find R-squared and give a sentence interpreting this statistic.
12. Refer to the Beer and Hot Wings Case Study in Section 1.8.
    (a) Create a scatter plot of beer consumed against hot wings eaten and find the correlation between these two variables.
    (b) Find the equation of the least-squares regression line (take `hotwings` as the independent variable) and give a sentence interpreting the slope.
    (c) Compute R-squared and state the interpretation of this statistic.
13. Figure 9.26 contains residuals plot for several least-squares regression models. Describe the unusual features.
14. Is there a relationship between female literacy and birth rate? The data set `Illiteracy` contains data on a sample of countries where female illiteracy is more that 5%. The variable `Illit` is the percentage of women over 15 years of age who are illiterate (2003) and the variable `Births` is the number of births per woman in that country (2005).[4]

[4] `http://www.unesco.org`, `www.data.worldbank.org`.

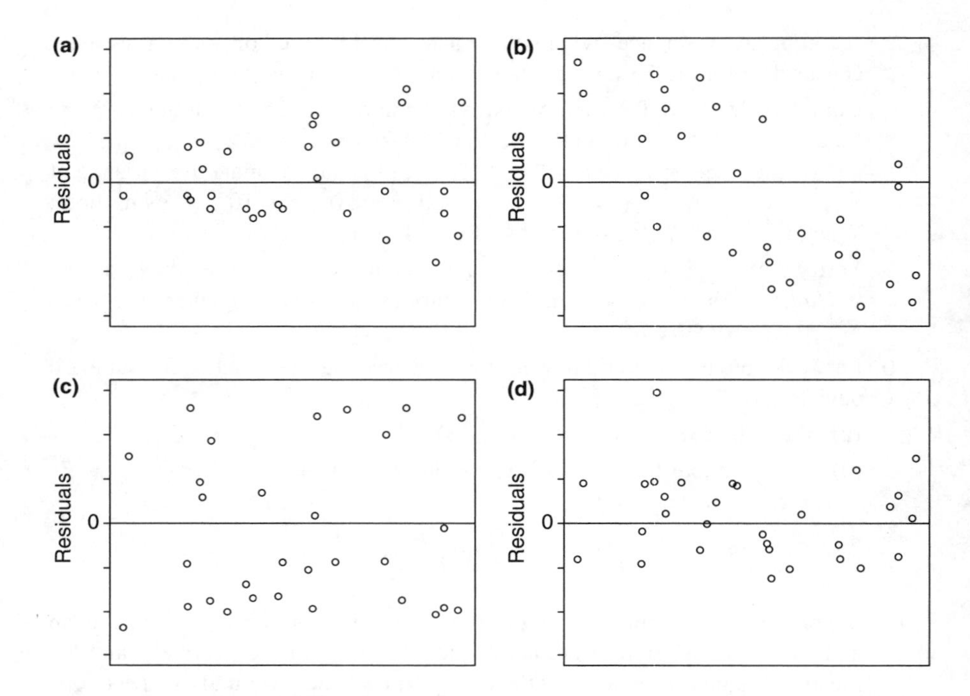

FIGURE 9.26 Residuals plot for regression models.

(a) Create a scatter plot of birth rate against illiteracy and comment on the relationship.

(b) Find the equation of the least-squares line and interpret the slope and r^2.

(c) Create a residuals plot and comment on the appropriateness of a straight line model.

(d) Can we say that improving literacy (reducing illiteracy) will cause the birth rate to go down? Explain.

15. The data set `Volleyball2009` contains data on 30 Division I women volleyball teams from the 2009 season (from `http://web1.ncaa.org/stats/StatsSrv/rankings`).[5]

    (a) Create a scatter plot of the number of kills per set (`Kills`) against assists per set (`Assts`) and describe the relationship.

    (b) Find the least-squares equation for the line and interpret the slope and r^2.

    (c) Create the residuals plot and comment on the appropriateness of a straight line model.

16. Refer to Exercise 15.

    (a) Find a 95% confidence interval for the slope.

    (b) Suppose a team not listed records 12.4 assists per set. Find the predicted number of kills per set and a 95% prediction interval.

[5]© 2009 American Volleyball Coaches Association; © 2009 National Collegiate Athletic Association.

17. Refer to the North Carolina Births Case Study in Section 1.2.
    (a) Create a scatter plot of weight against gestation period and compute the correlation between the two variables.
    (b) Find the least-squares regression line.
    (c) Give a sentence with the interpretation of the slope and the R-squared.
    (d) Create the residuals plot and comment on any unusual features. Is a linear model appropriate for the relationship between gestation period and weight?
18. North Carolina Births (continued)
    (a) What is the estimate of σ for the linear model of weight against gestation period?
    (b) Find a 95% confidence interval for the true slope β.
19. In the ice cream example (Example 9.6), the sum of squares for the sugar variable (ss_x) is 324.446 and the residual standard error is 2.069. Find a 95% confidence interval for the true slope (use the model including all the data).
20. Suppose a company offers tutoring to students interested in improving their math SAT scores. From a random sample of 100 previous customers, they find that $\widehat{\texttt{Score}} = 502.7 + 1.45 \cdot \texttt{Hours}$, where `Score` is the SAT math score and `Hours` is the number of hours the student was tutored. The R-squared and residual standard error for this model is 0.855 and 16.54, respectively.
    (a) For every additional 10 h of tutoring, what is the corresponding change in the test score?
    (b) If the standard deviation of the hours variable is 27.5, what is the standard deviation of the score variable?
    (c) Find a 95% confidence interval for the true slope.
    (d) Find a 95% confidence interval for the mean score for students who are tutored 50 h. Assume the mean number of hours tutored is 55 h.
21. The Mauna Loa Observatory, located on the Hawaiian islands of Mauna Loa, specializes in research in the atmospheric sciences. The facility has been collecting data on carbon dioxide levels since 1950s. The data file `Maunaloa` contains data on average CO_2 levels (ppm) for the month of May from 1990 to 2010 (from `www.esrl.noaa.gov/gmd/ccgg/trends/`).
    (a) Create a scatter plot of CO_2 levels against year and describe the relationship.
    (b) Find the equation for the least-squares equation.
    (c) Plot the residuals against year. Is a straight line model appropriate? Discuss.
22. The data set `Walleye` from the Minnesota Pollution Control Agency contains data on length (inches) and weight (pound) measurements for a sample of 60 walleye caught in Minnesota lakes during 1990s (Monson (2010)).
    (a) Create a scatter plot of weight against length. Does the relationship appear linear?

In fact, biologists have determined that the relationship between length and weight of fish is given by $W = aL^b$, where a and b depend on the species (Ricker (1973, 1975)). We will consider $\log(W) = \log(a) + b\ \log(L)$ (base 10).

(b) Transform the weight and length variables by log base 10 and then create a scatter plot of log(`Weight`) against log(`Length`). Describe the relationship.

(c) Use least-squares to find estimates of a and b based on this sample.

(d) What is the 95% confidence interval for b?

23. Import from the data set `Alelager` the data on alcohol content (per volume) and calories for a sample of beers (12 ounces). Find the correlation between alcohol and calories and then compute a 95% bootstrap percentile confidence interval for the true correlation.

24. Using the data set `Illiteracy`,

    (a) find the correlation between illiteracy rates and births via the bootstrap and also find a 95% bootstrap percentile interval; and

    (b) using a permutation test, find if illiteracy rates and births are independent.

25. Prove Proposition 9.2

26. Prove the second part of Proposition 9.4.

27. Prove Proposition 9.5.

28. Prove the results about $\bar{Y}$ in Theorem 9.6.

29. In Theorem 9.3, we stated without proof that $\hat{\beta}$ and $\bar{Y}$ are independent. Instead, prove that they are uncorrelated.

30. A campaign manager conducts a survey to gauge voter support for his candidate Lopez. He gathers data on the age of a registered voter (x) and whether this person supports Lopez ($Y = 1$) or somebody else ($Y = 0$). An analysis yields the following logistic equation:

$$\ln\left(\hat{p}/(1-\hat{p})\right) = -0.324 + 0.012x,$$

    where p is the probability of a vote for Lopez.

    (a) Find the estimated probability that a 21-year-old voter will vote for Lopez.

    (b) Compare the odds of support for Lopez between two people who are 10 years apart in age and give your answer in a complete sentence.

    (c) At what age is the expected response equal to 0.5?

31. On January 28, 1986, the space shuttle Challenger exploded during lift-off, killing all seven astronauts aboard.[6] In the follow-up investigation, attention was focused on the rubber O-rings that sealed the booster rockets. Engineers

[6] `http://history.nasa.gov/sts51l.html`.

had concerns earlier that the ambient temperature at the time of lift-off could affect the integrity of the O-ring. The data set `Challenger` contains data on 23 Challenger flights before the January 21 flight. The binary variable `Incident` records 1 if one of the O-rings on one of the booster rockets was damaged on this flight. The variable `Temperature` records the temperature (Fahrenheit) at the time of lift-off.

(a) Find the logistic regression equation modeling the log-odds of an O-ring incident against temperature. Plot the graph of the predicted probabilities against temperature and add the observed incidents also.

(b) How does a 10°F degrees decrease in temperature affect the odds of an incident? State your answer in a complete sentence.

(c) On the day of the Challenger accident, the temperature was 33°F. What is the predicted probability of an O-ring incident?

(d) Some would argue that it is not appropriate to use this model to predict an O-ring incident at 33°F. Why not?

32. The biologist in the Black Spruce Seedings Case Study in Section 1.9 was also interested in the relationship between seedling growth and water table depth. The data set `Watertable` contains data on a sample of the seedlings. The variable binary `Alive` indicates 1 if the seedling was alive at the end of the second year of the study and 0 otherwise. The variable `Depth` gives the depth of the water table (centimeter).

(a) Find the logistic equation modeling the log-odds of a seedling being alive against water table depth. Plot the graph of the predicted probabilities against depth and add the observed data points.

(b) Interpret the slope of the regression equation (in terms of odds).

(c) For a seedling growing in soil with a water table depth of 15 cm, find the predicted probability of being alive at the end of the second year.

33. Import from the data set `Phillies2009` the data on the Philadelphia Phillies baseball team.

(a) Find the logistic equation modeling the log-odds of the team winning (`Outcome`) against the number of hits in a game (`Hits`). (`R` will automatically convert `Lose` to 0 and `Win` to 1).

(b) Interpret the slope of the equation (in terms of odds).

(c) Find a 95% bootstrap percentile interval for the slope.

(d) Predict the probability of winning if the team has 17 hits and then find a 95% bootstrap percentile interval.

(e) For inference, we need to assume that the observations are independent. Is that condition met here?

34. Import from the data set `Titanic` the data on male passengers on the Titanic. The variable `Survived` is 1 if the passenger survived the sinking and 0 if the passenger died.

(a) Find the logistic equation modeling the log-odds of a male passenger surviving against age.
(b) Compare the odds of survival for a 30-year-old male with a 40 year old male.
(c) Find a 95% bootstrap percentile interval for the slope.
(d) Estimate the probability of a 69-year-old male surviving, and find a 95% bootstrap percentile interval for the probability.

35. Verify that Equation 9.20 is equivalent to $p_i = 1/\left(1 + e^{-(\alpha+\beta x)}\right)$.

10

BAYESIAN METHODS

You may have opened your email to find an inbox filled with offers for cheap Viagra, discount software, and alluring (?) new companions. And the problem used to be far worse. Many email providers have made a huge dent using Bayesian spam filtering.

Since certain words are more likely to occur in spam mail than legitimate mail, spam filters assign messages containing these words a higher probability of being spam. Certain senders get lower or higher probabilities. The Bayesian filtering combines information from all the words and other characteristics of the email to assign a probability that each message is spam. If that probability exceeds a certain threshold, the message may be sent directly to the trash.

More generally, Bayes offers a mechanism for combining information from multiple sources. We do this all the time in our real lives (though we may not do it well). What is the probability that our team will win the next game? We combine what we know about the strengths of the two teams, any injuries, where the game is played, how hot key players have been, and so on to come up with an estimate.

In statistics, Bayes offers a way to combine *prior information* with information provided by the data. The Bayesian approach contrasts to the *frequentist* approach. The confidence intervals and significance tests we have done up to this point are frequentist techniques, based on what would happen under repeated sampling from the population.

Bayesian answers are often easier to understand. For example, a 95% confidence interval means that if the sampling procedure and interval calculation procedure were repeated many times, 95% of the intervals would include the true parameter. In

Mathematical Statistics with Resampling and R, First Edition. By Laura Chihara and Tim Hesterberg.

contrast, using Bayes we produce an interval that has a 95% probability of including the parameter. Similarly, in hypothesis testing, a P-value is the probability that repeated the sampling and test statistic calculation would give a result as extreme as the actual data. In contrast, using Bayes we compute the probability that each hypothesis is true.

10.1 BAYES' THEOREM

The starting point for Bayesian methods is Bayes' theorem,

$$P(\theta \mid X) = \frac{P(\theta)P(X \mid \theta)}{P(X)}, \tag{10.1}$$

where θ is the parameter and X the data. On the face of it, this is trivial, following from two applications of the definition of conditional probability, $P(A \mid B) = P(AB)/P(B)$ or equivalently (also called the multiplication rule) $P(AB) = P(B)P(A \mid B)$:

$$P(\theta \mid X) = \frac{P(\theta X)}{P(X)} = \frac{P(X\theta)}{P(X)} = \frac{P(\theta)P(X \mid \theta)}{P(X)}.$$

Yet this result has some far-reaching implications.

First, Bayesian methods treat θ as random, rather than a fixed (but unknown) parameter, implying that θ has a probability distribution. This is both a major strength and a major weakness of the approach. It allows a scientist to assign his or her own probabilities, a *prior distribution*, to θ, to take advantage of past experience. For instance, in drug discovery companies screening thousands of compounds for efficacy against a particular cancer may use any prior information they have about which compounds are more or less likely to be effective in deciding which compounds to study further. But conversely, allowing a scientist to assign his or her own probabilities means that different people may get different results when analyzing the same data—just as different people have different estimates of the home team winning. Hence, Bayesian methods are viewed with suspicion in some settings. For instance, in Phase III clinical trials, the final phase testing new drugs, the results should stand on their own, based just on the data rather than on prior beliefs of the sponsor. One counter to that disadvantage is the use of "noninformative" prior information, a type of artificial prior information. In some cases these methods yield the same answers as frequentist methods. Bayesian methods have some other advantages. They are not subject to the same multiple testing issues as frequentist approaches, and they offer a logical self-consistent framework.

10.2 BINOMIAL DATA, DISCRETE PRIOR DISTRIBUTIONS

We begin with cases where both θ and X are discrete, in particular where the data are binomial and θ is the probability of success. In Bayes' theorem, Equation 10.1, we refer to $P(\theta)$ as the *prior*, $P(X \mid \theta)$ as the *likelihood*, and $P(\theta \mid X)$ as the *posterior*. We may restate the equation as

$$\text{Posterior} = \frac{\text{Prior} \times \text{Likelihood}}{\text{Data}}. \tag{10.2}$$

Suppose that θ has k possible values, $\theta_1, \theta_2, \ldots, \theta_k$, and let A_j be the event that $\theta = \theta_j$; these are mutually exclusive events whose union is the whole sample space. By the law of total probability,

$$\begin{aligned} P(X) &= P(XA_1) + P(XA_2) + \cdots + P(XA_k) \\ &= P(A_1)P(X \mid A_1) + P(A_2)P(X \mid A_2) + \cdots + P(A_k)P(X \mid A_k) \\ &= \sum_{j=1}^{k} P(A_j)P(X \mid A_j). \end{aligned} \tag{10.3}$$

Thus, Equation 10.1 can be expressed as

$$P(\theta = \theta_j \mid X) = \frac{P(\theta = \theta_j)P(X \mid \theta = \theta_j)}{\sum_{i=1}^{k} P(A_i)P(X \mid A_i)}$$

$$= \frac{P(\theta = \theta_j)P(X \mid \theta = \theta_j)}{\sum_{i=1}^{k} P(\theta = \theta_i)P(X \mid \theta = \theta_i)} \tag{10.4}$$

$$= \frac{\text{Prior} \times \text{Likelihood}}{\sum \text{Prior} \times \text{Likelihood}}. \tag{10.5}$$

This means that the posterior is proportional to the product of the prior and the likelihood, and that the denominator is obtained by adding the numerator across all possible values for θ. We call the denominator a *normalizing constant*. In general, a normalizing constant makes the distribution add or integrate to 1.

Example 10.1 Suppose you are playing with someone new in tennis and you do not know who is stronger, but you suspect you are. Let θ be the probability that you win a single game. For now we will treat θ as discrete, with possible values $0, 0.1, \ldots, 1.0$, and suppose you guess that the corresponding probabilities are

θ:	0.0	0.1	0.2	0.3	0.4	0.5	0.6	0.7	0.8	0.9	1.0
Prior $P(\theta)$:	0.00	0.02	0.03	0.05	0.10	0.15	0.20	0.25	0.15	0.05	0.00

In other words, before you play, you believe that there is a 2% chance that your probability of winning any given game is 10%, a 3% chance that your winning probability is 20%, and so on. Suppose you win the first time (and you are feeling pretty smug), but then you lose twice (ouch). How does this information change your belief about the probabilities for θ? For example, what is the posterior probability that $\theta = 0.2$?

The likelihood of one win and two losses (WLL) given $\theta = 0.2$ is $(0.2)^1(0.8)^2 = 0.128$. So, the posterior probability is

$$P(\theta = 0.2 |\ \mathrm{WLL}) = \frac{P(\theta)P(WLL \mid \theta)}{P(WLL)} = \frac{(0.03)(0.127)}{P(WLL)}.$$

To complete this calculation, we need the normalizing constant, the marginal probability $P(WLL)$. The following table shows the calculations:

θ	Prior	Likelihood $\theta(1-\theta)^2$	Prior × Likelihood	Posterior Previous/**0.0862**
0.0	0.00	0.000	0.0000	0.0000
0.1	0.02	0.081	0.0016	0.0188
0.2	0.03	0.128	0.0038	0.0446
0.3	0.05	0.147	0.0073	0.0853
0.4	0.10	0.144	0.0144	0.1671
0.5	0.15	0.125	0.0188	0.2176
0.6	0.20	0.096	0.0192	0.2228
0.7	0.25	0.063	0.0158	0.1827
0.8	0.15	0.032	0.0048	0.0557
0.9	0.05	0.009	0.0004	0.0052
1.0	0.00	0.000	0.0000	0.0000
Sum	1		**0.0862**	1

By the law of total probability,

$$\begin{aligned} P(WLL) &= P(\theta = 0.1)P(WLL \mid \theta = 0.1) + P(\theta = 0.2)P(WLL \mid \theta = 0.2) \\ &\quad + \cdots + P(\theta = 0.9)P(WLL \mid \theta = 0.9) \\ &= 0.0862, \end{aligned}$$

which is the sum of the entries in the fourth column (Prior × Likelihood). So, the posterior probability for $\theta = 0.2$ is

$$P(\theta = 0.2 \mid \mathrm{WLL}) = \frac{(0.03)(0.128)}{0.0862} = 0.0445.$$

In other words, prior to seeing any data, you thought there was a 3% chance that your long-term winning proportion is 0.2. After winning just one of the three games, you now believe that there is a 4.46% chance that the proportion is 0.2.

Similarly, you thought that $P(\theta = 0.7) = 0.25$, that is, there was a 25% chance that in the long run you would win an average of 70% of the games.

$$\begin{aligned} P(\theta = 0.7 \mid WLL) &= \frac{\text{Prior} \times \text{Likelihood}}{P(WLL)} \\ &= \frac{0.25 \times 0.063}{0.0862} \\ &= 0.1827, \end{aligned}$$

so now you think there is an 18.28% chance of winning 70% of games.

Before the games, you thought your overall chances of winning were

$$\mathrm{E}\,[\theta] = \sum \theta \times \text{Prior} = \sum_{i=0}^{10} \theta_i \times P(\theta_i) = 0.598,$$

almost 60%. Now, you think they are

$$\mathrm{E}\,[\theta \mid WLL] = \sum \theta \times \text{Posterior} = \sum_{i=0}^{10} \theta_i \times P(\theta_i \mid WLL) = 0.523;$$

you still think you are better!

R Note:

```
> theta <- seq(0, 1, by = .1)
> prior <- c(0, .02, .03, .05, .1, .15, .2, .25, .15, .05, 0)
> likelihood <- theta * (1 - theta)^2
> constant <- sum(prior * likelihood)
> posterior <- prior * likelihood / constant
> posterior
 [1] 0.000000000 0.018802228 0.044568245 0.085306407 0.167130919
 [6] 0.217618384 0.222841226 0.182799443 0.055710306 0.005222841
[11] 0.000000000
> sum(theta * prior)              # prior mean
[1] 0.598
> sum(theta * posterior)          # posterior mean
[1] 0.5229805
```

But what if you play five more times and lose three more? There are two different ways to calculate this—either start from scratch and use all the data, or use the current posterior as a new prior and work with just the new data. They are equivalent. Results are shown in Figure 10.1, using the following R code. Your posterior mean is now 0.49—you are not so cocky any more!

R Note:

```
> likelihood2 <- theta^3 * (1 - theta)^5    # 3 success, 5 fail
> constant2 <- sum(prior * likelihood2)
> posterior2 <- prior * likelihood2 / constant2
> posterior2
[1] 0.0000000000 0.0056870025 0.0378705884 0.1092609179
[5] 0.2396498173 0.2821578712 0.2130220598 0.1003416593
[9] 0.0118345589 0.0001755248 0.0000000000
```

Now, using the previous posterior as a prior, we add 2 wins and 3 losses:

```
> likelihood3 <- theta^2 * (1 - theta)^3
> constant3 <- sum(posterior * likelihood3)
> posterior3 <- posterior * likelihood3 / constant3
> posterior3                          # not shown, matches posterior2
> sum(theta*posterior2)               # posterior mean
[1] 0.485538
```

This code reproduces Figure 10.1. The `type="b"` argument gives **b**oth lines and points.

```
> plot(theta, prior, type = "b", ylim = c(0, max(posterior3)),
     ylab = "probability")
> lines(theta, posterior, type = "b", lty = 2)
> lines(theta, posterior2, type = "b", lty = 3)
> legend("topleft", legend = c("prior", "posterior1", "posterior2"),
     lty = 1:3)
```

□

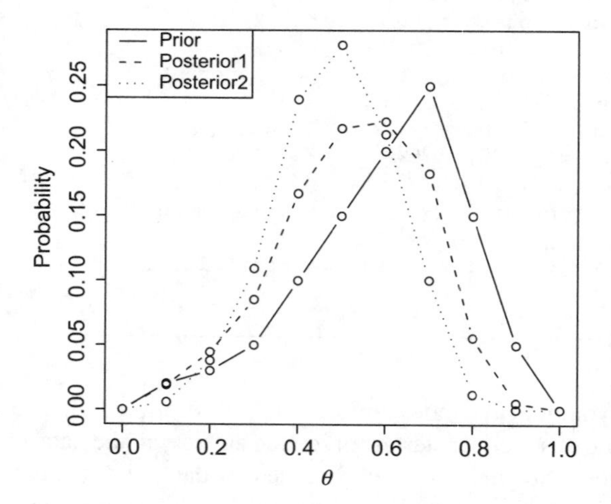

FIGURE 10.1 Prior distribution, and posterior distributions after three games (one win) and eight games (three wins), for tennis example.

TABLE 10.1 Calculations for Example

θ	Prior	Likelihood $\theta^2(1-\theta)$	Prior × Likelihood	Posterior Previous/**0.1177**
0.4	0.35	0.096	0.0336	0.2855
0.5	0.50	0.125	0.0625	0.5310
0.6	0.15	0.144	0.0216	0.1835
Sum	1		0.1177	1

Example 10.2 Suppose you are conducting an exit poll during a state governors election. You survey every fifth voter leaving the polling station and ask him whether he voted for candidates Cobb or Moore. Before the election there appears to be a slight edge for Cobb, so you decide to assume that there is a 50% chance that Cobb will get 50% of the votes, but a 35% chance that he will get 40% and a 15% chance of getting 60% of the votes. Let θ be Cobb's percentage of the votes. If the first three voters that you survey tell you that they voted for Cobb, Cobb, and Moore, respectively, what is the posterior probability of $\theta = 0.4$?

Solution Given $\theta = 0.4$, the likelihood of CCM is $0.4^2(1-0.4)^1 = 0.096$. Thus, the posterior probability is

$$P(\theta = 0.4 \mid CCM) = \frac{P(\theta = 0.4)P(CCM \mid \theta = 0.4))}{P(CCM)} = \frac{(0.30)(0.096)}{P(CCM)}.$$

Again, $P(CCM)$ is found by summing the entries in the column labeled Prior × Likelihood (see Table 10.1), so the posterior probability is $P(\theta = 0.4 \mid CCM) = 0.2855$. Thus, after surveying three voters, your belief that Cobb will get only 40% of the vote drops from 35% to 28.6%.

Figure 10.2 shows two Venn diagrams for the governor's exit poll (Example 10.2). In both sides, the area of a box equals the joint probability $P(\theta, X)$ for the values of θ and X (the number of people who voted for Cobb) shown in the box. Figure 10.2a has box heights equal to the prior probabilities and widths equal to conditional probabilities for X given θ. To find $P(X = 2)$, we add the areas for boxes with $X = 2$. Now look at Figure 10.2b at the column of boxes above the "2" at the bottom. These are versions of those $X = 2$ boxes from Figure 10.2a—same area, but different shapes. Their width is $P(X = 2)$ and areas are $P(\theta = \theta_j, X = 2)$, so their heights must be the posterior probabilities $P(\theta = \theta_j \mid X = 2)$.

Compare the heights in Figure 10.2a (the prior) to the heights above $X = 2$ in Figure 10.2b (the posterior given two votes for Cobb). We see that two votes for Cobb makes the small values of θ less likely and the large values of θ more likely. □

Remark

- The posterior must always add to 1 (or integrate to 1, for densities).
- Multiplying all likelihoods by the same constant does not change the posterior.

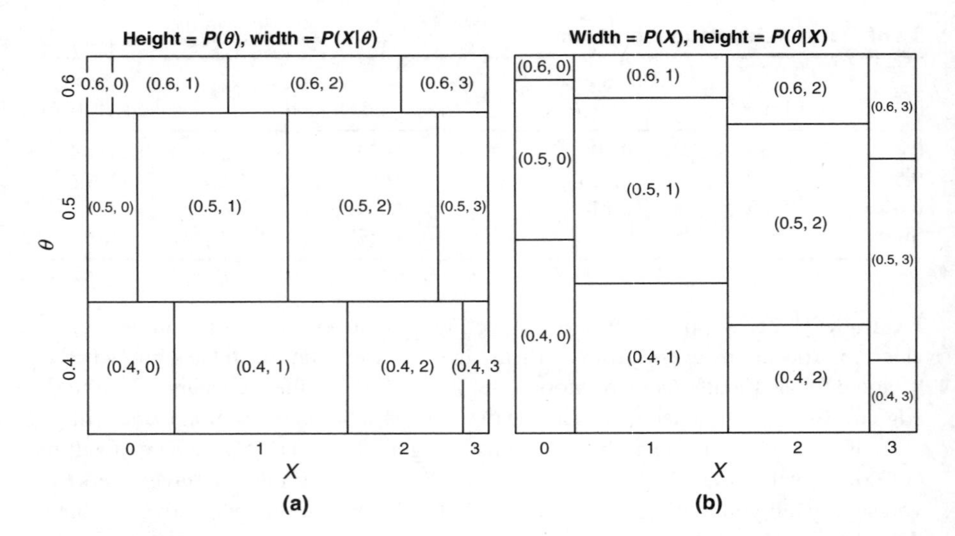

FIGURE 10.2 Venn diagrams with box areas showing joint probabilities for θ and X. Each box has area equal to the joint probability $P(\theta = \theta_j, X = x)$. (a) Box heights are $P(\theta)$ and widths are $P(X \mid \theta)$. (b) Box widths are $P(X)$ and heights are $P(\theta \mid X)$, the result of Bayes' theorem.

For instance, the calculations in the previous example were done assuming that order matters—you recorded voter preferences in the order they came out of the polling station. Suppose instead that we just know that out of the first three voters surveyed, two voted for Cobb. Then, if we assume that $\theta = 0.40$, then the likelihood is $\binom{3}{2}0.4^2(1-0.4)^1$. So all the likelihoods in Table 10.1 would have an extra factor of $c = \binom{3}{2}$. However, this extra factor would carry over to the next column and to the calculation of the marginal probability $P(CCM)$ and thus be canceled out in the end.

		Likelihood		Posterior
θ	Prior	$\theta^2(1-\theta)$	Prior × Likelihood	Previous/**0.1177*c***
0.4	0.35	0.096*c*	0.0336*c*	0.2855
0.5	0.50	0.125*c*	0.0625*c*	0.5310
0.6	0.15	0.144*c*	0.0216*c*	0.1835
Sum	1		0.1177*c*	1

- Multiplying all the priors by the same constant also does not change the posterior probability. Hence, we can be careless about specifying priors, they need not add to 1. We may use an *improper prior*. This is particularly useful in some calculations with continuous distributions below, where we let the prior be a function that integrates to ∞. So the posterior is unaffected by constants in the prior or likelihood; what is important is that it is proportional to their product:

POSTERIOR $\propto$ PRIOR $\times$ LIKELIHOOD

The prior distribution is proportional to the prior times the likelihood, $P(\theta \mid X) \propto P(\theta)P(X \mid \theta)$.

- We can be careless about prior distributions in another way—they can include some impossible values, for example, negatives values of the rate parameter λ for an exponential distribution. The likelihood is zero for negative λ, so it will not affect the posterior. Still, including such negative values adds nothing but confusion and may be a sign of an error. ||

Example 10.2 (Continued). Suppose after having observed CCM, you poll four more voters and their preferences are CMCC. What is the new posterior distribution?

Solution After having observed CCM, you have updated your beliefs about the probabilities for θ—they are the values in the posterior column of Table 10.1. These values become your new *priors* and the likelihoods are computed via $\theta^3(1-\theta)$.

θ	Prior	Likelihood $\theta^3(1-\theta)$	Prior $\times$ Likelihood	Posterior Previous/**0.0600**
0.4	0.2855	0.0384	0.0109	0.1827
0.5	0.5310	0.0625	0.0332	0.5531
0.6	0.1835	0.0864	0.0159	0.2642
Sum	1		0.0600	1

Thus, we now think there is a 18.3% chance that Cobb will receive 40% of the votes. Alternately, we could start over, work with the original prior, and analyze all the data at once, using the sequence CCMCMCC. Then, we can fill out the table using the likelihood $\theta^5(1-\theta)^2$:

θ	Prior	Likelihood $\theta^5(1-\theta)^2$	Prior$\times$Likelihood	Posterior Previous/**0.0071**
0.4	0.35	0.0037	0.0013	0.1827
0.5	0.50	0.0078	0.0039	0.5531
0.6	0.15	0.0124	0.0019	0.2642
Sum	1		0.0071	1

Note that we obtain exactly the same posterior probabilities whether we analyze the data sequentially or all together. We come back to this point in Section 10.5. □

10.3 BINOMIAL DATA, CONTINUOUS PRIOR DISTRIBUTIONS

In the above examples, we used discrete prior distributions. This may be fine as a rough approximation, but is bad in the longer term. We essentially claimed that it is impossible for the probability θ to be 0.45 or any other value between 0.4 and 0.5, for example. That is just wrong.

To avoid this, we need to use a continuous prior distribution. Instead of writing $P(\theta)$, we will write $\pi(\theta)$, where π is some density (or an improper prior), and write $p(\theta \mid X)$ for the posterior density.

For the data, we will work with the binomial model $X \sim \text{Binom}(n, \theta)$, so the likelihood is

$$\begin{aligned} P(X = x \mid \theta) &= \binom{n}{x} \theta^x (1-\theta)^{n-x} \\ &\propto \theta^x (1-\theta)^{n-x}. \end{aligned}$$

Then the density for the posterior distribution is

$$p(\theta \mid X = x) \propto \pi(\theta) \times \theta^x (1-\theta)^{n-x}. \tag{10.6}$$

To get rid of the $\propto$ and have an exact formula, we need to divide by the integral of the product of the prior and likelihood

$$p(\theta \mid X = x) = \frac{\pi(\theta) \times \theta^x (1-\theta)^{n-x}}{\int_0^1 \pi(\theta) \times \theta^x (1-\theta)^{n-x}\, d\theta},$$

a continuous analog of Equation 10.5. Depending on what we choose for $\pi(\theta)$, this integral can get nasty. However, it is relatively simple if $\pi(\theta) \propto \theta^c (1-\theta)^d$ for some constants c and d. In this case, we can combine terms to obtain $p(\theta \mid x) \propto \theta^{x+c}$ $(1-\theta)^{n-x+d}$. There is one family of distributions with that form, the beta distributions (see Section B.12.) Some examples of the densities corresponding to different values of the parameters are shown in Figure 10.3. This is a flexible family of distributions, able to accommodate not only expected values—the prior E $[\theta]$ estimates of your overall chance of winning—but also levels of uncertainty.

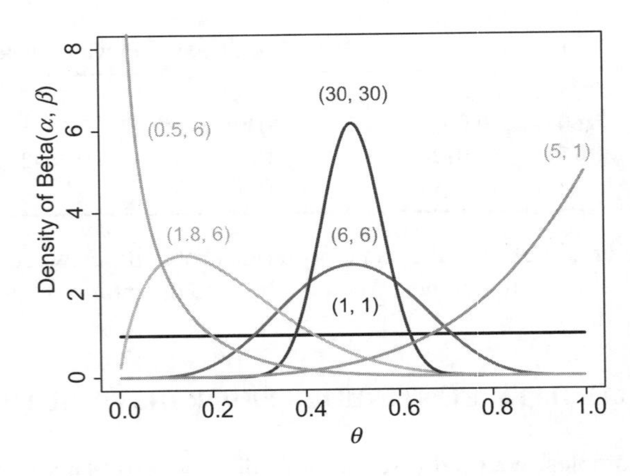

FIGURE 10.3 Beta(α, β) with different α's and β's.

For example, compare the curves for (30, 30) and (6, 6)—they both have a mean of 50%, but the narrower curve expresses a much stronger prior belief that the long-term probability must be close to 50%, while the wider curve expresses much more uncertainty. The curves may be pessimistic (Beta(0.5, 6) has a vertical asymptote at $\theta = 0$) or optimistic (e.g., Beta(5, 1)). This flexibility is useful for specifying prior distributions.

Now using the beta family for prior distributions, we have $\theta \sim \text{Beta}(\alpha, \beta)$, so the prior density is

$$\pi(\theta) \propto \theta^{\alpha-1}(1-\theta)^{\beta-1},$$

the likelihood is

$$P(X = x \mid \theta) \propto \theta^x (1-\theta)^{n-x},$$

and hence the posterior density is

$$\begin{aligned} p(\theta \mid X = x) &\propto \pi(\theta) \times P(X = x \mid \theta) \\ &\propto \theta^{\alpha+x-1}(1-\theta)^{\beta+n-x-1}, \end{aligned}$$

which is the density for $\text{Beta}(\alpha + x, \beta + n - x)$.

The posterior distribution looks similar to the prior distribution, but with the number of successes x added to α and the number of failures $(n - x)$ added to β. We can interpret α and β as giving prior successes and prior failures, with the posterior parameters giving total numbers of successes and failures. From Theorem B.20, we have

BINOMIAL DATA, BETA PRIOR

For the binomial model $X \sim \text{Binom}(n, \theta)$, suppose we place the continuous prior distribution $\text{Beta}(\alpha, \beta)$ on θ. Then the prior mean and the variance for θ are

$$\text{E}[\theta] = \frac{\alpha}{\alpha + \beta}, \tag{10.7}$$

$$\text{Var}[\theta] = \frac{\alpha\beta}{(\alpha+\beta)^2(\alpha+\beta+1)} \tag{10.8}$$

and the posterior distribution of θ given $X = x$ is

$$\theta \mid x \sim \text{Beta}(\alpha + x, \beta + n - x),$$

with posterior mean and variance

$$\text{E}[\theta \mid x] = \frac{\alpha + x}{\alpha + \beta + n}, \tag{10.9}$$

$$\text{Var}[\theta \mid x] = \frac{(\alpha + x)(\beta + n - x)}{(\alpha+\beta+n)^2(\alpha+\beta+n+1)}. \tag{10.10}$$

Remark

- The prior Beta(0, 0) is known as a *noninformative* prior with a posterior mean x/n determined solely by the data. This is an example of an improper prior that integrates to infinity.
- The prior Beta(1, 1) is the standard uniform distribution and is referred to as a *flat prior*.
- We say that the beta prior is a *conjugate family* to the binomial likelihood. A conjugate family is one for which the posterior distribution belongs to the same family as the prior distribution. ||

Example 10.3 You conduct a poll of students and ask "Do you watch the Daily Show with Jon Stewart?" Of the $n = 200$ randomly chosen students you survey, $x = 110$ say "yes." Let θ be the true proportion of students at your institution who would say yes (and assume that 200 is a negligible fraction of the population).

Let us compare the frequentist approach with two Bayesian approaches to this problem.

Frequentist The estimated proportion of students who watch is $\hat{\theta} = x/n = 110/200 = 0.55$. A 95% confidence interval, using the Agresti–Coull interval on page 193, is $\tilde{\theta} \pm 1.96\sqrt{\tilde{\theta}(1-\tilde{\theta})/\tilde{n}}$, where $\tilde{n} = n + 4 = 204$ and $\tilde{\theta} = (x+2)/\tilde{n} = 112/204$. Thus, the 95% confidence interval is $(0.4807, 0.6174)$.

Bayesian A One Bayesian analyst has no information about θ and uses a flat prior. The prior density $\pi(\theta)$ is from Beta(1, 1) with $\alpha = 1, \beta = 1$. Hence, the posterior density $p(\theta \mid X = 110)$ is from Beta$(1 + 110, 1 + 200 - 110) = $ Beta(111, 91). This density is shown in Figure 10.4. This distribution is much narrower than the prior because the data provide a lot of information about θ.

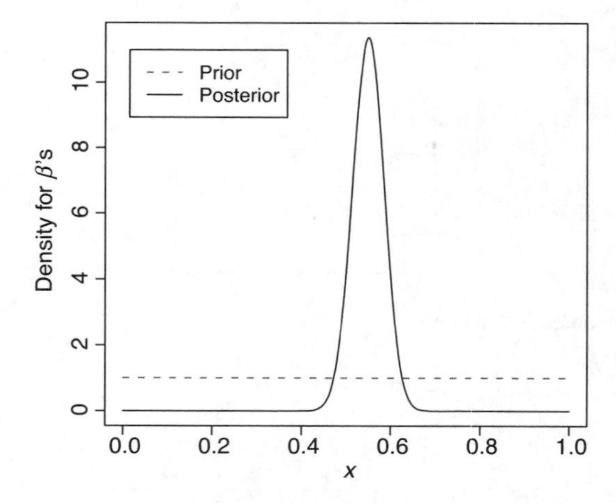

FIGURE 10.4 Pdf for Beta(111, 91), resulting from a uniform prior and 110 successes in 200 observations. The posterior distribution is concentrated near 0.55.

The posterior mean for θ is

$$\mathrm{E}[\theta \mid x = 110] = \frac{1 + 110}{1 + 1 + 200} = 111/202 = 0.5495.$$

This corresponds to a sample proportion with one artificial success and one failure added to the data. The middle 95% of the posterior distribution is (0.4807, 0.6174), nearly identical to the frequentist interval. There is a 95% probability that θ is within this interval. This is known as a *credible interval*, rather than a confidence interval.

CREDIBLE INTERVAL

The Bayesian analog of a confidence interval is a credible interval. The range of the middle 95% of a posterior distribution, between the 0.025 and 0.975 quantiles, is a 95% credible interval.

Since we have the pdf for the posterior distribution, technically, we can actually find any probability we wish, evaluating the integral using statistical software or otherwise.

For instance,

$$P(\theta \geq 0.5 \mid X = 110) = \int_{0.5}^{1} \frac{\Gamma(202)}{\Gamma(111)\Gamma(91)} \theta^{110}(1-\theta)^{90}\, d\theta = 0.9209,$$

where software is used to evaluate the integral. There is a 92% probability that at least 50% of the students at this school like the show.

R Note:

```
> qbeta(.025, 111, 91)
[1] 0.4806705
> qbeta(.975, 111, 91)
[1] 0.6174106
> 1-pbeta(.5, 111, 91)
[1] 0.9209173
```

To create the densities for the prior and posterior distributions in Figure 10.4,

```
curve(dbeta(x, 111, 91), from = 0, to = 1)
curve(dunif(x), add = TRUE, col = "blue", lty = 2)
legend(.05, 11, legend = c("Prior", "Posterior"), lty = c(2, 1),
  col = c("blue", "black"))
```

Bayesian B On the basis of previous polls at this or other institutions, someone expects that 58% of students watch the Daily Show with Jon Stewart, with an uncertainty quantified by a standard deviation of 0.03.

Again, we will use a beta prior. As is common in practice, the prior is given in a form that makes sense for the person stating the prior, rather than in a form convenient for the analysis. Here, we need to find parameters for the beta prior that match the given mean and standard deviation. Using Equations 10.7 and 10.8, we solve the following:

$$\mathrm{E}[\theta] = \frac{\alpha}{(\alpha+\beta)} = 0.58,$$

$$\mathrm{Var}[\theta] = \frac{\alpha\beta}{(\alpha+\beta)^2(\alpha+\beta+1)} = (0.03)^2.$$

This yields $\alpha = 156.4$ and $\beta = 113.26$. This analyst has quite a bit of prior information, corresponding to about $156 + 113 = 269$ prior observations.

SOLVING TWO EQUATIONS IN TWO UNKNOWNS:

Computer algebra systems such as Mathematica™ or Maple™ can solve the system of two equations and two unknowns above. The web site Wolfram | Alpha™(`http://www.wolframalpha.com`) provides another means for students to solve two equations with two unknowns:

```
Solve[{a/(a+b)=.58, (a*b)/((a+b)^2*(a+b+1))=.03^2}, {a,b}]
```

The resulting posterior probability distribution (Equation 10.9) is

$$\mathrm{Beta}(110 + 156.4, 200 - 110 + 113.26) = \mathrm{Beta}(266.4, 203.26).$$

The moments of the posterior, using Equations 10.9 and 10.10, are the following:

$$\mathrm{E}[\theta \mid X = 110] = \frac{266.4}{266.4 + 203.26} = 0.567,$$

$$\mathrm{Var}[\theta \mid X = 110] = \frac{(266.4)(203.26)}{(266.4 + 2.3.26)^2(266.4 + 203.26 + 1)} = 0.00005.$$

The posterior mean for θ is 0.567, a bit lower than the prior mean, with a standard deviation of 0.0228, 24% smaller than the prior standard deviation of 0.03. The prior and posterior are shown in Figure 10.5.

The 95% credible interval is the range from the 0.025 and 0.975 quantiles of Beta(226.4, 203.26) or (0.5222, 0.6117). The interval is narrower than either of the previous intervals, reflecting the amount of the person's prior information. It is centered to the right of the previous intervals, reflecting this person's prior belief about the true probability.

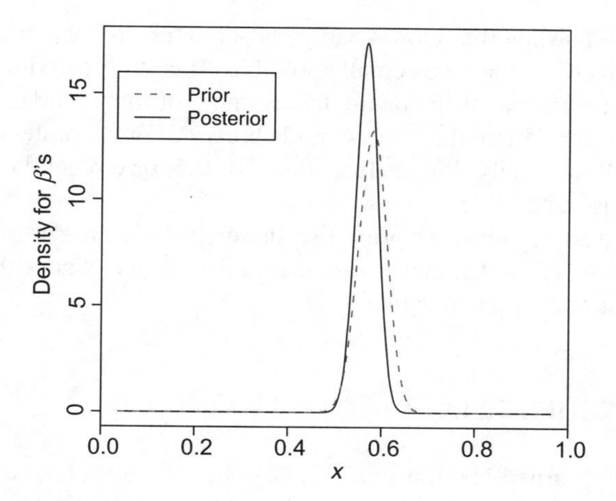

FIGURE 10.5 Pdf's for the prior Beta(156, 113.26) and posterior Beta(266.4, 203.26) for Bayesian B.

Figure 10.6 compares the prior and posterior densities for the two Bayesian approaches.

Even though the prior distributions are quite different, after observing the data, $x = 110$, the posterior distributions are similar. In practice, the data usually swamps the prior—with a reasonable amount of data, the posterior usually depends more on the data than on the prior, as long as the prior distribution is not unreasonable. □

Remark In the previous example, we saw that the posterior distribution depends more on the data than on the prior. This is usually the case in practice—the prior provides a framework to get an answer, but most of the information comes from the data.

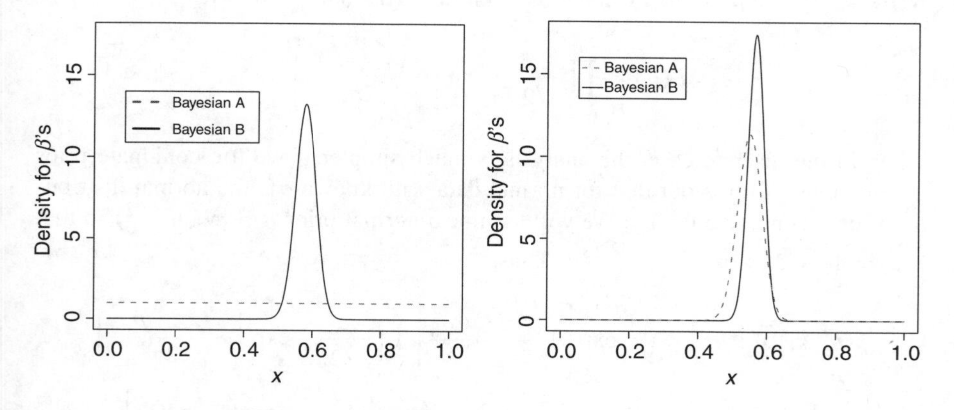

FIGURE 10.6 Comparison of pdf's for the priors of Bayesians A and B, as well as the posteriors of Bayesians A and B.

One exception is when the prior is badly chosen—for instance, the prior at some values of θ is zero when they are actually possible. Then the posterior is always zero at these values, regardless of the data. One example of this is the use of a discrete prior distribution for a binomial success probability θ. For example, our exit poll in Example 10.2 allowed only three values, $\theta = 0.4$, 0.5, or 0.6, so the posterior says that any other θ is impossible.

In the Jon Stewart example, if a prior distribution had been unreasonable, say indicating zero chance of $\theta > 0.1$, then no amount of data would change that conclusion. (Perhaps you know someone like that?) ||

10.4 CONTINUOUS DATA

Most of what we learned previously about Bayesian methods for discrete data also applies to continuous data, with one notational change—instead of working with probabilities for the data, we need to work with densities for the data. In this section, we also focus on continuous θ, the more common case.

Let $\pi(\theta)$ denotes the pdf for the prior distribution and f the density for the data's distribution. Then Bayes' theorem becomes

$$\begin{aligned} p(\theta \mid x) &= \frac{\pi(\theta) f(x \mid \theta)}{f(x)} \\ &= \frac{\pi(\theta) f(x \mid \theta)}{\int \pi(\theta) f(x \mid \theta)\, d\theta} \qquad (10.11) \\ &\propto \pi(\theta) f(x \mid \theta). \end{aligned}$$

We begin with the case of normal distributions. Suppose the data $x_1, x_2, \ldots, x_n \sim N(\mu, \sigma^2)$, where μ is unknown but σ^2 is known. The likelihood of μ is

$$L(\mu) = \prod_{i=1}^{n} \frac{1}{\sqrt{2\pi}\sigma} \exp\left[-\frac{(x_i - \mu)^2}{2\sigma^2}\right].$$

As in the discrete case, the analysis is much simpler if we use conjugate prior distributions. It turns out that for normal data with known σ^2, the normal distributions are a conjugate family. We will assume a normal prior $\mu \sim N(\mu_0, \sigma_0^2)$ so that $\pi(\theta) = 1/(\sqrt{2\pi}\sigma_0)e^{-(\mu-\mu_0)^2/(2\sigma_0^2)}$. Thus,

$$\begin{aligned} p(\mu \mid x_1, x_2, \ldots, x_n) &\propto \exp\left[-\frac{(\mu - \mu_0)^2}{2\sigma_0^2}\right] \prod_{i=1}^{n} \frac{1}{\sqrt{2\pi}\sigma} \exp\left[-\frac{(x_i - \mu)^2}{2\sigma^2}\right] \\ &= \frac{1}{(\sqrt{2\pi}\sigma)^n} \exp\left[-\frac{(\mu - \mu_0)^2}{2\sigma_0^2} - \sum_{i=1}^{n} \frac{(x_i - \mu)^2}{2\sigma^2}\right]. \end{aligned}$$

Now, a tedious algebraic calculation involving multiplying out the products in the exponents, collecting like terms, and then completing the square results in the posterior density simplifying to

$$p(\mu \mid x_1, x_2, \ldots, x_n) \propto \exp\left[-\frac{(\mu - \mu_1)^2}{2\sigma_1^2}\right], \tag{10.12}$$

where μ_1 and σ_1^2 are given in Equations 10.13 and 10.14. Thus, the posterior distribution of μ is normal, $N(\mu_1, \sigma_1^2)$.

NORMAL DATA, NORMAL PRIOR

Let $x_1, x_2, \ldots, x_n \sim N(\mu, \sigma^2)$, where μ is unknown, but σ^2 is known. Let $\bar{x}$ denote the sample mean.

If the prior distribution of μ is $N(\mu_0, \sigma_0^2)$, then the posterior distribution of μ is

$$\mu \mid x_1, x_2, \ldots, x_n \sim N(\mu_1, \sigma_1^2),$$

where

$$\mu_1 = \frac{1/\sigma_0^2}{\left(1/\sigma_0^2 + n/\sigma^2\right)}\,\mu_0 + \frac{n/\sigma^2}{\left(1/\sigma_0^2 + n/\sigma^2\right)}\,\bar{x} \tag{10.13}$$

and

$$\frac{1}{\sigma_1^2} = \frac{1}{\sigma_0^2} + \frac{n}{\sigma^2}. \tag{10.14}$$

The updated mean μ_1 is a weighted average of the prior mean and the observed data.

Definition 10.1 The reciprocal of the variance, $1/\sigma^2$, is called the *precision* for a normal distribution. ||

You can think of precision as information. The information of the posterior is the sum of the information provided by the prior and the data. Furthermore, the updated mean μ_1 is a weighted average, with weights proportional to the information provided by the prior and the data.

Example 10.4 A marine biologist is investigating a species of trout in a certain area of California. She assumes that the lengths of these fish are normally distributed with mean μ (cm) and variance 8^2. She obtains a random sample of 15 fish and records their lengths, $x_1, x_2, \ldots, x_{15} \sim N(\mu, 8^2)$. Based on her knowledge of this species at other locations, she assumes that the prior distribution is $\mu \sim N(50, 6^2)$. Suppose the mean of this random sample is $\bar{x} = 45$.

We have $n = 15$, $\sigma^2 = 8^2$, $\mu_0 = 50$, $\sigma_0^2 = 6^2$, and μ is unknown. Thus,

$$\begin{aligned}\mu_1 &= \frac{1/36}{1/36 + 15/64}(50) + \frac{15/64}{1/36 + 15/64}(45)\\ &= 0.106(50) + 0.894(45)\\ &= 45.53.\end{aligned}$$

$$\begin{aligned}\frac{1}{\sigma_1^2} &= \frac{1}{36} + \frac{15}{64}\\ &= 0.2622,\end{aligned}$$

which results in a variance of $\sigma_1^2 = 3.8146 = 1.953^2$. Her updated belief about μ is $\mu \mid x_1, x_2, \ldots, x_{15} \sim N(45.53, 1.953^2)$. In this example, the observed mean receives more weight in determining μ_1. The prior and posterior are shown in Figure 10.7.

A 95% credible interval for μ can be obtained by finding the 0.025 and 0.975 quantiles of $N(45.53, 0.1.953^2)$: the probability that the true μ lies in the interval (41.70, 49.36) is 0.95. □

Remark

- The noninformative prior for the normal distribution is $\pi(\theta) = 1$ for all θ (an improper prior).
- The case of normal distributions with unknown σ^2 is more complicated; the conjugate family has inverse gamma distributions for σ^2 (i.e., where the precision $1/\sigma^2$ has a gamma distribution). This is beyond the scope of this book; we

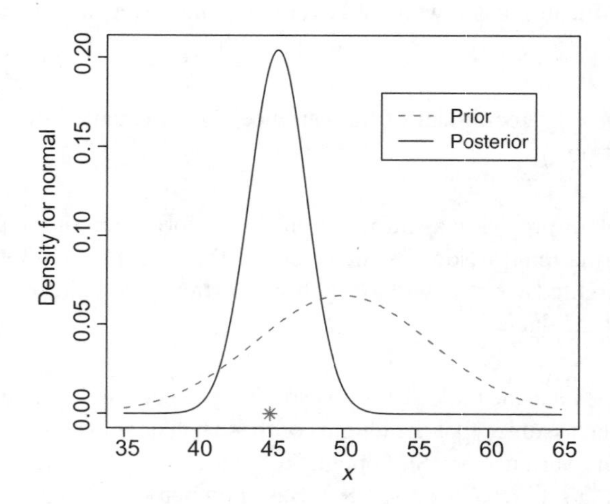

FIGURE 10.7 Prior $N(50, 6^2)$ and posterior $N(45.53, 1.953^2)$ distributions of μ. The observed mean $\bar{x} = 45$ has more weight in determining the mean of the posterior distribution than does the prior mean.

encourage interested students to see Carlin and Louis (2009) or Gelman et al. (2010) for more details. ||

10.5 SEQUENTIAL DATA

We commented at the end of Section 10.2 that when data arrive over time, we may either analyze all the data at once, using the original prior, or use the prior from an earlier step and the new data. This is easiest to show using the notation from continuous distributions (though everything that follows also applies to discrete data if we interpret f as being a density or a probability mass function for continuous and discrete distributions, respectively).

The posterior distribution after observing data x_1 is

$$p(\theta|x_1) = \frac{\pi(\theta)f(x_1|\theta)}{\int \pi(\theta)f(x_1|\theta)d\theta}.$$

The posterior distribution based on data x_2, using the first posterior as a prior, is

$$\begin{aligned}
p_2(\theta|x_1, x_2) &= \frac{p(\theta|x_1)f(x_2|\theta)}{\int p(\theta|x_1)f(x_2|\theta)d\theta} \\
&= \frac{\left(\pi(\theta)f(x_1|\theta)/\int \pi(\theta)f(x_1|\theta)d\theta\right)f(x_2|\theta)}{\left(\int \pi(\theta)f(x_1|\theta)/\int \pi(\theta)f(x_1|\theta)d\theta\right)f(x_2|\theta)d\theta} \\
&= \frac{\pi(\theta)f(x_1|\theta)f(x_2|\theta)}{\int \pi(\theta)f(x_1|\theta)f(x_2|\theta)d\theta} \\
&= \frac{\pi(\theta)f(x_1, x_2|\theta)}{\int \pi(\theta)f(x_1, x_2|\theta)d\theta}
\end{aligned}$$

where the last step assumes that x_1 and x_2 are independent, given θ. This is Bayes' formula for the combined data. In other words, we get the same posterior whether we analyze the data in one step or two. We can iterate this and find that the posterior distribution is the same whether we analyze data one observation at a time or all at once. This self-consistency of the answers is a valued property of Bayesian estimates.

Example 10.5 Suppose from a random sample of 38 students, 10 approve of the new graduation requirements. To estimate θ, the true proportion of students who approve, we will use a beta distribution and assume a flat prior. Thus, the posterior distribution, with $x = 10, n = 38$, and $\alpha = 1, \beta = 1$, is $P(\theta|x = 10) = \text{Beta}(10 + 1, 1 + 38 - 10) = \text{Beta}(11, 29)$. Thus, $\text{E}[\theta|x = 10] = 11/40 = 0.275$.

Suppose we poll an additional 15 students and find 5 approve of the new graduation requirements. We update the posterior using $x = 5, n = 15$, and $\alpha = 11, \beta = 29$ to find $P(\theta|x = 5) = \text{Beta}(5 + 1, 29 + 15 - 5) = \text{Beta}(16, 39)$ Thus, $\text{E}[\theta|x = 10] = 15/53 = 0.283$. On the other hand, suppose we wait until we have all the data. In this case, we have $x = 10 + 5 = 15$ out of $n = 38 + 15 = 53$ who approve of the graduation requirements. Then, starting with the flat prior, the posterior distribution is $P(\theta|x = 15) = \text{Beta}(15 + 1, 1 + 53 - 15) = \text{Beta}(16, 39)$, which matches the two-step answer. □

Sequential data are important in a wide range of applications, including the following:

- Clinical trials, where data from new patients arrive at different times. Rather than waiting for the end of a multiyear trial to analyze all data, the Food and Drug Administration mandates that the data be analyzed periodically. If a new drug or treatment appears to be harming patents, the trial may be stopped early. A trial may be stopped early because the new drug is convincingly better than anything else available, though this is rare because the FDA also wants enough data to look for adverse side effects.
- Google continuously updates its algorithms for producing search results.
- Google Web Optimizer (GWO), described in Section 8.3, reports results to web site owners on demand using the latest data.

Bayesian analysis offers substantial advantages in these situations. Frequentist testing suffers from multiple testing—if you do a significance test many times, each at the 5% level, and keep collecting new data, then eventually a test will come out significant, even when the null hypothesis is true. To compensate for that, frequentist clinical trials must be designed and analyzed with special software that calculates how to adjust critical values and *P*-values due to sequential testing. (One of the authors worked on such software, *S+SeqTrial*; it is complicated and expensive.)

In contrast, Bayesian analysis may be performed as often as desired. Say that we are interested in the difference between θ values for the treatment and control groups $(\theta_{\mathrm{t}} - \theta_{\mathrm{c}})$. Say that a difference of 0.03 would be important in practice. We might stop a trial whenever there is a high probability of an improvement that large, say, $P(\theta_{\mathrm{t}} - \theta_{\mathrm{c}} > 0.03) \geq 80\%$.

Bayesian analysis also offers advantages in *multiple-arm trials*,[1] when three or more things are being compared.

- In clinical trials, instead of just one treatment arm and a control arm, there may be multiple treatment arms at different levels of a drug, or perhaps using combinations of drugs.
- In Google Web Optimizer, web site owners can test multiple arms (different versions of their page) simultaneously, using different combinations of graphics and text content.

For example, a web site owner may choose to stop the experiment if there is a 60% probability that a particular version of the page is better than all other versions

[1] The terminology comes from casinos with *multi-armed bandits*, slot machines with more than one arm—you get to choose which arm to pull to lose your money.

being tested. If the current posterior distribution for each θ_j (for arm j) has a beta distribution with parameters $(\alpha_j + x_j, \beta_j + n_j - x_j)$ and the results from different arms are independent, then a simple simulation using values randomly generated from the posterior distributions can estimate that probability.

For another example, suppose a content publisher (a web site owner) is testing six versions of her home page (arms), and the numbers of impressions (visitors) on each arm and corresponding successes (e.g., purchases or donations) to date are the following:

n	1874	1867	1871	1868	1875	1875
X	52	41	55	49	39	39

(This is artificial data, motivated by GWO.) To estimate the probability that each arm is best, using prior consisting of independent noninformative beta priors, we use the following code:

R Note:

```
n <- c(1874, 1867, 1871, 1868, 1875, 1875)
X <- c(52, 41, 55, 49, 39, 39)
alpha <- X      # vector of posterior parameters
beta <- n - X   # vector of posterior parameters
N <- 10^5                    # replications
theta <- matrix(0.0, nrow = N, ncol = 6)
for (j in 1:6)
{
   theta[, j] <- rbeta(N, alpha[j], beta[j])
}
probBest <- numeric(6)       # vector for results
best <- apply(theta, 1, max) # maximum of each row
for (j in 1:6)
{
   probBest[j] <- mean(theta[, j] == best)
}
```

The vector `probBest` contains the probabilities of each of the six arms being best.

```
> probBest
[1] 0.29254 0.02027 0.49754 0.17027 0.00975 0.00963
```

These commands reproduce Figure 10.8.

```
plot(theta[1:10^4, 1], theta[1:10^4, 3], pch = ".")
abline(0, 1)
text(.037, .042, substitute(theta[3] > theta[1]))
text(.042, .037, substitute(theta[3] > theta[1]))
```

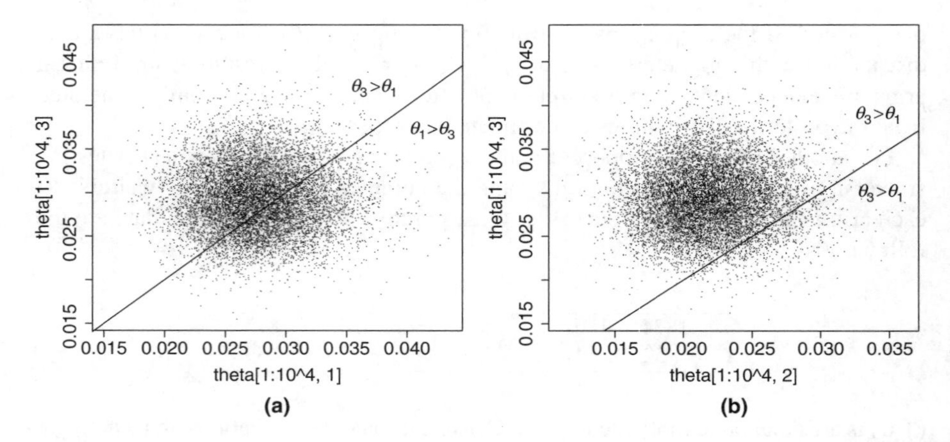

FIGURE 10.8 Posterior distribution for artificial GWO example. (**a**) Joint posterior for θ_1 and θ_3, with regions where each is better than the other. (**b**) θ_2 and θ_3.

Figure 10.8a shows the joint posterior distribution for two of the parameters θ_1 and θ_3. While θ_3 has a good chance of being the better of the two, there is a good chance that θ_1 is better, and possibly substantially so. In contrast, the right panel shows the joint posterior distribution for θ_2 and θ_3; θ_2 has only a small chance of being the better of the two, and even smaller chance of being substantially better.

We estimate that arm 3 has almost a 50% chance of being the best, arm 1 has almost 30%, and the other arms have smaller chances. Given this information, a webmaster might opt to discontinue the three or four worst performing arms, to avoid subjecting further site visitors to those arms and continue testing the better performing arms. After collecting more data, we may update the analysis and eventually eliminate all but one arm.

10.6 EXERCISES

1. We return to your favorite game. A friend thinks the possible values for θ, your long-term probability of winning, are $0.4, 0.5, 0.6$, with probabilities $1/5, 3/5, 1/5$, respectively. Suppose you win four games out of five.

θ	Prior	Likelihood	Prior × Likelihood	Posterior
0.4	1/5			
0.5	3/5			
0.6	1/5			
Sum	1			

   (a) Fill out the rest of this table.

   (b) Give a sentence interpreting the posterior probability for $\theta = 0.6$.

   (c) Find the expected value of θ and the posterior expected value of θ.

2. Another friend is positive that the long-term probabilities are one of 0.4, 0.5, 0.6, with probabilities 1/10, 3/10, 1/10, respectively. (He is not too good at math, these do not add to 1.) The corresponding table, with $c = 1/2$, is

θ	Prior	Likelihood of WLL	Prior × Likelihood	Posterior
0.4	$c/5$	0.144	$0.0288c$	
0.5	$3c/5$	0.125	$0.0750c$	
0.6	$c/5$k	0.096	$0.0192c$	
Sum				

Another friend insists on using binomial probabilities with this table (with $d = 3$):

θ	Prior	Likelihood of One Win	Prior × Likelihood	Posterior
0.4	c/5	0.144d	0.0288cd	
0.5	3c/5	0.125d	0.0750cd	
0.6	c/5k	0.096d	0.0192cd	
Sum				

Verify that the posterior probabilities are the same.

3. Suppose the possible values of θ for your winning probabilities in tennis are $0, 0.1, \ldots, 1$, with equal probabilities.

θ	Prior	Likelihood	Prior × Likelihood	Posterior
0 0	1/11			
0.1	1/11			
0.2	1/11			
0.3	1/11			
0.4	1/11			
0.5	1/11			
0.6	1/11			
0.7	1/11			
0.8	1/11			
0.9	1/11			
1	1/11			
Sum	1			

   (a) Fill out the rest of this table, if you win four game out of seven.
   Hint: Adapt the `R` code on page 305.

   (b) Give a sentence interpreting the posterior probability for $\theta = 0.2$.

   (c) Find the expected value of θ and the posterior expected value of θ.

4. (Exercise 3 continued) Suppose you play another six games and you win five of them.

   (a) Compute the new posterior probabilities using the posteriors from Exercise 3 as your new priors.

(b) Compute the posterior probabilities by considering the data all together that is, by considering 13 games of which you won 9.

5. According the 2002 General Social Survey (Case Study in Section 1.6), 295 out of 1324 surveyed thinks that the U.S. government spends too little on the military.
   (a) Find a 90% confidence interval for the true proportion of people who think that the government spends too little on the military.
   (b) A Bayesian analyst thinks that the true proportion has a beta distribution with a mean of 0.3 and standard deviation 0.02. Find a 90% credible interval for the true proportion.
   (c) For this Bayesian, find the probability that $\theta < 0.23$, given the data.
6. The Pew Research Center conducted a survey in fall 2005 on attitudes toward pets. For those who owned dogs, they asked, "Do you think of your dog as a member of your family?" Of the 178 people who were between 18 and 29 years old, 160 responded yes (see `http://pewresearch.org/pubs/303/gauging-family-intimacy` for full report.[2]).
   (a) Find a 95% confidence interval for the true proportion of 18–29 years old who responded yes.
   (b) A Bayesian statistician at this polling company suspects that the prior for θ, the true proportion, is a beta distribution with mean θ of 0.85 and a variance of 0.0025. A second statistician uses a flat prior. A third statistician thinks the prior is a Beta(6, 4). Find the estimates for θ (the posterior means) and 95% credible intervals.
   (c) For each statistician, plot their prior and posterior distributions on one graph.
   (d) For each statistician, find the probability, given the data, that $\theta > 0.90$.
7. Analyze the Jon Stewart question (page 312) using the noninformative prior.
8. Suppose you want to find the mean μ of the math SAT scores for the class of 2008 high school seniors in your town. You know the distribution of scores is normal with a standard deviation of 116, which is the standard deviation for the national distribution of scores. On the basis of the information on previous classes, you put a normal prior on μ, say $\mu \sim N(600, 25^2)$. If a sample of size 60 yields a mean score of 538,
   (a) find the posterior distribution of μ;
   (b) find a 95% credible interval for the true μ; and
   (c) find the probability that the posterior mean math SAT score is greater than 600.
9. Bone fractures are a common medical problem as we grow older, so there is much interest in determining risk factors. One assessment tool is bone mineral density (BMD), a measure of the amount of certain minerals such as calcium in the bone (in g/cm^2). The lower the BMD, the higher the risk for bone fractures.

[2] © 2006 Pew Research Center: Social and Demographic Trends Project.

Suppose a researcher wants to determine the mean BMD μ in female vegetarians between 30 and 39 years old. On the basis of other research, he believes $\mu \sim N(0.72, 0.08)$. He measures the BMD in the lumbar spine of 18 female vegetarians aged 30 to 39 years old and finds a mean BMD of $\bar{x} = 0.85$ g/cm^2. He will assume that BMD measurements for this population come from a normal distribution with unknown mean μ but known standard deviation $\sigma = 0.15$.

(a) Find the posterior distribution of μ.

(b) Find a 99% credible interval for μ.

(c) Find the posterior probability that μ is less than or equal to 0.8 g/cm^2.

10. A biologist is trying to determine the true mean weight μ of a certain species of fish in Lyman Lake. She is certain that the distribution of weights is normal with known standard deviation $\sigma = 100$ g. Based on prior work, her belief about μ is $\mu \sim N(900, 160^2)$. A random sample of $n = 10$ fish yields a sample mean of $\bar{x} = 970$ g.

(a) Find the posterior distribution of μ.

(b) Find a 95% credible interval for μ.

(c) Suppose she goes out and gets another sample of $n = 15$ fish and computes $\bar{x} = 940$ g. Find the new posterior distribution and a 95% credible interval for μ.

11. Suppose $x_1, x_2, \ldots, x_n \sim N(\mu, 6^2)$ with mean $\bar{x} = 19$.

(a) If $n = 15$ and the prior is $\mu \sim N(25, 5^2)$, find the posterior distribution (mean, standard deviation) of μ and the posterior precision.

(b) If $n = 50$ and the prior is $\mu \sim N(25, 5^2)$, find the posterior distribution of μ and the posterior precision.

(c) If $n = 15$ and the prior is $\mu \sim N(25, 10^2)$, find the posterior distribution of μ and the posterior precision.

(d) Compare the above three outcomes and discuss the impact of sample size and the standard deviation of the prior distribution on the posterior mean, standard deviation, and precision.

12. Suppose $x_1, x_2, \ldots, x_{15}$ are data from a normal distribution and $\bar{x} = 30$ and the prior distribution is normal with mean $\mu_0 = 40$ and $\sigma_0^2 = 5^2$.

(a) Compute the posterior distribution and posterior precision if the data distribution is $N(\mu, 3^2)$.

(b) Compute the posterior distribution and posterior precision if the data distribution is $N(\mu, 10^2)$.

(c) Compare the two outcomes and discuss the impact of the standard deviation of the data's distribution.

13. Name two disadvantages for using a normal distribution as the prior distribution for binomial data.

14. Show that the posterior distribution resulting from the noninformative prior $\pi(\theta) = 1$ for a normal mean (with known σ^2) is equal to the limit as $\sigma_0^2 \to \infty$ of the posterior resulting from the informative prior $\mu \sim N(\mu_0, \sigma_0^2)$.

15. Let $X_1, X_2, \ldots, X_N \overset{i.i.d.}{\sim} \text{Unif}[0, \theta]$.
    (a) Write down the likelihood function $f(\theta)$.
    (b) Suppose the prior distribution for θ is from a Pareto distribution with pdf $\pi(\theta \mid \alpha, \beta) = \alpha\beta^\alpha/\theta^{\alpha+1}$ for $\theta \geq \beta > 0$, $\alpha > 0$. Write down the posterior density and conclude that the Pareto family of distributions is a conjugate family to the uniform distribution.
    (c) Suppose the random sample 6, 6, 8, 10 is drawn from Unif$[0, \theta]$ and you use the Pareto distribution with $\alpha = 0.3$, $\beta = 5$ as your prior distribution. Find the probability that $\theta \geq 15$.
16. Let $X_1, X_2, \ldots, X_N$ be a random sample from the Poisson distribution with pdf $f(x) = \theta^x e^{-\theta}/x!$, $x = 0, 1, 2, \ldots$.
    (a) Write down the likelihood function $f(\theta)$.
    (b) Suppose the prior for θ is the gamma distribution with parameters r, λ. Let $\pi(\theta)$ denote the pdf for this prior. Find the posterior density $f(\theta)\pi(\theta)$.
    (c) Recognize this posterior density as the pdf for what known distribution with what parameters?
    (d) Suppose you observe the values 6, 7, 9, 9, 16 and you believe the prior is gamma with parameters $r = 15$, $\lambda = 3$. Find the posterior density.
    (e) Find a 95% credible interval for the θ.
17. Let $X_1, X_2, \ldots, X_N$ be a random sample from the exponential distribution with pdf $f(x) = \theta e^{-\theta x}$.
    (a) Write down the likelihood function $L(\theta)$.
    (b) Suppose the prior for θ is the gamma distribution with parameters r, λ. Let $\pi(\theta)$ denote the pdf for this prior. Find the posterior density $L(\theta)\pi(\theta)$.
    (c) Recognize this posterior density as the pdf for what distribution with what parameters?
    (d) Suppose you observe the values 1, 2, 4, 6 and you believe the prior is gamma with parameters $r = 10$, $\lambda = 4$. Find the posterior density.
    (e) Find a 95% credible interval for the θ.
18. In the Google Website Optimizer example at the end of Section 10.5, the number of visitors assigned to each arm were similar: 1874, 1867, 1871, 1868, 1875, 1875. These seem almost too good to be true—closer together than we would expect from random chance. Perform a two-sided goodness-of-fit test for the hypothesis that the customers were assigned randomly with equal probabilities. What is your conclusion?
    If the test fails on the lower end, this indicates that the numbers are closer together than would reasonable occur by random chance, and we conclude that assignment was not purely random. (In practice, the assignment is semirandom; assignment is performed by a number of different data centers operating independently, and each data center does a systematic round-robin.)

11

ADDITIONAL TOPICS

This chapter includes a variety of topics. We will begin by considering the fundamental bootstrap principle—plugging in an estimate for the population in place of the population and looking at two new possibilities for what to plug in—in Sections 11.1 (smoothed bootstrap) and 11.2 (parametric bootstrap). We will then consider some computational methods. One motivation is for use in Bayesian analysis in nonconjugate situations, but the methods are useful in general. We will discuss the delta method for finding standard errors, stratified sampling, Monte Carlo integration, and importance sampling.

11.1 SMOOTHED BOOTSTRAP

Recall that a sampling distribution is the distribution of a statistic when drawing random samples from a population. In practice, drawing thousands or millions of repeated samples from the population is impossible, so the fundamental bootstrap idea is to draw samples from an estimate of the population.

In earlier chapters, we sampled from the observed data, with empirical cumulative distribution function $\hat{F}_n(x) = (1/n)\#\{x_i \leq x\}$, where $\#\{x_i \leq x\}$ is the number of x_i values that are below x.

But that is not always a good choice. Recall Section 5.8, where we found that the ordinary bootstrap does not work well for the median; the bootstrap distributions in Figure 5.20 on page 128 look nothing like the sampling distribution.

Mathematical Statistics with Resampling and R, First Edition. By Laura Chihara and Tim Hesterberg.

Figure 11.1 shows another view of why that happened. The population and sampling distribution are continuous, but the empirical cdfs are discrete, so the bootstrap distributions for the median are also discrete.

For comparison, the mean is less sensitive to whether the distribution is discrete or continuous. The right column of Figure 11.1 displays the bootstrap distributions for the mean from the same empirical cdfs. Even though the empirical cdfs are discrete, the bootstrap distributions are nearly continuous. In fact, the theoretical (exhaustive) bootstrap distribution has $\binom{29}{15} = 77,558,760$ jumps (see Exercise 5).

11.1.1 Kernel Density Estimate

We return to the case of estimating the median; we need to draw samples from something other than the empirical distribution. Instead, we will draw samples from a smoothed version of the empirical distribution, using what is known as a *kernel density estimate*. Figure 11.2a illustrates the idea. Think of this as a *cow-pie estimate*—drop a cow-pie at every data point and see how high they pile up. Where there are data points close together, such as four observations beginning with the fourth smallest, the pile gets high. More formally, at each observation we center $1/n$ times a normal density with a small standard deviation σ_K and then add those up to obtain the density estimate,

$$\hat{f}(x) = \frac{1}{n}\sum_{i=1}^{n} g(x - x_i), \tag{11.1}$$

where g is the normal density with mean 0 and standard deviation σ_K (the "kernel"). The cdf estimate is

$$\hat{F}(x) = \frac{1}{n}\sum_{i=1}^{n} G(x - x_i) = \frac{1}{n}\sum_{i=1}^{n} \Phi((x - x_i)/\sigma_K), \tag{11.2}$$

where G is the corresponding normal cdf.

Figure 11.2a is jarring, with a number of wiggles. We might prefer to use a wider kernel, as shown in Figure 11.2c. There is a trade-off—a wider kernel reduces the variability in the density estimate (i.e., fewer wiggles)—but at the cost of increased bias—the resulting density estimate gets flatter and wider, with increasing standard deviation, eventually beyond what could be justified by the data. As the sample size increases, we typically want σ_K to decrease, but not too fast. The choice $\sigma_K = s/\sqrt{n}$ goes to zero at a reasonable rate, is easy to remember, and has another nice property that we will see in the following sections.

Remark The kernel density estimate is useful in its own right, as a way to look at data. It is an alternative to a histogram. The appearance of histograms can vary dramatically even if the width of the bars changes only slightly. Kernel density estimates are more stable as the kernel standard deviation varies. ||

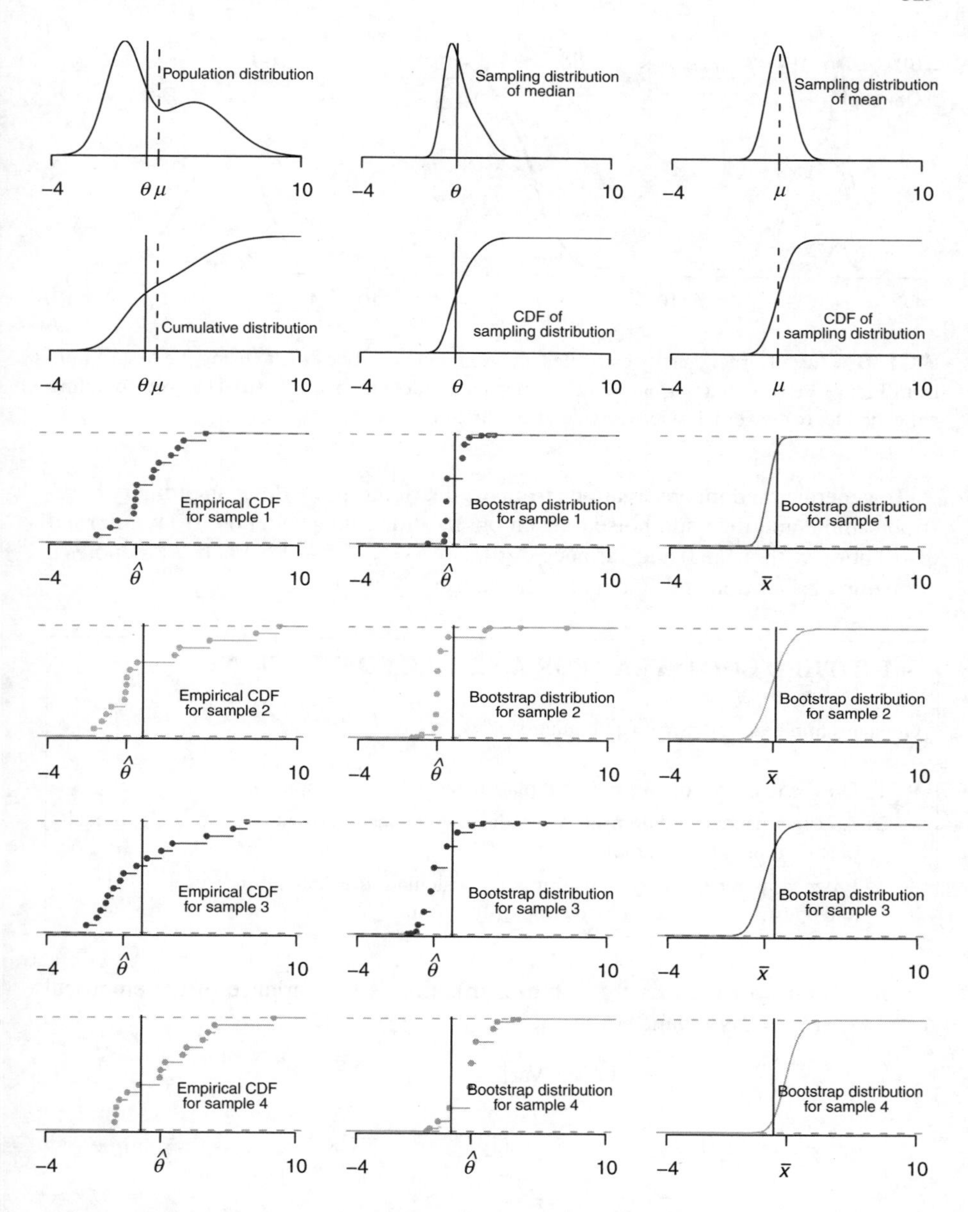

FIGURE 11.1 Ordinary Bootstrap distributions for the median and mean, $n = 15$. The left column shows the population density and cdf, as well as the cdfs for four samples. The middle and right columns are for the mean and the median, respectively, in each case showing the density and cdf for the sampling distribution, as well as bootstrap distributions for the four samples.

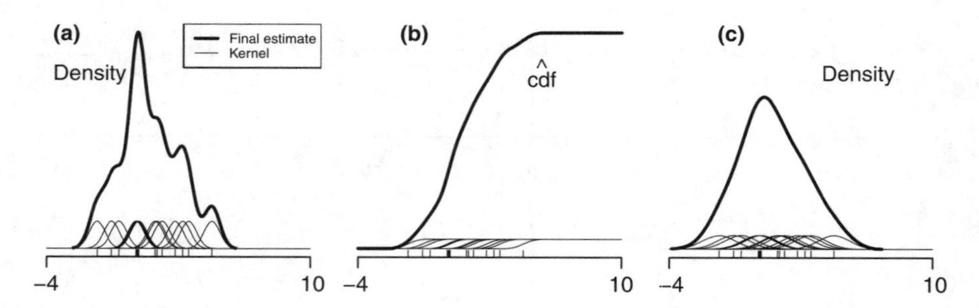

FIGURE 11.2 Kernel density estimates. (a) A kernel density estimate using the first sample from Figure 11.1, with a normal kernel with standard deviation $s/\sqrt{n}$. (b) The corresponding cdf estimate. (c) A kernel twice as wide as the first one.

To generate random observations from the distribution, we draw an ordinary bootstrap sample and then add noise to each observation, independently, from a normal distribution with mean 0 and variance σ_{K}; that is, $Y_i = X_i^* + V_i$, where X_i^* denotes a bootstrap observation and V_i denotes the "noise."

SMOOTHED BOOTSTRAP FOR A SINGLE POPULATION

Given a sample of size n from a population,

1. Draw a resample of size n with replacement from the sample.
2. Add independent random normal values with mean zero and variance σ_{K}^2 to each observation in the resample.
3. Proceed as for ordinary bootstrapping—calculate the statistic, repeat many times, and construct and use the bootstrap distribution.

The variance of the kernel smooth distribution is the variance of the empirical distribution plus the variance of the noise,

$$\begin{aligned}\mathrm{Var}[Y] &= \mathrm{Var}[X^* + V] \\ &= \frac{1}{n}\sum_{i=1}^{n}(x_i - \bar{x})^2 + \sigma_{\mathrm{K}}^2 \\ &= \frac{n-1}{n}s^2 + \sigma_{\mathrm{K}}^2.\end{aligned}$$

The choice

$$\sigma_{\mathrm{K}} = s/\sqrt{n} \tag{11.3}$$

makes $\mathrm{Var}[Y] = s^2$. This is particularly useful for bootstrapping. One problem with bootstrapping using the empirical distribution is that the bootstrap distributions tend to be too narrow—in the case that the statistic is the sample mean, the exhaustive (non-Monte Carlo) bootstrap standard error is $\sqrt{(n-1)/n}\, s/\sqrt{n}$. It is narrower than the common formula standard error by $\sqrt{(n-1)/n}$, because empirical distributions tend to be narrower than the population. Adding the right amount of noise by using $\sigma_K = s/\sqrt{n}$ corrects for that.

Figure 11.3 shows the smoothed bootstrap estimates for the mean and the median. The bootstrap distributions are much improved for the median; they are now continuous and the spreads are closer to the spread of the sampling distribution. As always, the centers are centered at the sample median $\hat{\theta}$ rather than the population median θ. And the spreads do still vary noticeably because with samples this small, the corresponding samples vary substantially. In contrast, smoothing does not make as much difference for the mean; the biggest difference is that the bootstrap distributions are slightly wider, with standard error larger by a factor $\sqrt{15/14}$.

Remark There is a problem with applying kernel density estimates to nonnegative data like Arsenic levels or Verizon repair times; some of the density falls to the left of 0, and doing smoothed bootstrap sampling by adding noise may make some values negative.

A remedy is to transform the data, say $y = \log(x)$, draw bootstrap samples from the y values, add noise to the bootstrap y values using kernel standard deviation $s_y/\sqrt{n}$, and then transform back to the original scale. Other transformations may be used, for example, $y = \sqrt{x}$. If x has some zero values, then a transformation like $\log(x + 0.1)$ prevents taking the log of zero. An example of `R` code for this is available at `https://sites.google.com/site/ChiharaHesterberg`. ||

11.2 PARAMETRIC BOOTSTRAP

In the ordinary bootstrap, we assume essentially nothing about the underlying population and sample from the empirical distribution. In the smoothed bootstrap, we assume that the population has a density and we sample from an estimate of this density. Suppose we are willing to make an even stronger assumption—that the underlying population has some parametric distribution. Then, we can use the *parametric bootstrap*: we estimate parameters based on the data, for instance, by maximum likelihood and then draw bootstrap samples from the corresponding parametric distribution.

Example 11.1 Recall the Wind Energy Case Study in Section 6.1.3, where we modeled wind speed using a Weibull distribution. We previously estimated the parameters to be $\hat{k} = 3.169$ and $\hat{\lambda} = 7.661$. Figure 11.4 shows the cdf of the data and the cdf of a Weibull distribution with these parameters.

We will compare two bootstraps, the ordinary bootstrap and a parametric bootstrap using bootstrap samples from the Weibull distribution.

The statistic we will use is the 10% sample quantile, $\hat{\eta}_1$ in Figure 11.4. We saw in Section 5.8 that the ordinary bootstrap does not work well for the median in small samples. We may expect similar problems with the 10% quantile. We are looking at

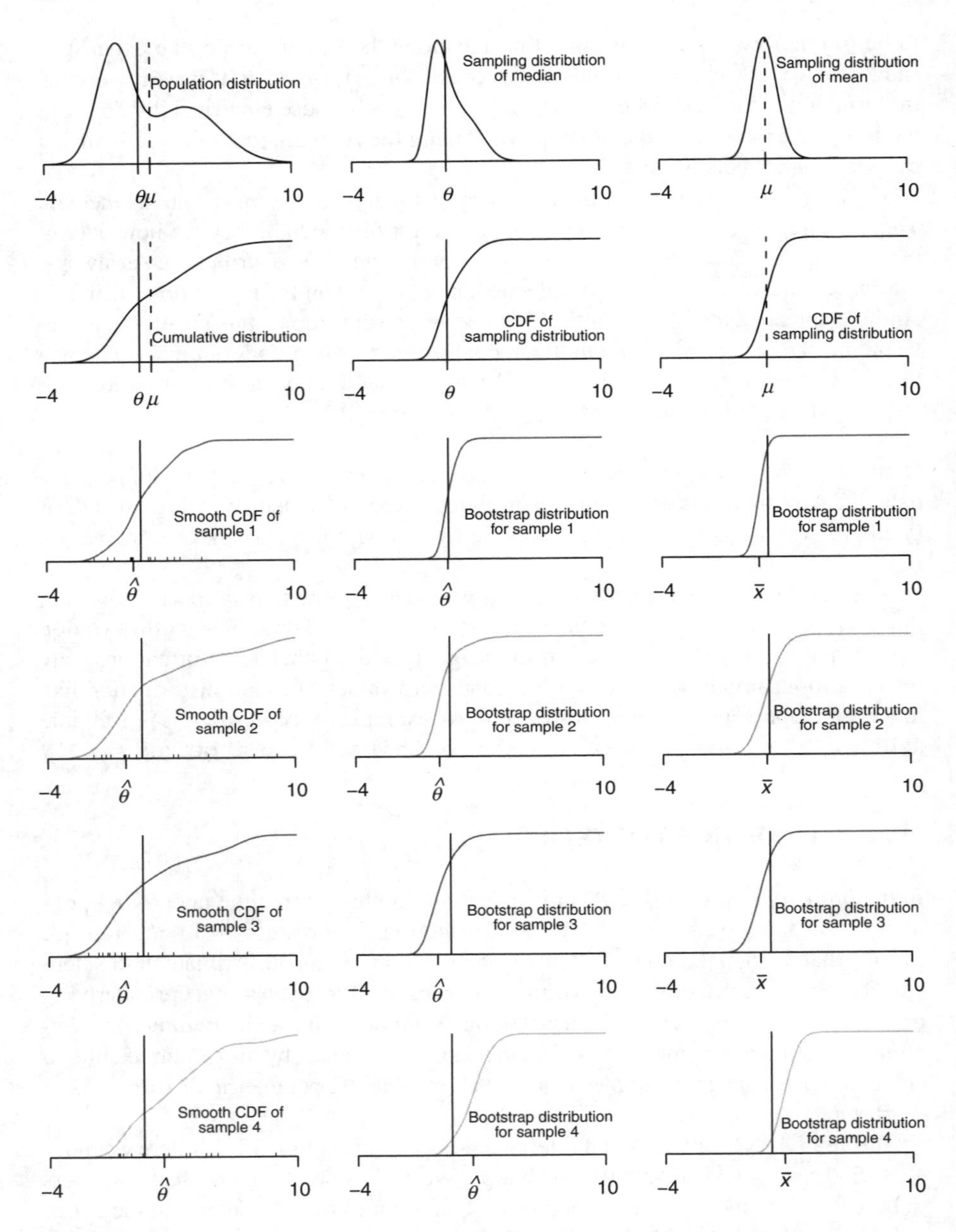

FIGURE 11.3 Smoothed bootstrap distributions for the median and the mean, $n = 15$. The left column shows the population density and cdf, as well as cdfs for four samples. The middle and right columns are for the mean and the median, respectively, in each case showing the density and cdf for the sampling distribution, as well as bootstrap distributions for the four samples.

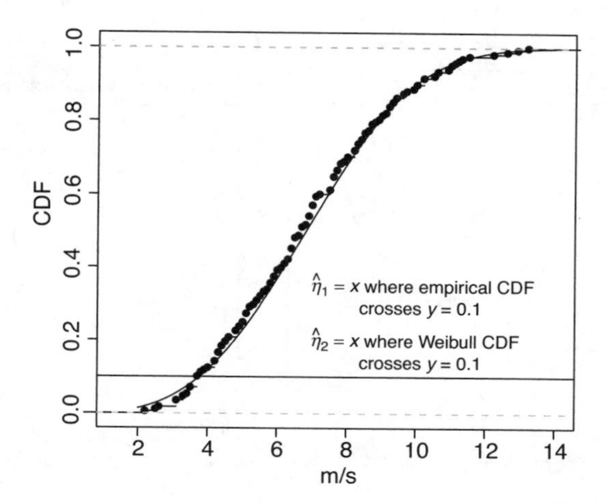

FIGURE 11.4 Empirical cumulative distribution function of wind speeds (m/s), with cdf for Weibull superimposed. $\hat{\eta}_1$ is the quantile where the empirical cdf crosses $y = 0.1$ and $\hat{\eta}_2$ is the quantile where the Weibull cdf crosses $y = 0.1$.

the 10% quantile rather than the median, because we are interested in a measure of reliability, the minimum amount of energy that the turbine will generate 90% of the time. While the sample size is 168, there are only 17 observations at or below the quantile, a small effective sample size, so the ordinary bootstrap may perform poorly. The two bootstrap distributions are shown in Figure 11.5. The ordinary bootstrap does badly. The parametric distribution seems more reasonable.

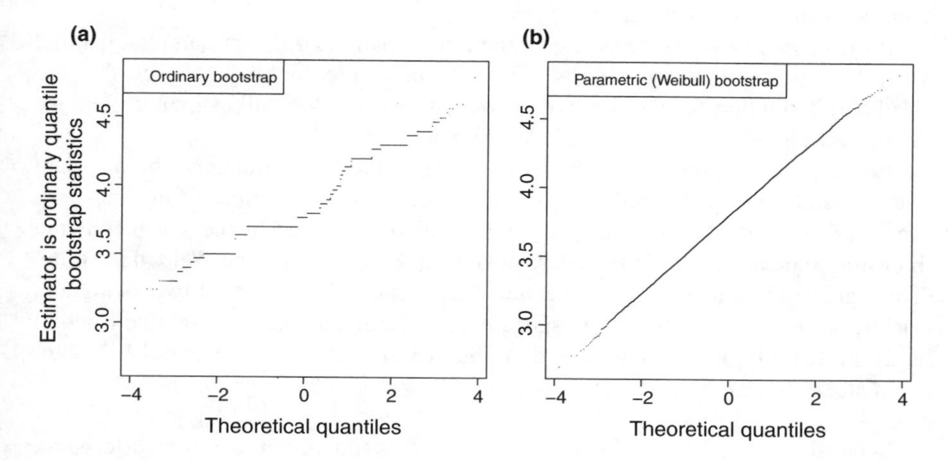

FIGURE 11.5 Bootstrap distributions for 10% quantile of wind speed. Wind speed data. (a) The ordinary bootstrap. (b) The Weibull parametric bootstrap. The estimator is the 10% empirical quantile (i.e., of each bootstrap sample). Each panel contains a normal quantile plot.

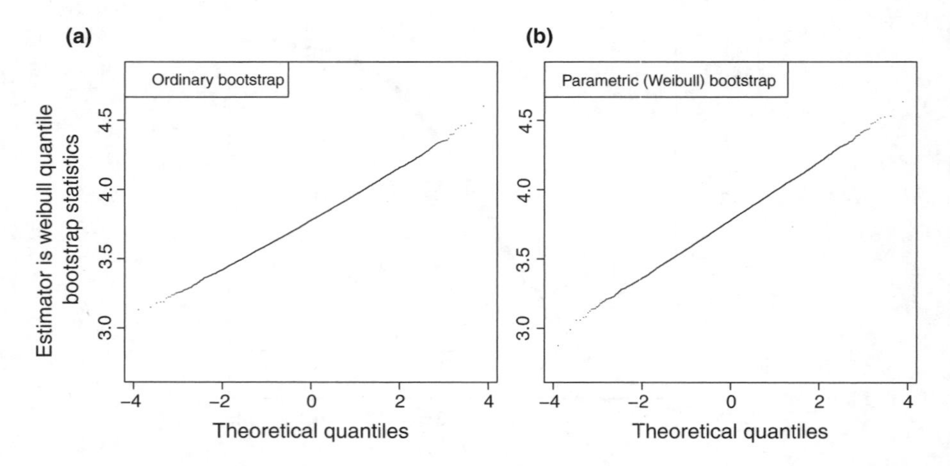

FIGURE 11.6 Bootstrap distributions for 10% quantile of wind speed, estimated from a Weibull distribution. Wind speed data. (a) The ordinary bootstrap. (b) The Weibull parametric bootstrap. The estimator is the 10% quantile of a Weibull distribution fitted to the data (i.e., of each bootstrap sample).

For comparison, we will also consider an alternative estimator for the 10% quantile of the distribution—given a set of data (original data or bootstrap data), fit a Weibull distribution to the data and then calculate the 10% quantile of that Weibull distribution. This is indicated as $\hat{\eta}_2$ in Figure 11.4. Again, we will do both the ordinary and Weibull bootstraps. The results are shown in Figure 11.6. For this estimator the ordinary bootstrap works just fine, there is little difference between the ordinary and Weibull bootstraps. So the problem with the ordinary bootstrap was limited to the sample quantile as an estimator.

If the estimator is the Weibull quantile rather than the ordinary sample quantile, then the ordinary bootstrap works fine. Still, if we are confident enough that the Weibull distribution fits the data well enough to use the Weibull estimator, then we might as well also do the parametric Weibull bootstrap.

It is also worth comparing the performance of the two estimators, the raw data quantile and the fitted Weibull quantile. Compare the vertical spread of the bootstrap statistics in Figures 11.5b and 11.6b; the standard deviation of the y values is the bootstrap standard error. The standard error is much smaller for the Weibull quantile. This agrees with our findings in Section 6.3.2, where we compared two estimators for the fraction of time that wind speed exceed 5 m/s, the ordinary sample fraction and a probability based on estimated Weibull parameters. In both cases the Weibull estimator is more efficient. □

When doing a parametric bootstrap, computing standard errors is straightforward: we use the sample standard deviation of the bootstrap values as we did for the ordinary bootstrap. Computing bias has a twist—we use the mean of the bootstrap values minus the parameter calculated from the parametric distribution we draw samples from. Recall that the definition of bias is

$$\text{bias} = \mathrm{E}\left[\hat{\theta}\right] - \theta = \mathrm{E}_F[\hat{\theta}] - \theta_F,$$

the expected value of a statistic when sampling from a population F minus the corresponding parameter for that population. We estimate this using the bootstrap by plugging in an estimate for F,

$$\mathrm{E}_{\hat{F}}[\hat{\theta}] - \theta_{\hat{F}},$$

and we use the same $\hat{F}$ in both places, whether that $\hat{F}$ is the empirical data or a parametric distribution.

11.3 THE DELTA METHOD

In many instances, we can calculate standard errors for certain quantities, but what we really need are the standard errors for related quantities.

- In Example 5.7, we are interested in the relative risk of cardiovascular disease, that is, the ratio of two proportions. We can compute standard errors for $\hat{p}_1$ and $\hat{p}_2$, but for a t interval we need a standard error for $\hat{p}_1/\hat{p}_2$.
- In the Verizon example in Section 1.3, let $\bar{X}$ be an estimate for μ_X, the mean repair time for the ILEC distribution, and let $\bar{Y}$ be an estimate for μ_Y, for the CLEC distribution. We may be interested in quantities such the average number of repairs that are completed in 1 h ($1/\mu_X$ and $1/\mu_Y$) or the ratio μ_Y/μ_X of the two average repair times. We would naturally estimate these quantities by $1/\bar{X}$, $1/\bar{Y}$, and $\bar{Y}/\bar{X}$, but need a way to compute standard errors for these quantities.
- In analyzing traffic deaths related to alcohol in Section 9.6, we may have estimates and standard errors for the logistic regression parameters α and β, but want to compute standard errors for the predicted probability of alcohol involvement in the automobile death of a 20-year old, $\exp(\hat{\alpha} + 20\hat{\beta})/(1 + \exp(\hat{\alpha} + 20\hat{\beta}))$.

We could use bootstrapping to estimate these standard errors. Here, we describe another method, the *delta method*.

Consider first the case involving a single estimate, such as $1/\mu_X$. Write

$$\eta = g(\theta),$$

where θ is a quantity whose standard error is known, η is the quantity of interest, and g is a differentiable function. The corresponding estimator is

$$\hat{\eta} = g(\hat{\theta}).$$

The core of the delta method is a linear Taylor series approximation of $g(\hat{\theta})$ about θ,

$$\begin{aligned}\hat{\eta} &= g(\hat{\theta}) \\ &= g(\theta) + g'(\theta)(\hat{\theta} - \theta) + \cdots .\end{aligned}$$

Now $g(\theta)$ and $g'(\theta)$ are constants, so

$$\mathrm{Var}[\hat{\eta}] \approx g'(\theta)^2 \mathrm{Var}[\hat{\theta}],$$

with corresponding standard deviations

$$\sigma_{\hat{\eta}} \approx |g'(\theta)| \sigma_{\hat{\theta}}.$$

Since $g'(\theta)$ is unknown, we use the statistician's favorite trick of plugging in an estimator to obtain standard errors

$$s_{\hat{\eta}} \approx |g'(\hat{\theta})| s_{\hat{\theta}}. \tag{11.4}$$

For example, if $g(\mu) = 1/\mu$, then $g'(\mu) = -1/\mu^2$ and the standard error for $1/\bar{x}$ is $s_{\bar{x}}/\bar{x}^2 = s/(\sqrt{n}\bar{x}^2)$. The mean Verizon ILEC repair time is 8.4 h with standard deviation 14.7 and standard error for the mean of 0.36; the estimated mean repairs per hour is 0.12, with standard error $0.12/8.4^2 = 0.005$.

The name comes from writing $\Delta = \hat{\theta} - \theta$ and then expanding $\hat{\eta} = g(\theta + \Delta)$ as a function of Δ.

In the case of a function of two parameters,

$$\eta = g(\theta_1, \theta_2)$$

and

$$\hat{\eta} = g(\hat{\theta}_1, \hat{\theta}_2),$$

with first-order Taylor series approximation

$$\hat{\eta} = g(\hat{\theta}_1, \hat{\theta}_2) \approx g(\theta_1, \theta_2) + \frac{\partial g(\theta_1, \theta_2)}{\partial \theta_1}(\hat{\theta}_1 - \theta_1) + \frac{\partial g(\theta_1, \theta_2)}{\partial \theta_2}(\hat{\theta}_2 - \theta_2).$$

Hence,

$$\mathrm{Var}[\hat{\eta}] \approx g_1^2 \mathrm{Var}[\hat{\theta}_1] + g_2^2 \mathrm{Var}[\hat{\theta}_2] + 2 g_1 g_2 \mathrm{Cov}[\hat{\theta}_1, \hat{\theta}_2],$$

where g_1 and g_2 indicate the partial derivatives with respect to θ_1 and θ_2. As before, we plug in estimates to obtain standard errors.

An important special case is $\bar{Y}/\bar{X}$ as an estimator for $r = \mu_X/\mu_Y$. In this case, we will use some algebra to obtain a similar expansion:

$$\begin{aligned}
\frac{\bar{Y}}{\bar{X}} &= \frac{\mu_Y + (\bar{Y} - \mu_Y)}{\mu_X + (\bar{X} - \mu_X)} \\
&= r\frac{1 + (\bar{Y} - \mu_Y)/\mu_Y}{1 + (\bar{X} - \mu_X)/\mu_X} \\
&= r\frac{1 + \Delta_Y}{1 + \Delta_X} \\
&= r(1 + \Delta_Y)(1 - \Delta_X + \Delta_X^2 + \cdots) \\
&= r(1 + (\Delta_Y - \Delta_X) + (-\Delta_Y\Delta_X + \Delta_w^2) + \cdots),
\end{aligned} \tag{11.5}$$

where $\Delta_Y = (\bar{Y} - \mu_Y)/\mu_Y$ and $\Delta_X = (\bar{X} - \mu_X)/\mu_X$ measure relative differences between the sample means and their expected values. The last line of Equation 11.5 groups terms by their size. As long as both Y and X have finite variance, $\bar{Y}$ and $\bar{X}$ have standard deviations that decrease at the rate $1/\sqrt{n}$; so Δ_Y and Δ_X tend to decrease at the rate $1/\sqrt{n}$ (more formally, they are $O_P(1/\sqrt{n})$ of "order $1/\sqrt{n}$ in probability," i.e., for any probability $\epsilon > 0$, there exists a constant z and integer m such that $P(|\Delta| > z/\sqrt{n}) < \epsilon$ for all $n \geq m$).

Using only the first approximation in Equation 11.5, $\bar{Y}/\bar{X} \approx \mu_Y$, suggests that the estimate is consistent.

The second approximation,

$$\bar{Y}/\bar{X} \approx r(1 + \Delta_Y - \Delta_X) = r + (\bar{Y} - r\bar{X})/\mu_X, \tag{11.6}$$

is a constant plus a sample average $1/n\sum(Y_i - rX_i)/\mu_X$. This suggests that the estimate is asymptotically normal with mean r and variance $\mathrm{Var}[Y - rX]/(n\mu_X^2)$.

The third approximation can be used to estimate the bias of the estimate. We do not do so here, but do note that the asymptotic bias decreases at the same rate $1/n$ as the terms $\Delta_Y\Delta_X$ and Δ_w^2; so for large n, the bias is negligible compared to the standard error.

Theorem 11.1 *If (X_i, Y_i) are i.i.d. from a joint distribution with $0 < \mathrm{E}[X] < \infty$ and $\mathrm{Var}_g[Y - \mu_Y X] < \infty$, then $\bar{Y}/\bar{X}$ is asymptotically normal with mean $r = \mu_Y/\mu_X$ and variance*

$$\mathrm{Var}[Y - rX]/(n\mu_X^2). \tag{11.7}$$

Proof See (Hesterberg (1991)). □

If $\bar{X}$ and $\bar{Y}$ are independent samples from two populations, then the theorem does not apply, but the expansions 11.5 and 11.6 still hold, $\hat{r} \approx r + (\bar{Y} - r\bar{X})/\mu_X$, so

$$\mathrm{Var}[\hat{r}] = \frac{\mathrm{Var}[\bar{Y}] + r^2\mathrm{Var}[\bar{X}]}{\mu_X^2}.$$

Plugging in estimates for unknown quantities, we obtain a squared standard error,

$$s_{\hat{r}}^2 = \frac{s_Y^2/n_y + \hat{r}^2 s_X^2/n_x}{\bar{x}^2}. \tag{11.8}$$

Example 11.2 In Example 5.7 (relative risk), 55 of 3338 smokers and 21 of 2676 nonsmokers died of cardiovascular disease, with a relative risk of $\hat{r} = \hat{p}_1/\hat{p}_2 = 0.0165/0.0078 = 2.1$. This is a ratio of means (a proportion is an average of Bernoulli variables), with independent numerator and denominator; so we use Equation 11.8 with $\hat{p}_1(1-\hat{p}_1)$ in place of s_y^2, and similarly for s_x^2.

$$\begin{aligned} s_{\hat{r}}^2 &= \frac{\hat{p}_1(1-\hat{p}_1)/n_1 + \hat{r}^2 \hat{p}_2(1-\hat{p}_2)/n_2}{\hat{p}_2^2} \\ &= \frac{0.016/3338 + 2.1^2(0.00778595)/2676}{0.0078^2} = 0.54^2. \end{aligned}$$

The estimated relative risk of 2.1 has a standard error of 0.54. A 95% confidence interval is $\hat{r} \pm 1.96 s_{\hat{r}} = (1.05, 3.15)$. Hence, there is substantial uncertainty in the estimate.

If we flip the numerator and denominator, $\hat{p}_2/\hat{p}_1$, the resulting 95% confidence interval is (0.2380.714). Inverting that gives a corresponding confidence interval for $\hat{p}_1/\hat{p}_2$, $(1/0.714, 1/.0238) = (1.34, 4.20)$. This is very different from $(1.05, 3.15)$. □

This is a weakness of standard intervals of the form $\hat{\theta} \pm se_{\hat{\theta}}$: different analysts can get nonequivalent confidence intervals depending on how they express their problem.

To partially address this issue, analysts should use standard transformations, where such transformations exist. For ratios, it is standard to use a log transformation, for example, to work with

$$\hat{\eta} = \log(\bar{Y}/\bar{X}) = \log(\bar{Y}) - \log(\bar{X}).$$

Example 11.3 Referring again to the relative risk example (Example 5.7), we continue with a log transformation,

$$\hat{\eta} = \log(\hat{p}_1/\hat{p}_2) = \log(\hat{p}_1) - \log(\hat{p}_2).$$

We again need the delta method. The partial derivatives are $\partial\eta/\partial p_1 = 1/p_1$ and $\partial\eta/\partial p_2 = 1/p_2$, and

$$\hat{\eta} \approx \eta + (\hat{p}_1 - p_1)/p_1 - (\hat{p}_2 - p_2)/p_2,$$

with variance (in this case, with independent samples)

$$\text{Var}[\hat{\eta}] \approx \frac{p_1(1-p_1)/n_1}{p_1^2} + \frac{p_2(1-p_2)/n_2}{p_2^2} = \frac{(1-p_1)}{n_1 p_1} + \frac{(1-p_2)}{n_2 p_2}.$$

The squared standard error is obtained by plugging in estimates. The resulting 95% confidence interval is (0.24, 1.24). Transforming back to the original scale by exponentiating gives a 95% confidence interval for $\hat{p}_1/\hat{p}_2$ of (1.27, 3.46), intermediate between the intervals obtained using the two fractions, and probably more accurate than either. □

Remark A confidence interval procedure is *transformation invariant* if it yields equivalent confidence intervals for any monotone transformation of the parameter of interest. t intervals and z intervals are not transformation invariant. Bootstrap percentile intervals are. Bootstrap t intervals (Section 7.5) are approximately transformation invariant. ||

11.4 STRATIFIED SAMPLING

We first described stratified sampling in the context of sample surveys given on page 7. Suppose we are sampling from a population with 50% men; we can stratify and survey exactly 50% women and 50% men or draw observations randomly from everyone and possibly get an unbalanced sample.

Stratified sampling is useful in computer simulations. For example, suppose we want to estimate $\mathrm{E}\left[e^X\right]$, where X has a standard normal distribution. We can divide the population (the normal distribution) into two strata, say $X < 0$ and $X > 0$, and draw exactly half the observations from each.

In any case, sampling from a finite population or infinite population, real data or computer simulation, stratified sampling can reduce the variance of the result. In this section, we will quantify this.

Consider the general case where a population has J *strata* (subpopulations), and that stratum j represents a fraction λ_j of the population, with $\sum \lambda_j = 1$. For now assume that $n\lambda_j$ is a whole number for each j, so we could draw exactly the "right" number of observations from each stratum.

Let μ_j and σ_j^2 be the mean and variance for stratum j and let μ and σ^2 be the mean and variance for an observation chosen from the whole population; then $\mu = \sum_j \lambda_j \mu_j$ and

$$\sigma^2 = \sum_j \lambda_j(\sigma_j^2 + (\mu_j - \mu)^2). \tag{11.9}$$

If observations are drawn from the whole population without regard to strata, then it is as if strata did not exist and

$$\mathrm{Var}[\bar{Y}] = \sigma^2/n. \tag{11.10}$$

In contrast, when stratifying, if exactly $n_j = n\lambda_j$ observations are chosen for stratum j, Y_{ji} for $i = 1, \ldots, n_j$, then

$$\bar{Y} = (1/n)\sum_{j=1}^{J}\sum_{i=1}^{n_j} Y_{ji}$$

and the variance is

$$\mathrm{Var}[\bar{Y}] = (1/n^2)\sum_{j=1}^{J} n_j\sigma_j^2 = (1/n)\sum_{j=1}^{J}\lambda_j\sigma_j^2. \tag{11.11}$$

Comparing Equations 11.10 and 11.11, with a view to Equation 11.9, we see that the variance of the stratified estimate is smaller than that of the unstratified estimate by $(1/N)\sum_j \lambda_j(\mu_j - \mu)^2$. As long as the strata means differ, the stratified estimate has smaller variance.

The standard error for the stratified estimate is obtained by plugging in sample variances s_j^2 for σ_j^2 in Equation 11.11.

Stratification can also be used in delta method situations with similar gains. For example, for a ratio estimate $\bar{Y}/\bar{X}$ for bivariate data, the asymptotic variance in the unstratified case is $\mathrm{Var}[(Y - rX)/(n\mu_X^2)]$, and in the stratified case

$$\mathrm{Var}[\bar{Y}/\bar{X}] \approx \frac{1}{n\mu_X^2}\sum_{j=1}^{J}\lambda_j \mathrm{Var}_j[Y - rX].$$

Standard errors are obtained by substituting sample estimates where needed. In particular, for $\mathrm{Var}_j[Y - rX]$, we substitute the sample standard deviation values of $Y_{ji} - \hat{r}X_{ji}$.

Another option is to bootstrap to obtain standard errors from a stratified sample, drawing separately from each stratum.

11.5 COMPUTATIONAL ISSUES IN BAYESIAN ANALYSIS

The examples in Chapter 10, involving beta and normal distributions, use conjugate priors. In these cases, the posterior distributions of the parameters are from known distributions and computations of expected values and credible intervals are straightforward, using `R` or other software packages.

In practice, many applications are too complicated for conjugate priors, or the analyst may have prior knowledge that suggests using another prior. In most nonconjugate applications and in some conjugate applications, closed-form solution are not available, so approximations must be found.

Bayesian calculations typically involve integrals. Recall from Equation 10.11 that the posterior density may be written $p(\theta \mid x) = (\pi(\theta)f(x \mid \theta))/(\int_\Theta \pi(\theta)f(x \mid \theta)\,d\theta)$, where Θ is the set of possible values for θ. The denominator is the normalizing constant. To compute an expected value, say $\mathrm{E}_{p(\theta \mid x)}[\theta]$, requires

$$\int_\Theta \theta p(\theta \mid x)\,d\theta = \int_\Theta \theta\pi(\theta)f(x \mid \theta)\,d\theta$$

and the normalizing constant. Similarly, to find a probability, say $P(\theta \in A \mid x)$, requires

$$\int_A \pi(\theta) f(x \mid \theta)\, d\theta$$

and the normalizing constant. Finding a credible interval requires finding values θ_1 and θ_2 that satisfy $\int_{-\infty}^{\theta_1} \theta\pi(\theta) f(x \mid \theta) d\theta = \int_{\theta_1}^{\infty} \theta\pi(\theta) f(x \mid \theta) d\theta = C\alpha/2$, where C is the normalizing constant. In the content publisher example in Example 10.5, θ is six dimensional, and to find the probability that arm 3 is the best, given the current data, requires

$$\int_{\Theta_1} \int_{\Theta_2} \int_{\Theta_4} \int_{\Theta_5} \int_{\Theta_6} \int_{\max \theta_1, \theta_2, \theta_4, \theta_5, \theta_6}^{\infty} \pi(\theta) f(x \mid \theta)\, d\theta_2 d\theta_6 d\theta_5 d\theta_4 d\theta_2 d\theta_1$$

and the normalizing constant. In all cases, calculations require evaluating integrals with respect to θ.

For univariate θ, the integrals may be evaluated with techniques you may remember from calculus, such as Simpson's rule or the trapezoid rule. In R, the `integrate` function uses an *adaptive* version of such methods: for a finite region, it starts by evaluating points on a grid and then splits the grid into finer grids until it estimates that the accuracy satisfies a level specified by the user. For an infinite region, it first does a transformation (a "u-substitution," in calculus terms) to express the problem as an integral over a finite region before proceeding with the methods for a finite region.

Though these methods can be extended to functions of more than one variable, the number of evaluations required to achieve a desired level of accuracy increases exponentially in the number of dimensions: this is the "curse of dimensionality," a term coined by applied mathematician Richard Bellman (1966). For Simpson's method, for example, we can improve the accuracy by cutting the grid spacing in half, but in d dimensions this requires approximately 2^d times as many points; in practice, d is often large enough to make this infeasible.

Alternatively, we may use simulation. We will describe some simulation methods next. One approach, known as *Monte Carlo integration*, uses simulation to estimate definite integrals. A variation of this approach, *importance sampling*, can improve efficiency and solve some otherwise intractable problems. We describe these methods next. Another approach, *Markov Chain Monte Carlo* (known as MCMC or MC^2), involves drawing sequences of dependent observations; for more information, see Gilks et al. (1996), Robert and Casella (2004, 2010), and Suess and Trumbo (2010).

11.6 MONTE CARLO INTEGRATION

Consider the problem of estimating a univariate integral over a finite region, $\int_a^b h(x)\, dx$, with $b > a$. Let $X_i \overset{\text{i.i.d.}}{\sim} U(a, b)$ for $i = 1, 2 \ldots$, and let $Y_i = (b - a)h(X_i)$.

By Theorem A.1,

$$\mathrm{E}\left[Y_i\right] = \mathrm{E}\left[h(X_i)\right] = \mu = \int_a^b (b-a)h(x)\frac{1}{b-a}\,dx = \int_a^b h(x)\,dx.$$

By the strong law of large numbers,

$$\lim_{N\to\infty} \bar{Y}_N = \mu = \int_a^b h(x)\,dx$$

with probability 1, where $\bar{Y}_N = (1/N)(Y_1 + Y_2 + \cdots + Y_N)$ is the average after N replications. Thus, for large N,

$$\bar{Y}_N \approx \int_a^b h(x)\,dx.$$

In other words, to approximate $\int_a^b h(x)\,dx$, we randomly draw N points from the interval $[a, b]$, evaluate h at each of these points, and compute the average times the length of the interval (since $Y_i = (b-a)h(X_i)$).

If $\mathrm{Var}[Y] < \infty$ (the usual case), we can compute the standard error for $\bar{Y}$ as $s/\sqrt{n}$ and calculate a confidence interval for μ.

Example 11.4 Estimate $\displaystyle\int_1^3 e^{-x^2}\,dx$.

Solution The following R code draws random numbers from the uniform distribution, evaluates $h(x) = e^{-x^2}/f(x)$ at each random point, and averages.

R Note:

```
> N <- 10^5
> x <- runif(N, 1, 3)     # draw random values from Unif[1, 3]
> out <- 2*exp(-x^2)      # evaluate h at each random x (times (b-a))
> mean(out)
[1] 0.1401142
> sd(out) / sqrt(N)       # standard error
[1] 0.0006160375
```

The command `integrate` computes definite integrals. The first argument must be an R function.

```
> integrate(function(x) exp(-x^2), 1, 3)   # adaptive
0.1393832 absolute error < 1.5e-15
```

One run of this simulation gives an estimate of 0.1401 with standard error 0.00062. Thus, a 95% confidence interval for the true value of the integral is $0.1401 \pm 1.96 \times 0.00062 = (0.1395, 0.1407)$. For comparison, the value obtained by adaptive numerical integration is 0.1394. □

Example 11.5 According to a poll conducted by the Pew Research Center for the People and the Press (October 2010), 66% of 325 people, between 18 and 29 years of age, surveyed believe that the average temperature on earth has been getting warmer over the past few decades (`http://people-press.org/report/669/`).

We will consider the binomial model for the data and assume $X \sim \text{Binom}(n, \theta)$ with $X = 215, n = 325, \hat{\theta} = 0.66$.

Consider the following prior density for θ:

$$\pi(\theta) = c\, e^{-25|\theta-0.5|}, \quad 0 \leq \theta \leq 1,$$

where c is a normalizing constant. Then by Equation 10.3, the posterior density is

$$\begin{aligned} p(\theta \mid X = 215) &= \frac{c\, e^{-25|\theta-0.5|}\, \theta^{215}(1-\theta)^{110}}{\int_0^1 c\, e^{-25|u-0.5|}\, u^{215}(1-u)^{110}\, du} \\ &= e^{-25|\theta-0.5|}\, \theta^{215}(1-\theta)^{110}/K, \end{aligned} \tag{11.12}$$

where K is another normalizing constant.

To compute the expected value of θ, given the data,

$$\begin{aligned} \mathrm{E}\,[\theta \mid X = 215] &= \int_0^1 \theta\, p(\theta \mid X = 215)\, d\theta \\ &= \int_0^1 \theta\, e^{-25|\theta-0.5|}\, \theta^{215}(1-\theta)^{110}\, d\theta/K. \end{aligned}$$

Thus, we have two integrals to compute: one for the numerator and another for the normalizing constant K.

R Note:

We define three functions in R, draw random numbers from the uniform distribution on [0, 1], and then evaluate each of these functions at these random numbers.

```
f0 <- function(u) exp(-25 * abs(u-0.5))     # prior, except constant
f1 <- function(u) f0(u) * u^215 * (1-u)^110# prior*likelihood
f2 <- function(u) u * f1(u)                 # for expected value
x <- runif(10^6)
a <- mean(f0(x))          # constant for prior
K <- mean(f1(x))          # constant in denom. of E[theta|x=215]
b <- mean(f2(x))          # numerator of E[theta|x=215]
c(a, K, b, b/K)           # output all

# Plot the prior and posterior
curve(f0(x)/a, from=0, to=1, ylim=c(0, 15))
curve(f1(x)/K, col="blue", lty = 2, add=TRUE)
legend(.05, 14, legend=c("Prior","Posterior"),
       col=c("black", "blue"), lty=c(1, 2))
```

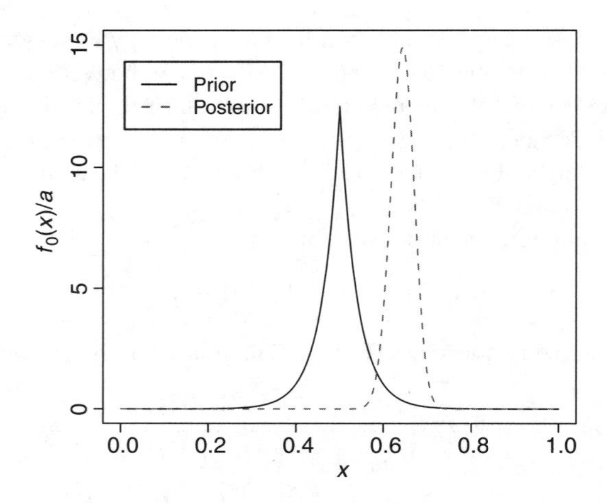

FIGURE 11.7 Prior and posterior densities for Pew survey example.

In one run of this simulation, we find that the estimated mean of the posterior distribution is 0.643, with density shown in Figure 11.7. □

We may draw random variables from any distribution, not just the uniform distribution.

> **GENERAL MONTE CARLO INTEGRATION**
>
> Suppose that an integral can be written in the form
>
> $$\int_C h(x) f(x)\, dx = \mathrm{E}\left[h(X)\right], \tag{11.13}$$
>
> where f is a density with $f(x) > 0$ over C and zero elsewhere, and that the integral exists and is finite. Then, we can estimate $\mathrm{E}\left[h(X)\right]$ by generating $X_1, X_2, \ldots, X_N$ i.i.d. from a distribution with pdf f and using the sample mean
>
> $$\frac{1}{N} \sum_{i=1}^{N} h(X_i). \tag{11.14}$$

We leave the verification of this as an exercise (Exercise 9).

We may need to do a bit of manipulation to express a problem in the form required by Equation 11.13. For example, to integrate $\int_C r(x)\, dx$, choose a density f with the proper domain and, then let $h(x) = r(x)/f(x)$. For example, to integrate $\int_0^\infty \sin(x^2) \exp(-x^2)\, dx$, we may choose an exponential distribution $f(x) = \exp(-x)$ and then let $h(x) = \sin(x^2) \exp(-x^2 + 1)$.

Or, suppose the integral appears to be in the proper form, but with the wrong domain, for example, $\int_0^5 x^2 \exp(-x)$ looks like $h(x) = x^2$ and $f = \exp(-x)I(x > 0)$

(a standard exponential distribution) would work, except that the domain is wrong. One option is to let f be a uniform distribution over $(0, 5)$ and define h accordingly. Another option is to let f be the standard exponential and incorporate the domain restriction into h, $h(x) = x^2 I(0 < x < 5)$.

In Bayesian applications, two natural choices for f are the prior and posterior distributions, if these are tractable.

Example 11.6 Refer to Example 10.4. Based on the data, the biologist's posterior distribution for the mean lengths (cm) of trout is $\mu \sim N(45.53, 1.953^2)$. Biologists have determined that the relationship between the weight (W) and length (X) of a fish is approximately $W = aX^b$ for constant a and b that are determined empirically for any give species (Ricker (1973, 1975)). Suppose for this species, $W = 0.088 \times X^{3.069}$ g. If $X \sim N(45.53, 1.953^2)$, find the expected value of the weight of the trout.

Solution We want $\mathrm{E}[W] = \mathrm{E}\left[0.088X^{3.069}\right] = \int_{-\infty}^{\infty} 0.088x^{3.069} f(x)\,dx$, where $f(x)$ is the pdf for the normal distribution with mean 45.53 and standard deviation 1.953.

R Note:

```
x <- rnorm(10^6, 45.53, 1.953)
out <- 0.088*x^(3.069)
mean(out)
```

One run of the simulation gives an estimated mean of 10872.55 g (23.9 pounds). □

For functions of a single variable, estimating integrals using Monte Carlo integration is generally less effective than deterministic methods such as Simpson's rule. But such deterministic methods suffer from the curse of dimensionality, whereas Monte Carlo does not; doubling the accuracy requires four times as many points, regardless of the dimension.

Monte Carlo can have another advantage—we can use it to obtain not only the expected value but also the whole distribution. In the previous example, where f is the posterior distribution for fish lengths, the values $h(X_i)$ are a sample from the posterior distribution of fish weights $h(X)$.

But in many applications it is not possible to draw samples from the posterior. We will use a technique called *importance sampling* to draw observations from some other distribution, but still obtain answers based on the posterior. We will start with basic importance sampling for estimating integrals or expected values and then turn to estimating distributions in Sections 11.7.1 and 11.7.2.

11.7 IMPORTANCE SAMPLING

The Monte Carlo procedure may be inefficient if $h(X)$ is skewed, for example, if $h(X)$ is small, most of the time and only rarely takes on large values. Then, the estimate of an integral may be largely determined by a relatively small number of points with large $h(X_i)$.

We are not limited to sampling from the prior, posterior, or uniform distributions. The original motivation behind *importance sampling* is to choose the distribution to emphasize important regions, to oversample these regions, and to obtain more observations there, increasing the effective sample size.

Importance sampling turns out to be useful in a variety of applications, Bayesian and otherwise, including many cases where it is the best way to get answers, regardless of efficiency.

Example 11.7 My (Hesterberg) first career job was for Pacific Gas and Electric Co., estimating how much fuel oil PG&E should carry in inventory at the start of a winter. PG&E had a diverse system, with electricity generated by hydroelectric power in California and the northwest, natural gas fired generators, nuclear, geothermal, wind, and burning oil. The amount of power available from various sources is random, depending on factors such as temperature, rainfall, and breakdowns. There were about 900 input random variables.

Burning oil was the fuel of last resort, more expensive and dirtier than other options, and PG&E would rarely burn oil. Oil inventories were very expensive (about $200 million in stock), so we did not want to hold more than necessary. However, a combination of high demand and low generation from other sources, over a sustained period, could require burning enough oil to exhaust the oil stocks and result in a potentially major, long-term shortage, where the company could not generate sufficient electricity. Naturally, this would be most likely when it would be most harmful, in the middle of a cold winter.

Oil has to be in inventory at the start of winter. PG&E burned low-sulfur crude that must be specially ordered from Indonesia; this takes two months for delivery, and delivery in mid-winter may be impossible due to winter storms.

The ratio between the marginal cost of running one barrel short and the marginal cost of holding an extra barrel was about 300 to 1, so an optimal inventory would result in around 99.7% reliability—holding enough oil to run even one barrel short only about once every 300 years.

The model was large and complex—I would often exceed quota and be kicked off the company mainframe, a multimillion dollar machine. Early versions of the model generated random data in a way that mimicked real life, generating synthetic years of temperature, rainfall, breakdowns, and so on. But only 1 of every 300 such replicates yielded useful information.

We switched to importance sampling. We oversampled colder drier weather with more breakdowns, so that more replicates yielded useful information. We then used importance sampling formulas to adjust for the sampling bias. The work is described in greater detail in Hesterberg (1988, 1995). □

We will start with the classical importance sampling approach and then see what is wrong with that and discuss an alternative. Let f be a pdf and h a function. We write

$$\mathrm{E}[h(X)] = \mu = \int h(x)f(x)\,dx = \int h(x)\frac{f(x)}{g(x)}g(x)\,dx = \int h(x)w(x)g(x)\,dx, \tag{11.15}$$

where g is another density with $g(x) > 0$ when $f(x)h(x) \neq 0$, and the ratio $w(x) = f(x)/g(x)$ is a weighting function.

Instead of drawing observations from f, we may instead draw observations $X_1, X_2, \ldots, X_N$ i.i.d. from g and, referring to General Monte Carlo Integration, compute the classical importance sampling estimate

$$\hat{\mu}_{\mathrm{IS}} = \frac{1}{N}\sum_{i=1}^{N} h(X_i)w(X_i). \tag{11.16}$$

f is called the *target distribution* and g the *design distribution* or *importance function*. The target is what we want answers for; we design our simulation as a means to get there. In the case of PG&E, we choose g with $g(x) > f(x)$ for cold, dry years with more breakdowns. But that sampling is biased, and if we then just took a sample average of the resulting costs, it would be biased. The weighting factor w counteracts the sampling bias by downweighting the oversampled cases and upweighting the undersampled cases.

Write $Y_i = h(X_i)w(X_i)$; then $\hat{\mu}_{\mathrm{IS}} = \bar{Y}$ is just a sample average, with standard error $s_Y/\sqrt{n}$. Thus, we may use the standard error or confidence intervals to gauge the accuracy of this estimate.

Example 11.8 In a *European Option*, an investor purchases the right to buy a stock on a certain future date at a given price K, the "strike price." Suppose the stock price on that date is X; the value of the option is then $h(X) = \max(X - K, 0)$. Suppose the strike price is $K = 700$ and that $X \sim N(500, 120^2)$, based on historic volatility and current market conditions. What is the expected value of $h(X)$? What is the distribution of the payout?

Solution Here the target distribution f is the normal density with mean 500 and standard deviation 120. This implies that X could be negative; in practice, it could not, but we ignore the discrepancy because the probability is small. The probability that the option has any value is $P(X > 700) = 1 - \Phi((700 - 500)/120) = 0.048$, slightly under 5%.

The expected value is

$$\mathrm{E}[h(X)] = \mu = \int_{-\infty}^{\infty} h(x)f(x)\,dx = \int_{700}^{\infty}(x - 700)f(x)\,dx = 2.38. \tag{11.17}$$

The average payout is \$2.38, even though 95% of the time the payout is zero; hence, the average payout given that there is a payout is $2.38/0.048 = 49.8$.

Now consider solving this problem using simulation, without importance sampling. We generate $X_i \overset{\text{i.i.d.}}{\sim} N(500, 120^2)$, compute the payout $h(X_i)$ for each, and calculate the sample average, as well as standard errors and confidence intervals. Using 10^5 replications, we estimate a mean payout of 2.33 with standard error 0.04.

As a bonus, we get an estimate of the whole distribution of the payout—the values of $h(X_i)$ are a sample from that distribution. This makes it easy to compute such quantities as the mean payout, given that the payout is nonzero (49.5) or the probability that the payout exceeds \$100 (0.0056)—the person selling the option may be interested in that or similar quantities.

As a side note, the payout distribution is neither discrete nor continuous, but has elements of both.

Now consider using importance sampling. Instead of sampling from $N(500, 120^2)$, we will sample from $N(700, 120^2)$: that is, we set g to be the density for $N(700, 120^2)$. With this design, half the observations have nonzero payout. The weighting factor $w(x) = f(x)/g(x)$ is then much less than zero for those observations with positive payout. Thus, drawing $X_i \overset{\text{i.i.d.}}{\sim} N(700, 120^2)$, our estimate is of the form

$$\hat{\mu}_{\text{IS}} = \frac{1}{N}\sum_{i=1}^{N} \max(X_i - 700, 0)w(X_i).$$

With 10^5 replications, we estimate a mean of 2.36, with standard error 0.0085. The standard error is about five times smaller than for simple Monte Carlo; in other words, we achieve comparable accuracy with about 5^2 times fewer replications (actually 27.6). This is the *relative efficiency*, or variance under simple Monte Carlo divided by variance with this importance sampling distribution.

R Note:

```
K <- 700      # strike price
mu <- 500     # mean of the stock price at option date
sigma <- 120 # sd

# P(option has value)
pnorm(K, mu, sigma, lower.tail = FALSE)

# expected value, and expected value given has value
integrate(function(x) (x - K) * dnorm(x, mu, sigma), K, Inf)
(integrate(function(x) (x - K) * dnorm(x, mu, sigma), K, Inf)$value/
  pnorm(K, mu, sigma, lower.tail = FALSE))
```

The option `lower.tail = FALSE` to `pnorm` gives `1 - pnorm(K, mu, sigma)`.

We next estimate the integral using Monte Carlo integration without importance sampling. The command pmax computes the coordinate-wise maximum of two vectors.

```
# Simulation, no ImpSamp
N <- 10^5
X <- rnorm(N, mu, sigma)
payout <- pmax(X-K, 0)
mean(payout)
sd(payout) / sqrt(N)
mean(payout[payout > 0])
mean(payout > 100)
```

payout > 0 returns a vector of TRUE or FALSE's (depending on whether or not a particular value is greater than 0. Thus, payout(payout > 0) returns those payout values corresponding to the TRUE's.
payout > 100 return a vector of TRUE's or FALSE's depending on whether or not a particular value is greater than 100. Thus, mean(payout > 100) returns the proportion of values greater than 100.
Now we integrate using importance sampling:

```
# ImpSamp, normal with mean = K
X2 <- rnorm(N, K, sigma)                           # drawing from g
w2 <- dnorm(X2, mu, sigma)/dnorm(X2, K, sigma)     # w2(x) = f(x)/g(x)
Y2 <- pmax(0, X2-K) * w2                           # h(x) * w(x)
mean(Y2)
sd(Y2) / sqrt(N)
sd(Y2) / sd(payout)
var(payout) / var(Y2)                              # relative efficiency
```

□

Definition 11.1 Let $\hat{\theta}_1$ and $\hat{\theta}_2$ be estimates of θ under two Monte Carlo methods. The *relative efficiency* of these two methods is the ratio of the variances under these methods: $\text{Var}[\hat{\theta}_1]/\text{Var}[\hat{\theta}_2]$. ||

Choosing to sample from $N(700, 120^2)$ in the previous example was a bit ad hoc. Let us look a bit deeper and figure out what makes a good importance function g.

We may try to choose g to minimize the variance of the estimate, σ_Y^2/N.

Theorem 11.2 *Let* $\text{E}_g[Y]$ *and* $\text{Var}_g[Y]$ *denote the expected value and variance of* Y *when sampling from* g*. Then,* $\text{Var}_g[Y]$ *is minimized when*

$$g^*_{\text{IS}}(x) \propto |h(x)f(x)|. \tag{11.18}$$

Proof We give here a nonrigorous proof based on calculus of variations; readers not familiar with this method may skip the proof.

$$\begin{aligned}\mathrm{Var}_g[Y] &= \mathrm{E}_g[Y^2] - \mu^2 \\ &= \int h(x)^2 w(x)^2 g(x)\, dx - \mu^2 \\ &= \int h(x)^2 f(x)^2 / g(x)\, dx - \mu^2.\end{aligned}$$

Write $H(x) = h(x) f(x)$; the objective is to minimize $\int H(x)^2/g(x)\, dx$ subject to $\int g(x)\, dx = 1$ and $g(x) \geq 0$ for all x. Consider minimizing

$$\int \frac{H(x)^2}{g(x)} + \lambda g(x)\, dx$$

and setting the derivative with respect to g in the integral equal to zero,

$$\frac{-H(x)^2}{g(x)^2} + \lambda = 0,$$

the solution has $g(x) = \lambda |H(x)|$ for some λ. □

In a case like this, where $h(x)$ is nonnegative for all x, Equation 11.18 reduces to

$$g^*(x) = \frac{h(x) f(x)}{\int_{-\infty}^{\infty} h(u) f(u)\, du}, \tag{11.19}$$

which makes

$$Y = h(X) w(X) = h(x) \frac{f(x)}{g(x)} = \int_{-\infty}^{\infty} h(u) f(u)\, du.$$

In other words, the optimal g makes Y a constant, so the estimate has zero variance. Unfortunately, this distribution requires the normalizing constant $\int_{-\infty}^{\infty} h(u) f(u)\, du$, which happens to be the answer we are looking for (Equation 11.17). In other words, we have to know the answer in order to get the answer. This is a problem!

Even though it is impossible to generate values from the optimal g^*, maybe we can come close. We can use Equation 11.18 as a guide—we would like g to be roughly proportional to $|h(x) f(x)|$. So g should be bigger than f when h is large—the sampling should be biased to produce more of the important cases where h is large. Conversely, g can be smaller than f where h is small.

Example 11.8 (Continued) In the European option example, f can even be zero where h is zero, for $x < 700$. Let us try a *translated exponential distribution*, an exponential distribution translated right 700 units:

$$g(x) = \begin{cases} \lambda \exp(-\lambda x), & x \geq 700, \\ 0, & x < 700. \end{cases} \tag{11.20}$$

This "wastes" no observations for values of x where $h(x) = 0$. To use this, we must choose λ. For a start, we will try using an exponential distribution with the same standard deviation as the normal distribution f. The relative efficiency is about 102, about four times better than using a normal distribution for importance sampling.

R Note (European Option Continued):

```
lambda <- 1 / sigma   # same standard deviation as the normal
X3 <- K + rexp(N, lambda)    # g ~ shifted exponential
w3 <- dnorm(X3, mu, sigma) / dexp(X3 - K, lambda)
Y3 <- pmax(0, X3-K) *w3
mean(Y3)
sd(Y3) / sqrt(N)
sd(Y3) / sd(payout)
var(payout) / var(Y3)

plot(X3[1:300], Y3[1:300], pch=".", xlab = "X",
     ylab = "Y = h(x) f(x)/g(x)")
```

A scatter plot of the resulting Y versus X is shown in Figure 11.8. Y is initially small, then large, then small, indicating that g was too large at both ends, relative to g^*. Changing λ could help at either end, at the cost of making the other end worse.

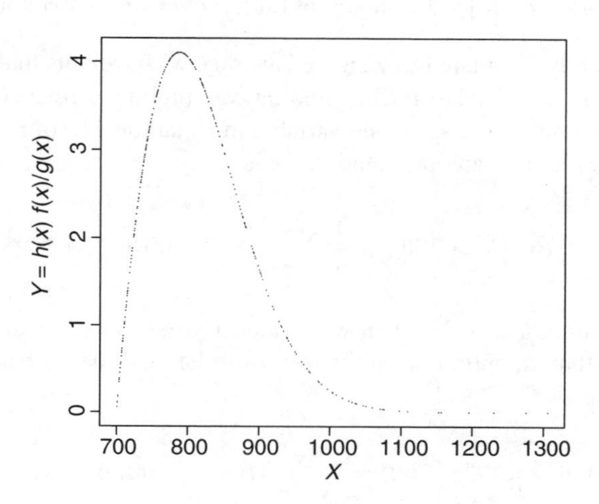

FIGURE 11.8 Importance sampling for the European option example, where $X \overset{\text{i.i.d.}}{\sim} g$ is exponential and $Y = h(X)f(X)/g(X)$.

We could put in more effort to find a better g. To reduce the number of replications, we would choose g to make $Y(x) = h(x)f(x)/g(x)$ as flat as possible. The next step might be to try a shifted gamma distribution with shape parameter $r = 2$ chosen to mimic the shape of $h \times f$ near 700, namely, $g(x) \approx c(x - 700)$ for some c, and scale parameter chosen to mimic $h \times f$ over a broader range. We would not do this here; in practice, there is always a trade-off between human and computer time, and for this application more effort is not needed. □

11.7.1 Ratio Estimate for Importance Sampling

We hinted above that there is something wrong with the classical importance sampling approach. Actually, there are a couple of problems. Let us see what they are and consider remedies.

In practice, we usually want to estimate more than one quantity. In the fuel inventory example, we want to estimate quantities such as the probability of running short, expected cost, probability of cost exceeding $300 million, probability of running short in each of several months, and the expected oil burn in each month, and the expected amount of oil left in inventory. For the European option, we want to estimate the expected value, probabilities of the value exceeding different levels, and the probability of a gain.

One problem with the classical approach is that a design distribution that works well for one quantity may be terrible for another; we will consider this below. The second problem is that its estimates for various quantities are inconsistent, for example,

$$\hat{P}(\text{run out of oil}) + \hat{P}(\text{no run out of oil}) \neq 1$$

and

$$\hat{\mathrm{E}}[\text{amount of oil burnt}] + \hat{\mathrm{E}}[\text{amount of oil left over}] \neq \text{total amount of oil.}$$

The reason is that the estimate is a weighted average with weights that do not sum to 1, so probabilities do not add to 1. The importance sampling estimate of a probability is obtained by letting h be a zero–one variable in Equation 11.16. For example, the estimated probability that an option has value is

$$\hat{P}_{\mathrm{IS}}(X > 700) = \frac{1}{N} \sum_{i=1}^{N} I(X_i > 700) w(X_i),$$

where $I(A)$ is the *indicator function*, with value 1 when A occurs and 0 otherwise. The probability that an option has no value is similar, and the probability of one or the other is

$$\begin{aligned}\hat{P}_{\mathrm{IS}}(X > 700) + \hat{P}_{\mathrm{IS}}(X \leq 700) &= \frac{1}{N} \sum_{i=1}^{N} (I(X_i > 700) + I(X_i \leq 700)) w(X_i) \\ &= \frac{1}{N} \sum_{i=1}^{N} w(X_i).\end{aligned}$$

In general, this is not equal to 1. In our simulation, the two probabilities add to 0.988 when using the normal design distribution and 0.048 when using the exponential design. In fact, with the latter, the estimated probability of the option having no value is zero!

The problem is that the classical estimate is a weighted average, with weights $w(X_i)/N$ that do not sum to 1. A remedy is to normalize the weights to sum to 1, using $w_i / \sum_{j=1}^{N} w_j$; this simplifies to the *importance sampling ratio estimate*:

$$\hat{\mu}_{\text{ratio}} = \sum Y_i / \sum w(X_i) = \bar{Y}/\bar{W}. \tag{11.21}$$

As long as $g(x) > 0$ when $f(x) > 0$, $\mathrm{E}\left[\bar{X}\right] = 1$ (the proof is an exercise), and this estimate is usually similar to the earlier estimate (Equation 11.16). The ratio estimate does require $g(x) > 0$ when $f(x) > 0$ to be consistent; the translated exponential distribution does not qualify.

One extra benefit of using the ratio estimate is that we get the whole distribution for free, as a weighted empirical distribution. For any output quantity h, the estimate of the distribution of $h(X)$ has cumulative distribution function

$$\hat{F}(v) = \frac{\sum_{i=1}^{N} w_i I(h(x_i) \le v)}{\sum_{i=1}^{N} w_i}. \tag{11.22}$$

A second extra benefit of the ratio estimate is that f need not be completely known, it may have an unknown normalizing constant. This is useful in Bayesian applications.

Remark Now a personal note. When I (Hesterberg) first applied importance sampling in the fuel inventory problem, I was not familiar with the importance sampling literature and did what made sense to me. Then, at one point my boss asked me, "You mean you normalize the weights to add to 1?" My answer was roughly, "Of course, doesn't everyone?" Well, actually no, they did not. It was bit embarrassing. However, this realization and the follow-up work turned into my PhD dissertation (Hesterberg (1988)).

One reason that people used the classical estimate instead of the ratio estimate is that the classical estimate is unbiased, while the ratio estimate is in general biased. However, the bias is small, $O(1/N)$,[1] so is negligible when the number of replications N is large; and when running computer simulations, we can make N large. This is a case where people were too concerned about unbiasedness. ||

As long as $g(x) > 0$ when $f(x) > 0$, $\mathrm{E}[W] = 1$, and the ratio estimate is approximately normal with mean $r = \mu_Y$ and variance $\mathrm{Var}[Y - \mu_Y W]/N$, by Theorem 11.1. The standard error is $s/\sqrt{N}$ where s is the sample standard deviation of the residuals $Y - \hat{\mu}_{\text{ratio}} W$.

[1]That is, for large N, the bias is bounded above by a constant multiple of $1/N$.

The asymptotically optimal design for the ratio estimate minimizes

$$\begin{aligned}\mathrm{Var}_g[Y - \mu_Y W] &= \mathrm{E}_g[(Y - \mu_Y W)^2] \\ &= \int (h(x) - \mu_Y)^2 w(x)^2 g(x)\,dx \\ &= \int (h(x) - \mu_Y)^2 f(x)^2 / g(x)\,dx\end{aligned}$$

and is

$$g^*_{\text{ratio}}(x) \propto |h(x) - \mu| f(x). \qquad (11.23)$$

This is often more reasonable than g^*_{IS} from Equation 11.18. g^*_{ratio} is large when $h(X)$ is far from its mean, whereas g^*_{IS} is large when $h(X)$ is far from zero. Consider two examples:

- To estimate $P(\text{option has value})$, $h(x) = I(x > 700)$, g^*_{IS} is zero for $x \le 700$; while to estimate $P(\text{option has no value})$, $h(x) = I(x \le 700)$, g^*_{IS} is zero for $x > 700$. Hence, to estimate equivalent quantities, the design would be the exact opposite. g^*_{ratio} is the same for estimating both probabilities and draws half of its observations from each region.
- Consider estimating the average amount of oil left in inventory at the end of the winter. This is usually identical to the initial amount. g^*_{IS} tries not to draw observations from the rare difficult cases (cold, dry years with power plant breakdowns) where a lot of oil is burned. In contrast, g^*_{ratio} tries to draw more observations from these cases, exactly as if we were estimating the average amount of oil burned.

A strategy that usually works well for both the classical and ratio estimates and is often nearly optimal for the ratio estimate is to use a a *defensive mixture distribution* (Hesterberg (1988, 1995)). Pick a distribution that emphasizes the rare cases and call it g_1. But we do not want to put all of our faith into g_1, so instead we use a mixture of 50% of samples from g_1 and 50% from f,

$$g(x) = 0.5 g_1(x) + 0.5 f(x). \qquad (11.24)$$

Taking half the replications from f defends against everything that can go wrong if g_1 happens to be bad, either for the main output or for a secondary output. It bounds the weight ratio $w(x)/g(x) \le 2$, so the estimate cannot be dominated by one or a few observations with huge weights.

For estimating probabilities, g^*_{ratio} puts 50% of its probability on the cases where the event occurs and 50% where it does not (the proof is an exercise). The 50% defensive mixture tends to approximate this.

To generate replications from Equation 11.24, it is best to stratify (Section 11.4) and draw exactly $N/2$ observations from each of the two mixture components.

11.7.2 Importance Sampling in Bayesian Applications

Now turn back to Bayesian applications. For importance sampling in a Bayesian context, we consider the posterior density for some parameter θ,

$$p(\theta \mid x) = \frac{\pi(\theta)L(\theta)}{\int \pi(\theta)L(\theta)\, d\theta}.$$

where $L(\theta)$ is the likelihood. The target distribution is $p(\theta \mid x)$. To compute the expected value of any function h of θ, we estimate

$$\mathrm{E}\left[h(\theta) \mid x\right] = \frac{\int h(\theta)\pi(\theta)L(\theta)\, d\theta}{\int \pi(\theta)L(\theta)\, d\theta} \tag{11.25}$$

$$= \frac{\int h(\theta)w(\theta)g(\theta)\, d\theta}{\int w(\theta)g(\theta)\, d\theta}, \tag{11.26}$$

where

$$w(\theta) = \frac{\pi(\theta)L(\theta)}{g(\theta)}.$$

This w is not equal to the ratio of the target divided by the design distribution, $p(\theta \mid x)/g(\theta)$, because it is missing a normalizing constant; that constant cancels out.

We can view Equation 11.25 as a ratio of two classical importance sampling estimates or as a single ratio importance sampling estimate. In either case we draw $\theta_1, \theta_2, \ldots, \theta_N$ from g and obtain the weighted empirical distribution with weights $w_i / \sum_{j=1}^{N} w_j$. The corresponding estimate of the mean of $h(\theta)$ is

$$\hat{\mathrm{E}}[h(\theta) \mid x] = \bar{Y}/\bar{w},$$

where $Y = hw$.

Example 11.9 Return to Example 10.3 (Jon Stewart), to the "Bayesian A" analysis with a flat (uniform) prior, $n = 200$ and $x = 1104$. Suppose that we did not know about conjugate families and we use importance sampling with observations drawn from the uniform distribution. In this case, the the prior $\pi(\theta)$ and design distribution g are both uniform and the posterior and w are both proportional to the likelihood $L(\theta) = \theta^{110}(1-\theta)^{90}$. The likelihood is quite close to 0 outside a narrow range around 110/200 (Figure 11.9a). Figure 11.9b shows the corresponding weighted empirical cdf. Most of the observations are almost wasted, with weights almost zero. We use a small $N = 180$ in this example, so the jumps in the empirical cdf are visible, but in practice N would be much larger, say $N = 10^6$.

We should be able to find a better design distribution than the uniform, one that is shaped more like the posterior, resulting in more even weights. The shape of L in Figure 11.9a appears approximately normal, and the posterior has the same shape as L here. Furthermore, in practice, posterior distributions are often approximately normal when sample sizes are large, so we will consider a normal distribution. We will find

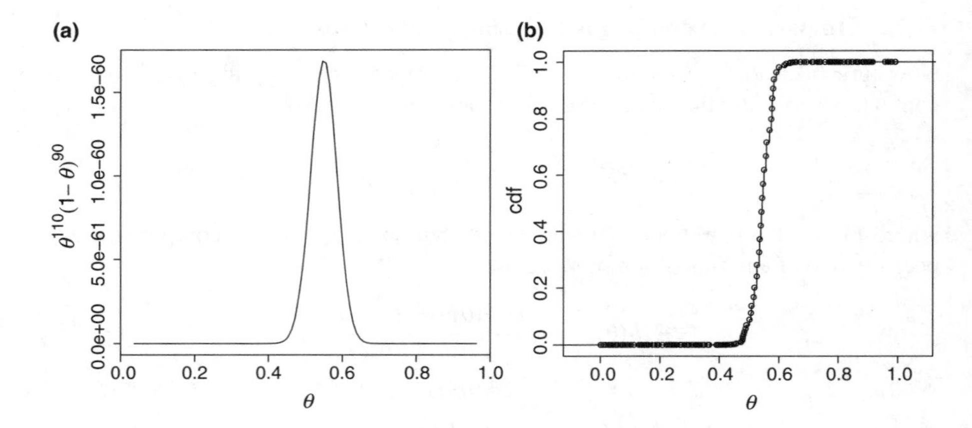

FIGURE 11.9 Importance sampling for the Jon Stewart example, $n = 200$ and $x = 110$, where the prior and design distributions are uniform. (a) The likelihood L. (b) The importance sampling weighted empirical cdf.

the location and value of the maximum of L and its second derivative at the maximum. We will pick the normal mean to match the maximum and standard deviation so that $g''/g = L''/L$ at the maximum. For the latter, it is easier to work with logs; we set $\log''(g(\theta)) = \log''(L(\theta))$. (In general, we would work with the product $\pi(\theta)L(\theta)$ rather than L; here they are the same.) This procedure of matching derivatives to choose the normal variance is the Laplace method (Carlin and Louis (2009)). The second derivative of the log of a normal density is $-1/\sigma^2$.

In this simple example, we can find the derivatives of $\log(L)$ using calculus. The maximum occurs at $\theta = 110/200$, and the second derivative of the log-likelihood at this value is $-200(1/110 + 1/90)$. In practice, we usually need to find the maximum numerically, for example, using the `optimize` function in `R`, and then evaluate the second derivative numerically.

The resulting weighted empirical cdf is shown in Figure 11.10a. Compare this with Figure 11.9a—in the middle there are many more steps, and they are smaller. This succeeded in sampling more in the important region in the middle. Figure 11.10b shows a plot of the weights; it is always a good idea to look at the weights, especially to ensure that the weights are not blowing up in the tails of the distribution. This appears not to be the case, the weights are nearly constant.

The normal design distribution does sometimes produce observations outside the range of (0, 1). This does not invalidate the analysis, but it is a waste of effort; any observation that occur there get a weight of zero. In this example, there are so few of them that we would not worry about it. In general, it is alright if $g(x) > 0$ where $f(x) = 0$, but not the reverse.

The normal design appears to work well here, but is often risky; the normal distribution drops to zero very quickly outside the ± 3 standard deviations; if the target distribution does not drop as quickly, there could be huge weights in the tails. In

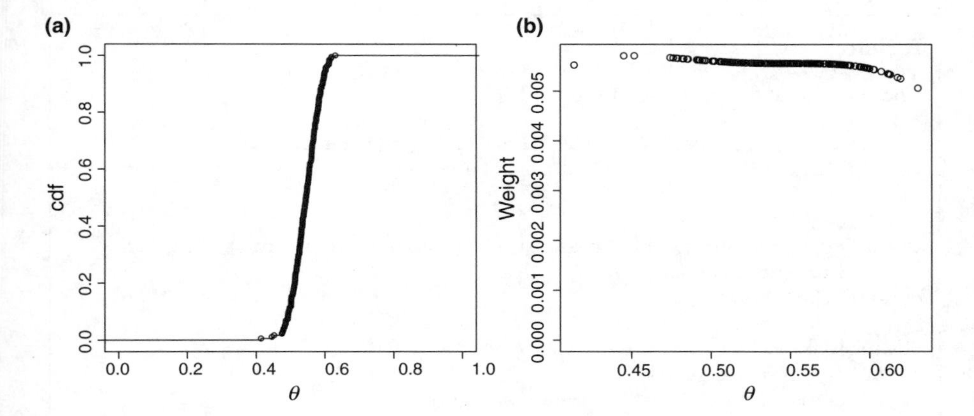

FIGURE 11.10 Weighted empirical cdf from importance sampling for the Jon Stewart example, using a normal design distribution. (a) The weighted cdf. (b) The corresponding weights.

practice, people often prevent this using a translated scaled t distribution or using a mixture distribution. Here, we will demonstrate using a mixture of 90% from the same normal distribution as before and 10% from a uniform distribution.

The resulting weighted empirical cdf is shown in Figure 11.11a. Compared to the pure normal, this has more observations in the tails, each receiving small weight. Figure 11.11b shows the weights; here they are smaller in the tails, which tends to be safe.

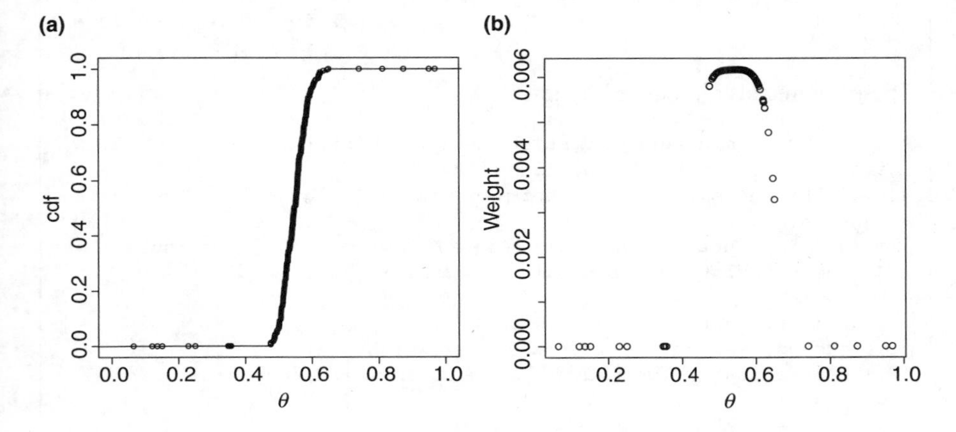

FIGURE 11.11 Weighted empirical cdf from importance sampling for the Jon Stewart example, prior distribution is uniform, with $n = 200$ and $x = 110$. (a) Based on a normal design distribution. (b) Based on a mixture of normal and uniform.

R Note:

These commands reproduce Figure 11.9.

```
likelihood <- function(theta) theta^110 * (1-theta)^90
logLikelihood <- function(theta) 110 * log(theta) + 90 *log(1-theta)

N <- 180
theta <- sort(runif(N))  # Sampling from a uniform distribution
w <- likelihood(theta)
curve(likelihood(x))

dev.new()
weight <- w / sum(w)
plot(stepfun(theta, c(0, cumsum(weight))),
     xlab = substitute(theta), ylab = "cdf", main = "")
```

Using calculus, we find that the maximum of the log-likelihood occurs at $\theta = 110/200$. The second derivative here is $-(110/\theta^2 + 90/(1-\theta)^2 = -200^2 \times (1/90 + 1/110) = -1/\sigma^2$.

```
mu <- 110/200
sigma <- 1/(200 * sqrt(1/90 + 1/110))
curve(dnorm(x, mu, sigma))
```

To find the maximum and second derivative numerically in R:

```
thetaMax <- optimize(logLikelihood, interval = 0:1, maximum = TRUE)
  $maximum
epsilon <- .001
(logLikelihood(thetaMax + epsilon) + logLikelihood(thetaMax -
 epsilon)
 - 2 * logLikelihood(thetaMax)) / epsilon^2  # second derivative
-200^2*(1/90 + 1/110)                        # for comparison,
                                               is very close
```

These commands reproduce Figure 11.10.

```
# Sampling from a normal distribution (no mixture)
theta2 <- sort(rnorm(N, mu, sigma))
w2 <- likelihood(theta2) / dnorm(theta2, mu, sigma)
weight2 <- w2 / sum(w2)
plot(stepfun(theta2, c(0, cumsum(weight2))), xlab = substitute
    (theta), ylab = "cdf", main = "", xlim = 0:1)

dev.new()
plot(theta2, weight2, ylim = c(0, max(weight2)))
# make sure that weights don't blow up anywhere, especially in
# the tails
```

Here is code to estimate a probability and standard error for the estimate:

```
y2 <- (theta2 > .6) * w2                       # h(theta) = (theta > .6)
```

```
r2 <- mean(y2) / mean(w2)                       # ratio estimate for
P(theta > .6)
r2
stdev(y2 - r2 * w2) / (mean(w2) * sqrt(N))   # standard error
```

This reproduces Figure 11.11:

```
# Mixture of 90% normal and 10% uniform
theta3 <- c(rnorm(.9*N, mu, sigma), runif(.1*N))
w3 <- likelihood(theta3) / (0.9 * dnorm(theta3, mu, sigma) + 0.1)
weight3 <- w3 / sum(w3)
out <- order(theta3)
plot(stepfun(theta3[out], c(0, cumsum(weight3[out]))),
     xlab = substitute(theta), ylab = "cdf", main = "", xlim = 0:1)

dev.new()

plot(theta3, weight3, ylim = c(0, max(weight2))) # check if weights
                                                 # blow up
```

To estimate a probability and standard error for the estimate based on stratified sampling:

```
y3 <- (theta3 > .6) * w3               # h(theta) = (theta > .6)
r3 <- mean(y3) / mean(w3)              # ratio estimate for P(theta > .6)
r3
sqrt(.9 * var((y3 - r3 * w3)[1:(.9*N)]) +
     .1 * var((y3 - r3 * w3)[(.9*N+1):N]))/(mean(w3) * sqrt(N)) # SE
```

□

11.8 EXERCISES

1. In Section 3.3, we performed a permutation test to determine if men and women consumed, on average, different amounts of hot wings.
   (a) Do smoothed bootstraps for the mean consumption for men and the mean consumption for women. How do the distributions differ?
   (b) Do a smoothed bootstrap for the difference in means.
   (c) Do an ordinary bootstrap for the difference in means.
   (d) How do the ordinary and smoothed bootstrap distributions differ?
2. Perform a smoothed bootstrap for the mean arsenic level from the `Bangladesh` data set (Example 5.3). Use a transformation to prevent the smooth from creating negative values.
3. Refer to Exercise 15 of Chapter 6 where, given the data, we used the method of moments to estimate λ and r for the Gamma distribution. It turns out that $\hat{r} = 22.88$ and $\hat{\lambda} = 3.138$. Now perform a parametric bootstrap for the mean.
4. Refer to Exercise 17 of Chapter 6 where we modeled the time between successive earthquakes using the Weibull distribution. From the given data, we

estimate $k = 0.917$ and $\lambda = 17.344$. Do a parametric bootstrap for the mean time between earthquakes and for the average number of earthquakes per year.

5. Suppose that $X_1, \ldots, X_n \overset{\text{i.i.d.}}{\sim} \text{Gamma}(r, \lambda)$.
   (a) Find the variance of $\bar{X}$.
   (b) Use the delta method to find the approximate variance of $\bar{X}^2$.
   (c) Use the delta method to find the approximate variance of $\log(\bar{X})$.
   (d) Use the delta method to find the approximate variance of $\sqrt{\bar{X}}$.
   (e) Check your answers using simulation.
6. Compute standard errors for the ratio of means $\bar{Y}/\bar{X}$ for the Verizon data, where $\bar{Y}$ and $\bar{X}$ are the mean of CLEC and ILEC repair times, respectively, using
   (a) the delta method for $\bar{Y}/\bar{X}$,
   (b) the delta method for $\log(\bar{Y}/\bar{X})$,
   (c) the ordinary bootstrap for $\bar{Y}/\bar{X}$, and
   (d) the ordinary bootstrap for $\log(\bar{Y}/\bar{X})$.

   Then compute confidence intervals,
   (e) t interval for $\bar{Y}/\bar{X}$ using the delta method SE,
   (f) t interval for $\log(\bar{Y}/\bar{X})$ using the delta method SE,
   (g) bootstrap percentile interval for $\bar{Y}/\bar{X}$, and
   (h) bootstrap percentile interval for $\log(\bar{Y}/\bar{X})$.

   For degrees of freedom for the t intervals, use the smaller sample size minus 1. Finally,
   (i) transform the two intervals on the log scale to the scale $\bar{Y}/\bar{X}$, and
   (j) compare the four intervals for μ_Y/μ_X.
7. Estimate $\int_0^1 \cos(x^2)\, dx$ by the following:
   (a) Using Monte Carlo integration.
   (b) Using the `integrate` function in **R**.
   (c) Expanding $\cos(x^2)$ as a power series out to degree 12. What is the error? *Hint*: This is an alternating series!
   (d) Using a computer algebra system.
8. Consider $4 \int_0^1 1/(1 + x^2)\, dx$.
   (a) Use calculus to evaluate the integral.
   (b) Use Monte Carlo integration to find an estimate for π.
9. Let $X_1, X_2, \ldots, X_N$ be i.i.d. from a distribution with density f and let h denote a real-valued function. Verify that Equation 11.14 holds. (Pick f and h, and verify using simulation that Equation 11.14 approximately equals Equation 11.13.)

10. Estimate $\int_{-\infty}^{\infty} e^{-x^4}/2\,dx$ using the design distribution $g(x) = (1/\sqrt{2\pi})e^{-x^2/2}$, the pdf for the standard normal distribution.

11. A computer algebra system gives the estimate $\int_0^{\infty} \ln(1+x)e^{-x}\,dx = 0.596347$.

    (a) Compute an estimate of the integral and a confidence interval for the estimate using the pdf of the chi-square distribution with 3 degrees of freedom as the importance function. Use the commands `rchisq` and `dchisq` for obtaining random values and density values, respectively, for the chi-square distribution.

    (b) Compute an estimate of the integral and a confidence interval for the estimate, using the pdf of the exponential distribution with $\lambda = 1/2$. Use the commands `rexp` and `dexp` for obtaining random values and density values, respectively, for the exponential distribution.

    (c) Write the integral in the form $\int_0^{\infty} h(x)f(x)dx$; graph both h functions on one set of axes over the range (0, 10).

    (d) Which of the two f gives more accurate estimates? Explain why based on the graph.

12. In importance sampling, $(1/N)\sum_{i=1}^{N} w(X_i)$ is in general not equal to 1. However, it may be equal to 1 on average.

    (a) Show that if $g(x) > 0$ whenever $f(x) > 0$, then $(1/N)\sum_{i=1}^{N} w(X_i) = 1$.

    (b) In one example in this chapter, this condition was not met; discuss what happened to $(1/N)\sum_{i=1}^{N} w(X_i)$ in that example.

13. For estimating a probability $P(A) = \int_A f(x)\,dx = \int h(x)f(x)\,dx$ with $h(x) = 1$ if $x \in A$ and 0 otherwise, show that if $g(x) \propto |h(x) - P(A)|f(x)$, then g has 50% of its probability on each of A and its complement.

    This g is g^*_{ratio}, the optimal design for the ratio estimate; this implies that the optimal design samples equally from successes and failures.

APPENDIX A

REVIEW OF PROBABILITY

This section contains a brief review of some definitions and results from probability that will be required in this book. Refer to a textbook on probability for more in-depth information, for example, Ghahramani (2004), Pitman (1993), Ross (2009), or Scheaffer and Young (2010).

A.1 BASIC PROBABILITY

Recall that the set of all possible outcomes of a random experiment is called a *sample space*, S. An event E is a subset of S.

Proposition A.1 (Law of Total Probability) *Let A denote an event in a sample space S and let $B_1, B_2, \ldots, B_n$ be a disjoint partition of A. Then*

$$P(A) = P(B_1)P(A \mid B_1) + P(B_2)P(A \mid B_2) + \cdots + P(B_n)P(A \mid B_n).$$

Definition A.1 A *discrete random variable* X is a function from S into the real numbers $\mathbf{R}$ with a range that is finite or countably infinite. That is, X: $S \to \{x_1, x_2, \ldots, x_m\}$, or X: $S \to \{x_1, x_2, \ldots\}$. ||

For instance, if we consider the experiment of rolling two dice, we can let X denote the sum of the two numbers that appear. Then X is a discrete random variable, X: $S \to \{2, 3, \ldots, 12\}$. The probability mass function (pmf) is a function $p : \mathbf{R} \to [0, 1]$

Mathematical Statistics with Resampling and R, First Edition. By Laura Chihara and Tim Hesterberg.

such that $p(x) = P(X = x)$, for all x in the range of X. Note then that $\sum_x p(x) = 1$, where the sum is over the range of X.

Definition A.2 A function X from S into the real numbers $\mathbf{R}$ is a *continuous random variable* if there exists a nonnegative function f such that for every subset C of $\mathbf{R}$, $P(X \in C) = \int_C f(x)\,dx$. In particular, for $a \leq b$, $P(a < X \leq b) = \int_a^b f(x)\,dx$. ||

The function f is called the *probability density function* (pdf) of X. Note that $\int_{-\infty}^{\infty} f(x)\,dx = 1$. The *(cumulative) distribution function* F of a random variable X is the function F: $\mathbf{R} \to [0, 1]$ that satisfies

$$F(x) = P(X \leq x), \quad -\infty < x < \infty.$$

F is a nondecreasing, right-continuous function with $\lim_{x \to -\infty} F(x) = 0$ and $\lim_{x \to \infty} F(x) = 1$. In the case that X is a continuous random variable,

$$F(x) = P(X \leq x) = \int_{-\infty}^{x} f(t)\,dt,$$

and thus at every point x at which $f(x)$ is continuous,

$$F'(x) = f(x)$$

by the fundamental theorem of calculus.

Example A.1 Recall that X is an exponential random variable with $\lambda > 0$ if its pdf is $f(x) = \lambda e^{-\lambda x}$, $x \geq 0$. Then the distribution function is

$$F(x) = P(X \leq x) = \int_0^x e^{-\lambda t}\,dt = 1 - e^{-\lambda x}.$$

We also have $\lim_{x \to -\infty} F(x) = 0$ and $\lim_{x \to \infty} F(x) = 1$. □

A.2 MEAN AND VARIANCE

Definition A.3 Let X: $S \to \mathbf{R}$ denote a random variable and f denote its density function. The *mean* of X, also known as the *expected value* of X, is

$$\mathrm{E}[X] = \mu = \int_{-\infty}^{\infty} x f(x)\,dx$$

and the *variance* is

$$\mathrm{Var}[X] = \sigma^2 = \int_{-\infty}^{\infty} (x - \mu)^2 f(x)\,dx.$$

||

The mean and variance may not exist if the integrals are not finite.

Definition A.4 The *standard deviation* of X is $\mathrm{SD}[X] = \sqrt{\mathrm{Var}[X]}$. ||

If $X\colon S \to \{x_1, x_2, \ldots\}$ is a discrete random variable with $P(X = x_i) = p_i$, then $\mathrm{E}[X] = \mu = \sum_i x_i \cdot p_i$ and $\mathrm{Var}[X] = \sum_i (x_i - \mu)^2 \cdot p_i$.

Example A.2 Suppose X is a finite discrete random variable, $X\colon S \to \{a_1, a_2, \ldots, a_n\}$, with $p(a_i) = 1/n$. That is, each outcome is equally likely. Then

$$\mathrm{E}[X] = \mu = \frac{1}{n}\sum_{i=1}^{n} a_i, \tag{A.1}$$

the "usual" average, and

$$\mathrm{Var}[X] = \mathrm{E}\left[(X-\mu)^2\right] = \sum_{i=1}^{n}(a_i - \mu)^2\frac{1}{n} = \frac{1}{n}\sum_{i=1}^{n}(a_i - \mu)^2. \tag{A.2}$$

□

Proposition A.2 $\mathrm{Var}[X] = \mathrm{E}\left[(X-\mu)^2\right] = \mathrm{E}\left[X^2\right] - \mu^2$.

Theorem A.1 *If X is a random variable with density f and g is any real-valued function, then*

$$\mathrm{E}\left[g(X)\right] = \int_{-\infty}^{\infty} g(x) f(x)\, dx.$$

We state a special case of Theorem A.1 as its own theorem:

Theorem A.2 *If X is a random variable and a and b are constants, then* $\mathrm{E}[a + bX] = a + b\mathrm{E}[X]$ *and* $\mathrm{Var}[a + bX] = b^2\mathrm{Var}[X]$.

Definition A.5 The random variables X and Y have a joint density if there exists a nonnegative function $f\colon \mathbf{R} \times \mathbf{R} \to \mathbf{R}$, such that for every subset C of the plane,

$$P((X, Y) \in C) = \int_C\int f(x, y)\, dx\, dy.$$

||

Definition A.6 Let X and Y be random variables. Then X and Y are *independent* if for any sets A and B,

$$P(X \in A, Y \in B) = P(X \in A)P(Y \in B).$$

||

Theorem A.3 *If X and Y are random variables, then* $\mathrm{E}[X + Y] = \mathrm{E}[X] + \mathrm{E}[Y]$.

Theorem A.4 *If X and Y are independent, then* $\mathrm{Var}[X+Y] = \mathrm{Var}[X] + \mathrm{Var}[Y]$.

Note that Theorem A.3 is true regardless of whether or not X and Y are independent, but Theorem A.4 is not. Failure to properly allow for the latter point is a common error, which was a prime contributor to the global financial meltdown of 2008. Financial institutions calculated the risk of their portfolios based on variability calculations without allowing the dependence between different assets. Variables that appear on the surface to be independent may in fact be dependent because of *lurking variables*, variables that affect both. For example, if X and Y are the amounts that two people default on their mortgages, then X and Y are dependent even if the people live in different states, because of economic conditions that affect both.

The combination of Theorem A.2 with Theorem A.3 or Theorem A.4 is very useful for working with two or more random variables. For example, for two variables, $\mathrm{E}[aX + bY] = a\mathrm{E}[X] + b\mathrm{E}[Y]$, and for two independent variables, $\mathrm{Var}[aX + bY] = a^2\mathrm{Var}[X] + b^2\mathrm{Var}[Y]$. For more variables, of central importance is the case of a mean of a set of variables.

A.3 THE MEAN OF A SAMPLE OF RANDOM VARIABLES

Definition A.7 Let $X_1, X_2, ..., X_n$ be identically distributed random variables with mean μ and variance σ^2. Then

$$\bar{X} = \frac{1}{n}\sum_{i=1}^{n} X_i \tag{A.3}$$

is called the *(sample) mean* of the $X_1, X_2, \ldots, X_n$. ||

Theorem A.5 *Let* $X_1, X_2, ..., X_n$ *be identically distributed random variables with mean* μ *and variance* σ^2.

Then $\mathrm{E}\left[\bar{X}\right] = \mu$.

If, in addition, $X_1, X_2, ..., X_n$ *are independent, then* $\mathrm{Var}[\bar{X}] = \sigma^2/n$.

Proof

$$\begin{aligned}
\mathrm{E}\left[\bar{X}\right] &= \mathrm{E}\left[\frac{1}{n}\sum_{i=1}^{n} X_i\right] \\
&= \frac{1}{n}\sum_{i=1}^{n} \mathrm{E}[X_i] \\
&= \frac{1}{n}\sum_{i=1}^{n} \mu \\
&= \mu
\end{aligned}$$

and

$$\begin{aligned}\operatorname{Var}[\bar{X}] &= \operatorname{Var}\left[\frac{1}{n}\sum_{i=1}^{n} X_i\right] \\ &= \frac{1}{n^2}\sum_{i=1}^{n}\operatorname{Var}[X_i] \\ &= \frac{1}{n^2}n\sigma^2 \\ &= \frac{\sigma^2}{n}.\end{aligned}$$

□

Example A.3 Let $X_1, X_2, \ldots, X_{20}$ be independent random variables, each normally distributed, $N(3, 4^2)$. Then $\mathrm{E}\left[\bar{X}\right] = 3$ and $\operatorname{Var}[\bar{X}] = 4^2/20$. □

Consider $X_1, X_2, \ldots, X_n$ independent Bernoulli random variables, each satisfying $X_i = 1$ with probability p. That is, $\mathrm{E}[X_i] = p$ and $\operatorname{Var}[X_i] = p(1-p)$ for $i = 1, 2, \ldots, n$. Then $\bar{X} = (1/n)\sum_{i=1}^{n} X_i =$ (number of 1's)$/n$ = proportion of 1's (sample proportion). Then $\mathrm{E}\left[\bar{X}\right] = p$ and $\operatorname{Var}[\bar{X}] = p(1-p)/n$. We will use the notation $\hat{p}$ to denote the sample proportion (i.e., $\hat{p} = \bar{X}$ in the case of Bernoulli random variables).

We summarize this:

Corollary A.1 *Let $X_1, X_2, \ldots, X_n$ be independent Bernoulli random variables with $\mathrm{E}[X_i] = p$ and $\operatorname{Var}[X_i] = p(1-p)$ for $i = 1, 2, \ldots, n$. Then $\mathrm{E}\left[\hat{p}\right] = p$ and $\operatorname{Var}[\hat{p}] = p(1-p)/n$.*

In this book, we will work with independent random variables drawn from a common distribution.

Definition A.8 Let $X_1, X_2, \ldots, X_n$ be independent random variables drawn from a common distribution. We say that the random variables are *independent and identically distributed* or i.i.d. Let F and f denote the cdf and pdf, respectively, of the common distribution. Then we may also denote this by

$$X_1, X_2, \ldots, X_n \overset{\text{i.i.d.}}{\sim} F \quad \text{or} \quad X_1, X_2, \ldots, X_n \overset{\text{i.i.d.}}{\sim} f.$$

||

A.4 THE LAW OF AVERAGES

You may be familiar with the "law of averages," the idea that in the long run the sample average (or sample proportion) gets closer and closer to the true mean (or proportion). The more rigorous version of this is the *strong law of large numbers*:

Theorem A.6 (Strong Law of Large Numbers (SLLN)) *Let $X_1, X_2, \dots$ be independent and identically distributed random variables with* $\mathrm{E}[X_i] = \mu < \infty$. *Then*

$$P\left(\lim_{N\to\infty} \frac{X_1 + X_2 + \cdots + X_N}{N} = \mu\right) = 1.$$

The SLLN is usually proved in a typical probability course (Casella and Berger (2001); Ghahramani (2004); Ross (2009)).

The SLLN works by *swamping*, not *compensation*. If you are flipping a fair coin and got 80 heads in the first 100 flips, then over the next 100 flips, the expected number of heads is 50—not 20. As you continue to flip more coins—say 10^8 more—the extra 30 heads become insignificant. The remarkable thing about the SLLN is that it works even if the variance of the distribution is infinite. On the other hand, the SLLN does not say anything about how close the sample mean is to the true mean. If the variance is finite, then we can obtain such estimates using Chebyshev's inequality, or a combination of the Central Limit Theorem (Section 4.3) and normal distributions.

Theorem A.7 *Let $X_1, X_2, \dots, X_n$ be i.i.d. random variables with* $\mathrm{E}[X_i] = \mu$ *and* $\mathrm{Var}[X_i] = \sigma^2 < \infty$. *Then by Chebyshev's inequality*

$$P(|\bar{X} - \mu| \geq \epsilon) \leq \frac{\sigma^2}{n\epsilon^2}.$$

Proof

$$\begin{aligned}\mathrm{Var}[\bar{X}] &= \sigma^2/n = \mathrm{E}\left[(\bar{X} - \mu)^2\right] \\ &= \int (y - \mu)^2 f_{\bar{X}}(y)\, dy \geq \int \epsilon^2 I(|y - \mu| > \epsilon) f_{\bar{X}}(y)\, dy.\end{aligned}$$

□

Chebyshev's inequality is typically quite conservative; a normal approximation is typically much more accurate.

A.5 THE NORMAL DISTRIBUTION

The random variable X has a normal distribution with parameters μ and σ if its pdf is

$$f(x) = \frac{1}{\sqrt{2\pi}\sigma} e^{-(x-\mu)^2/(2\sigma^2)}.$$

We denote this by $X \sim N(\mu, \sigma^2)$.

Theorem A.8 *Let $X \sim N(\mu, \sigma^2)$. Then* $\mathrm{E}[X] = \mu$ *and* $\mathrm{Var}[X] = \sigma^2$.

Theorem A.9 *Let $X \sim N(\mu, \sigma^2)$ and define $Z = (X - \mu)/\sigma$. Then $Z \sim N(0, 1)$.*

Subtracting the mean and dividing by the standard deviation is called *standardization* (for any random variable, not just normal). For normal random variables, it is a key step in performing probability calculations.

$$\begin{aligned} P(X \le x) &= \int_{-\infty}^{x} f(t)\,dt \\ &= P(Z \le (x-\mu)/\sigma) \\ &= \Phi((x-\mu)/\sigma), \end{aligned} \tag{A.4}$$

where

$$\Phi(z) = P(Z \le z) = \frac{1}{\sqrt{2\pi}} \int_{-\infty}^{z} e^{-t^2/2}\,dt \tag{A.5}$$

is the cumulative distribution function for the standard normal distribution.

Remark Recall the 68-95-99.7 rule: For a standard normal random variable $Z \sim N(0, 1)$,

$$\begin{aligned} P(-1 < Z < 1) &\approx 0.68, \\ P(-2 < Z < 2) &\approx 0.95, \\ P(-3 < Z < 3) &\approx 0.997. \end{aligned}$$

||

A.6 SUMS OF NORMAL RANDOM VARIABLES

Theorem A.10 *Let X be a normal random variable with mean μ_1 and variance σ_1^2, and let Y be a normal random variable with mean μ_2, and variance σ_2^2. Assume that X and Y are independent. Then $X \pm Y$ is a normal random variable with mean $\mu_1 \pm \mu_2$ and variance $\sigma_1^2 + \sigma_2^2$.*

Example A.4 In the town of Sodor, the weights of boys are normally distributed, $N(100, 5^2)$, while the weights of girls are normally distributed, $N(90, 6^2)$. If a boy and a girl are selected independently and at random from the population, what is the probability that the boy will weigh at least 6 pounds more than the girl?

Solution Let X and Y denote the weights of the selected boy and girl, respectively. We want $P(X \ge Y + 6) = P(X - Y \ge 6)$. By the theorem, $X - Y$ is normally distributed with mean $100 - 90 = 10$ and variance $5^2 + 6^2 = 61$. So,

$$\begin{aligned} P(X - Y \ge 6) &= P\left(\frac{X - Y - 10}{\sqrt{61}} \ge \frac{6 - 10}{\sqrt{61}}\right) \\ &= P(Z \ge -0.5121) \approx 0.6957. \end{aligned}$$

□

More generally,

Theorem A.11 *Let $X_1, X_2, \ldots, X_n$ be independent normal random variables with mean μ_i, variance σ_i^2, $i = 1, 2, \ldots, n$, respectively. Let $a_1, a_2, \ldots, a_n$ be arbitrary constants. Then $a_1X_1 + a_2X_2 + \cdots + a_nX_n$ is a normal random variable with mean $a_1\mu_1 + a_2\mu_2 + \cdots + a_n\mu_n$ and variance $a_1^2\sigma_1^2 + a_2^2\sigma_2^2 + \cdots + a_n^2\sigma_n^2$.*

Corollary A.2 *Let $X_1, X_2, \ldots, X_n$ be independent normal random variables with common mean μ and common variance σ^2. Let $\bar{X}$ denote the sample mean. Then $\bar{X}$ is normally distributed with mean μ and variance (σ^2/n).*

Example A.5 The amount of coffee a machine dispenses is normally distributed with mean 8 ounces and variance 0.47 ounces. If 10 cups of coffee dispensed from this machine are chosen at random, what is the probability that the average amount of coffee is more than 8.5 ounces?

Solution The amount of coffee in one cup is a random variable with mean $\mu = 8$ and standard deviation $\sigma = \sqrt{0.47}$. If $\bar{X}$ denotes the mean of 10 cups of coffee, then $\bar{X}$ is distributed normally with mean 8 ounces and standard deviation $\sqrt{0.47}/\sqrt{10} \approx 0.2168$. Thus,

$$
\begin{aligned}
P(\bar{X} > 8.5) &= P\left(\frac{\bar{X} - 8}{0.2168} > \frac{8.5 - 8}{0.2168}\right) \\
&\approx P(Z > 2.3062) \\
&= 1 - P(Z \leq 2.3062) = 0.0105.
\end{aligned}
$$

□

A.7 HIGHER MOMENTS AND THE MOMENT GENERATING FUNCTION

The material in this section is used in Sections 2.7, B.9, and B.10. The mean and standard deviation of a normal distribution describe its *center* and *spread*. Every normal distribution has the same basic shape, so the mean and standard deviation completely determine a normal distribution. To describe nonnormal variables, it is also useful to describe how asymmetrical the variable is and how peaked. All four characteristics, center, spread, asymmetry, and peakedness, are based on *moments* of a variable.

Definition A.9 Let X be a random variable with mean μ. For a positive integer n, the nth moment of X is

$$\mu_n' = \mathrm{E}\left[X^n\right], \tag{A.6}$$

and the nth central moment of X is

$$\mu_n = \mathrm{E}\left[(X - \mu)^n\right], \tag{A.7}$$

provided $\int_{-\infty}^{\infty} |x|^n f(x)\, dx < \infty$, where f denotes the pdf of X. ||

Thus, the mean of X is its first moment, and the variance of X is its second central moment. In many situations, it is often easier to characterize distributions using *moment generating functions (mfg)*.

Definition A.10 Let X be a random variable with cdf F. The *moment generating function (mfg) of X* is defined by

$$M(t) = \mathrm{E}\left[e^{tX}\right],$$

provided the expected value exists in a neighborhood of 0. ||

Example A.6 Let $X \sim \mathrm{Binom}(n, p)$. Then

$$\begin{aligned} M(t) &= \mathrm{E}\left[e^{tX}\right] \\ &= \sum_{x=0}^{n} e^{tx} \binom{n}{x} p^x (1-p)^{n-x} \\ &= \sum_{x=0}^{n} \binom{n}{x} (e^t p)^x (1-p)^{n-x} \\ &= (e^t p + 1 - p)^n, \end{aligned}$$

which holds for all t. □

Example A.7 Let $X \sim \mathrm{Exp}(\lambda)$, with $\lambda > 0$. Then

$$\begin{aligned} M(t) &= \mathrm{E}\left[e^{tX}\right] \\ &= \int_0^\infty e^{tx} \lambda e^{-\lambda x}\, dx \\ &= \int_0^\infty \lambda e^{(t-\lambda)x}\, dx \\ &= \int_0^\infty \lambda e^{-u} \frac{1}{\lambda - 1}\, du \qquad \text{where } u = (t-\lambda)x \\ &= \frac{\lambda}{\lambda - t}, \end{aligned}$$

where $t < \lambda$. □

We now state the result that explains the name given to $M(t)$.

Theorem A.12 *Let X be a random variable with mgf $M(t)$. Then*

$$\frac{d^n}{dt^n} M(t)|_{t=0} = \mathrm{E}\left[X^n\right].$$

That is, the nth *moment of X is the* nth *derivative of $M(t)$ evaluated at $t = 0$.*

Example A.8 For $X \sim \text{Binom}(n, p)$, we have $M(t) = (e^t p + 1 - p)^n$. Thus, $M'(t) = n(e^t p + 1 - p)^{n-1} e^t p$, so $M'(0) = np = \text{E}[X]$. □

The following theorem allows us to use moment generating functions to characterize distributions.

Theorem A.13 *Let X and Y be random variables with corresponding cdf's F_X and F_Y, respectively. Suppose all the moments of X and Y exist. If the moment generating functions for X and Y exist and $M_X(t) = M_Y(t)$ for all t in some neighborhood about 0, then $F_X(s) = F_Y(s)$ for all s.*

Moment generating functions for the sum of independent random variables can be found easily.

Theorem A.14 *Let $X_1, X_2, \ldots, X_n$ be independent random variables with moment generating functions $M_{X_1}(t), M_{X_2}(t), \ldots, M_{X_n}(t)$, respectively. Then*

$$M_{X_1+X_2+\cdots+X_n}(t) = M_{X_1}(t) M_{X_2}(t) \cdots M_{X_n}(t).$$

APPENDIX B

PROBABILITY DISTRIBUTIONS

In this section, we gather results about some special probability distributions. Many of these are typically covered in a first probability course, so the presentations for them are brief. A document with information on computing probabilities and quantiles in `R` is provided at `https://sites.google.com/site/ChiharaHesterberg`.

B.1 THE BERNOULLI AND BINOMIAL DISTRIBUTIONS

Consider a random experiment that has only two outcomes, tossing a coin and recording "heads" or "tails." This experiment is called a *Bernoulli trial*. We define a random variable X on the set of these two outcomes, letting X have the value 0 with probability p for one outcome, and value 1 with probability $1 - p$ for the other outcome. Then X is called a *Bernoulli* random variable. For instance, if we roll a die and consider seeing a 1 a "success" and anything else a "failure," we can let $X = 1$ for a success with probability $p = 1/6$ and $X = 0$ for a failure with probability $1 - 1/6 = 5/6$.

We denote a random variable X with probability p of success by $X \sim \text{Bern}(p)$.

$$\text{E}[X] = p, \quad \text{Var}[X] = p(1 - p). \tag{B.1}$$

Now, let $X_1, X_2, \ldots, X_n$ be n independent Bernoulli random variables, each with probability p of success, and let Y denote the number of successes. Then Y is

Mathematical Statistics with Resampling and R, First Edition. By Laura Chihara and Tim Hesterberg.

a *binomial* random variable denoted by $Y \sim \text{Binom}(n, p)$ with probability mass function

$$f(y) = P(Y = y) = \binom{n}{y} p^y (1-p)^{n-y}, \quad y = 0, 1, \ldots, n. \tag{B.2}$$

For instance, if we roll a die 10 times and consider seeing a 1 as a success, then $Y \sim \text{Binom}(10, 1/6)$. So the probability of seeing four 1's is $f(4) = \binom{10}{4}(1/6)^4(5/6)^6 \approx 0.0543$.

Theorem B.1 *Let* $Y \sim \text{Binom}(n, p)$. *Then*

$$\mathrm{E}[Y] = np, \quad \mathrm{Var}[Y] = np(1-p). \tag{B.3}$$

B.2 THE MULTINOMIAL DISTRIBUTION

We saw that the binomial distribution can be used to model the number of heads in n tosses of a coin. In particular, each coin toss has only two possible outcomes, heads or tails. What if an action has more than two outcomes, such as the tossing of a die? How might we model the distribution of the different possible outcomes in n tosses of the die?

First, we recall the multinomial coefficient. Suppose we have n objects of which x_1 are of type O_1, x_2 are of type O_2,..., x_r are of type O_r. Then the number of ways to arrange these $n = x_1 + x_2 + \cdots + x_r$ objects is

$$\binom{n}{x_1, x_2, \ldots, x_r} = \frac{n!}{x_1!\,x_2!\,\ldots, x_r!}. \tag{B.4}$$

To see why this is so, consider placing the objects in a row in positions 1, 2, ..., n. Then there are $\binom{n}{x_1}$ ways to chose the x_1 positions where the objects of type O_1 can be placed. Now, there are $n - x_1$ positions remaining, so there are $\binom{n-x_1}{x_2}$ ways to chose the positions where the objects of type O_2 can be placed. Continuing in this fashion, we have

$$\begin{aligned}
&\binom{n}{x_1}\binom{n-x_1}{x_2}\binom{n-x_1-x_2}{x_3}\cdots\binom{n-x_1-x_2-\cdots-x_{r-1}}{x_r} \\
&= \frac{n!}{x_1!(n-x_1)!} \cdot \frac{(n-x_1)!}{x_2!(n-x_1-x_2)!} \\
&\quad \cdot \frac{(n-x_1-x_2)!}{x_3!(n-x_1-x_2-x_3)!} \cdots \frac{(n-x_1-x_2-\cdots-x_{r-1})!}{x_r!0!} \\
&= \frac{n!}{x_1!\,x_2!\ldots x_r!}.
\end{aligned}$$

Suppose a random experiment consists of n independent, identical trials, each of which has r possible outcomes, say $O_1, O_2, \ldots, O_r$. Let X_i denote the number of times outcome O_i occurs and $p_i = P(X_i = 1)$. We say that $X_1, X_2, \ldots, X_r$ has a *multinomial distribution with parameters n and* $p_1, p_2, \ldots, p_r$. The joint probability mass function of $X_1, X_2, \ldots, X_r$ is

$$f(x_1, x_2, \ldots, x_n) = P(X_1 = x_1, X_2 = x_2, \ldots, X_n = x_n) = \binom{n}{x_1, x_2, \ldots, x_n} p_1^{x_1} p_2^{x_2} \cdots p_r^{x_r}, \tag{B.5}$$

where $n = \sum_{i=1}^{n} x_i = n$.

Example B.1 Toss a die 15 times. Find the probability of seeing three 1's, two 2's, five 3's, one 4, two 5's, and two 6's.

Solution If $p_i, i = 1, 2, \ldots, 6$, denotes the probability of seeing an i on the die, then $p_i = 1/6$, for $i = 1, 2, \ldots, 6$. Thus, the desired probability is

$$\binom{15}{3, 2, 5, 1, 2, 2} \left(\frac{1}{6}\right)^{3+2+5+1+2+2} = \frac{15!}{3!2!5!1!2!2!} \left(\frac{1}{6}\right)^{15} = 0.0005.$$

□

Example B.2 According to Mars Candy Company, the distribution of colors in the Milk Chocolate M& M^{TM}'s is blue 24%, orange 20%, green 16%, yellow 14%, red 13%, and brown 13%. Suppose in a well-mixed extremely large vat of M& M^{TM}'s, you draw out 30 pieces of candy. What is the probability that you get 8 blue, 5 orange, 6 green, 5 yellow, 4 red, and 2 brown candies?

Solution

$$\binom{30}{8, 5, 6, 5, 4, 2} (0.24)^8 (0.20)^5 (0.16)^6 (0.14)^5 (0.13)^4 (0.13)^2 = 0.0002.$$

□

Example B.3 Suppose five numbers are drawn at random from a distribution with pdf $f(x) = 3x^2$, $0 \le x \le 1$. What is the probability that one of the numbers lies in the interval $I_1 = [0, 1/3)$, two lie in the interval $I_2 = [1/3, 2/3)$, and two lie in the interval $I_3 = [2/3, 1]$?

Solution If X is a number drawn from this distribution, then

$$p_1 = P(0 \le X < 1/3) = \int_0^{1/3} 3x^2\, dx = 1/27,$$

$$p_2 = P(1/3 \le X < 2/3) = \int_{1/3}^{2/3} 3x^2\, dx = 7/27,$$

$$p_3 = P(2/3 \le X \le 1) = \int_{2/3}^{1} 3x^2\, dx = 19/27.$$

Thus, if X_i denotes the number of values that lie in interval I_i, $i = 1, 2, 3$, then

$$P(X_1 = 1, X_2 = 2, X_3 = 2) = \binom{5}{1, 2, 2} \left(\frac{1}{27}\right) \left(\frac{7}{27}\right)^2 \left(\frac{19}{27}\right)^2 = 0.0369.$$

□

Note that when there are only two types of outcomes ($r = 2$), then the multinomial distribution reduces to the binomial distribution since we can consider outcome O_1 as "heads" (or successes) and outcome O_2 as "tails" (failures). We can generalize this when we have more than two outcomes. If $X_1, X_2, \ldots, X_r$ has a multinomial distribution with parameters n and $p_1, p_2, \ldots, p_r$, consider X_i to be the number of occurrences of type O_i, each occurrence a "success" with probability p_i. Consider any occurrence of the other types to be a "failure." This then occurs with probability $1 - p_i$. Thus, X_i is a binomial random variable with parameters n and p_i, and

$$\mathrm{E}[X_i] = np_i, \quad \mathrm{Var}[X_i] = np_i(1 - p_i), i = 1, 2, \ldots, r. \tag{B.6}$$

Example B.4 Referring to Example B.2, suppose you draw out 38 candies at random from the vat of M&M^{TM}'s. What is the expected number of reds?

Solution Let X denote the number of red candies. Then, $X \sim \mathrm{Binom}(38 \times 0.13)$ and $\mathrm{E}[X] = 38 \times 0.13 = 4.94$, so about five red candies. □

B.3 THE GEOMETRIC DISTRIBUTION

Toss a fair six-sided die. What is the probability that the first occurrence of "2" will occur on the 15th toss? More generally, consider a sequence of independent Bernoulli trials X_i, each with probability p of success. If X denotes the number of trials up to and including the first success, then X has a *geometric* distribution. In order for the first success to occur on the kth trial, we must have $(k - 1)$ failures, each occurring with probability $1 - p$. Hence, the probabilities are given by

$$p(k) = P(X = k) = (1 - p)^{k-1} p, \quad k = 1, 2, 3, \ldots \tag{B.7}$$

Example B.5 The probability that the first occurrence of a "2" will occur on the 15th toss of a six-sided die is

$$P(X = 15) = (1 - 1/6)^{14}(1/6) = 0.01298.$$

□

We denote a geometric random variable X with probability of success p by $X \sim \mathrm{Geom}(p)$.

Theorem B.2 *Let* $X \sim \text{Geom}(p)$. *Then*

$$\mathrm{E}[X] = \frac{1}{p} \qquad \mathrm{Var}[X] = \frac{1}{p^2}. \tag{B.8}$$

B.4 THE NEGATIVE BINOMIAL DISTRIBUTION

Toss a fair six-sided die until the 8th appearance of "2". What is the probability that this will occur on the 30th toss? More generally, consider a sequence of independent Bernoulli trials with each trial having success probability p. What is the probability that the rth success will occur on the xth trial? Let X denote the trial at which the rth success occurs. Then X has a negative binomial distribution with parameters r and p, and its probability mass function is

$$P(X = x) = \binom{x-1}{r-1} p^r (1-p)^{x-r}, \quad x = r, r+1, r+2, \ldots. \tag{B.9}$$

If the rth success occurs on the x trial, then there must be $(r-1)$ successes in the first $(x-1)$ trials. There are $\binom{x-1}{r-1}$ ways to choose the trials where the success occurs; in addition, the probability of r successes and $(x-r)$ failures is $p^r(1-p)^{x-r}$.

Example B.6 The probability that in tossing a fair six-sided die, the 8th appearance of "2" occurs on the 30th toss is $P(X = 30) = \binom{30-1}{8-1}(1/6)^8(5/6)^22 = 0.0168$. □

We check that Equation B.9 is indeed a probability mass function:

$$\begin{aligned}
\sum_{x=r}^{\infty} \binom{x-1}{r-1} p^r (1-p)^{x-r} &= \sum_{y=0}^{\infty} \binom{y+r-1}{r-1} p^r (1-p)^y \quad \text{set } y = x - r \\
&= \frac{p^r}{(1-(1-p))^r} \\
&= 1,
\end{aligned}$$

where we use the identity

$$\frac{1}{(1-w)^m} = \sum_{k=0}^{\infty} \binom{k+m-1}{m-1} w^k.$$

Theorem B.3 *Let X be a negative binomial random variable with parameters r and p. Then*

$$\mathrm{E}[X] = \frac{r}{p} \qquad \mathrm{Var}[X] = \frac{r(1-p)}{p^2} \tag{B.10}$$

Proof Let X_1 denote the geometric random variable giving the number of trials until the first success, X_2 the geometric random variable giving the number of additional trials after the first success until the second success,..., X_r the number of additional trials after the $(r-1)$st success until the rth success. Then $X = \sum_{i=1}^{r} X_i$ and since $X_i, i = 1, 2, \ldots, r$, are independent, we have

$$\mathrm{E}[X] = \sum_{i=1}^{r} \mathrm{E}[X_i] = \sum_{i=1}^{r} \frac{1}{p} = \frac{r}{p}$$

$$\mathrm{Var}[X] = \sum_{i=1}^{r} \mathrm{Var}[X_i] = \sum_{i=1}^{r} \frac{1-p}{p^2} = \frac{r(1-p)}{p^2}.$$

□

B.5 THE HYPERGEOMETRIC DISTRIBUTION

Suppose there are M women and N men at a college. You form a committee of size $n, n \le \min\{M, N\}$. There are $\binom{M+N}{n}$ ways to choose a committee, of which $\binom{M}{x}\binom{N}{n-x}$ have x women and $(n-x)$ men. Suppose the committee is chosen randomly, and let X denote the number of women on the committee. Then X is called a *hypergeometric random variable*, with pmf

$$f(x) = P(X = x) = \frac{\binom{M}{x}\binom{N}{n-x}}{\binom{M+N}{n}}, \quad x = 0, 1, 2, \ldots, n; x \le \min\{M, n\}; x \ge n - M. \tag{B.11}$$

It is helpful to visualize this using Table B.1. None of the four cells in the table $(n, n-x, M-x, N-n+x)$ may be negative, leading to four constraints on the range of x. For example, if the committee is of size $n = 300$ and there are only $N = 250$ men at the college, there must be at least $n - N = 50$ women on the committee.

Since each committee of size n can have either $0, 1, 2, \ldots,$ or n women serving on it, $\sum_{x=0}^{n} \binom{M}{x}\binom{N}{n-x} = \binom{M+N}{n}$, the total number of committees of size n from $M + N$ people. Thus $\sum_{x=0}^{n} f(x) = 1$.

Theorem B.4 *For a hypergeometric variable X with parameters M, N, and n,*

$$\mathrm{E}[X] = \frac{nM}{M+N} \qquad \mathrm{Var}[X] = \frac{nMN(M+N-n)}{(M+N)^2(M+N-1)}$$

Proof In a college of M women and N men, select a committee of size n, $n \le \min\{M, N\}$. Let $X_i = 1$ if the ith person chosen is a woman, and $X_i = 0$ otherwise,

TABLE B.1 Table for Hypergeometric Distribution

	Women	Men	Total
On committee	x	$n-x$	n
Off committee	$M-x$	$N-n+x$	$M+N-n$
Total	M	N	$M+N$

$i = 0, 1, \ldots, n$. Then $P(X_i = 1) = M/(M+N)$, so $\mathrm{E}[X_i] = 1 \times P(X_i = 1) + 0 \cdot P(X_i = 0) = M/(M+N)$. Thus, $X = \sum_{i=1}^{n} X_i$ gives the number of women on the committee and

$$\mathrm{E}[X] = \sum_{i=1}^{n} \mathrm{E}[X_i] = \sum_{i=1}^{n} \frac{M}{M+N} = \frac{nM}{M+N}.$$

Now, from Corollary 9.2 and since, by symmetry, $\mathrm{Cov}[X_i, X_j] = \mathrm{Cov}[X_1, X_2]$, we have

$$\mathrm{Var}[X] = \sum_{i=1}^{n} \mathrm{Var}[X_i] + 2\sum_{i}\sum_{j} \mathrm{Cov}[X_i, Y_i] \tag{B.12}$$

$$= \frac{nNM}{(M+N)^2} + n(n-1)\mathrm{Cov}[X_1, X_2] \tag{B.13}$$

Now, consider the case that $n = M + N$. Then $X = M$ since *all* the women at the college must be selected. In this case, $\mathrm{Var}[X] = 0$. Thus, from Equation B.13, we have

$$\mathrm{Cov}[X_1, X_2] = -\frac{MN}{(M+N)^2(M+N-1)}.$$

Thus, in the general case (n not necessarily equal to $M + N$), Equation B.13 gives

$$\mathrm{Var}[X] = \frac{nMN}{(M+N)^2} \cdot \frac{M+N-n}{M+N-1},$$

which establishes the result. □

B.6 THE POISSON DISTRIBUTION

Definition B.1 The discrete random variable X with range $0, 1, 2, \ldots$ has a Poisson distribution with parameter $\lambda > 0$ if its pmf is

$$f(x) = P(X = x) = \frac{\lambda^x e^{-\lambda}}{x!}, \quad x = 0, 1, 2, \ldots. \tag{B.14}$$

||

We see that Equation B.14 is a probability mass function since

$$\sum_{x=0}^{\infty} \frac{\lambda^x e^{-\lambda}}{x!} = e^{-\lambda} \sum_{x=0}^{\infty} \frac{\lambda^x}{x!} = e^{-\lambda} e^{\lambda} = 1.$$

The Poisson distribution is often used to model counts.

Example B.7 Suppose the number of typographical errors on a web page for a certain company follows a Poisson random variable with $\lambda = 1.5$. Find the probability that there are at most two typographical errors on a given page.

Solution Let X denote the number of typographical errors on a web page. Then the probabilities are given by $f(x) = 1.5^x e^{-1.5}/x!$, $x = 0, 1, 2, \ldots$, so $F(2) = e^{-1.5}(1 + 1.5 + 1.5^2/2!) \approx 0.8088$. □

We leave as an exercise the proof of the following.

Theorem B.5 *Let X be a Poisson random variable with $\lambda > 0$. Then,*

$$\mathrm{E}[X] = \lambda \qquad \mathrm{Var}[X] = \lambda. \tag{B.15}$$

The following theorem allows us to work with sums of Poisson random variables.

Theorem B.6 *Let $X_1, X_2, \ldots, X_n$ be independent Poisson random variables with parameters $\lambda_1, \lambda_2, \cdots, \lambda_n$, respectively. Then $X = X_1 + X_2 + \cdots + X_n$ is Poisson with parameters $\lambda_1 + \lambda_2 + \cdots + \lambda_n$.*

Proof We will prove the result for $n = 2$.

$$\begin{aligned}
P(X_1 + X_2 = m) &= \sum_{j=0}^{m} P(X_1 = j, X_2 = m - j) \\
&= \sum_{j=0}^{m} P(X_1 = j)P(X_2 = m - j) \\
&= \sum_{j=0}^{m} \frac{\lambda_1^j e^{-\lambda_1}}{j!} \frac{\lambda_2^{m-j} e^{-\lambda_2}}{(m-j)!} \\
&= e^{-(\lambda_1+\lambda_2)} \sum_{j=0}^{m} \frac{\lambda_1^j \lambda_2^{m-j}}{j!(m-j)!} \\
&= \frac{e^{-(\lambda_1+\lambda_2)}}{m!} \sum_{j=0}^{m} \frac{m!}{j!(m-j)!} \lambda_1^j \lambda_2^{m-j} \\
&= \frac{e^{-(\lambda_1+\lambda_2)}}{m!} (\lambda_1 + \lambda_2)^m,
\end{aligned}$$

where the last equality is from the binomial theorem. Thus, $X_1 + X_2$ is Poisson with parameter $\lambda_1 + \lambda_2$.

A proof by induction can be used to extend to a sum of n Poisson random variables. □

B.7 THE UNIFORM DISTRIBUTION

Definition B.2 A random variable X has a uniform distribution on the interval $[a, b]$ (for $a < b$) if its pdf is

$$f(x) = \begin{cases} \frac{1}{b-a}, & a \leq x \leq b, \\ 0, & \text{otherwise} . \end{cases} \tag{B.16}$$

We denote this by $X \sim \text{Unif}[a, b]$. ||

We can also define the uniform distribution over the intervals (a, b), $(a, b]$, and $[a, b)$ by making the appropriate adjustment in the pdf.

Theorem B.7 *Let $X \sim \text{Unif}[a, b]$. Then*

$$\text{E}[X] = \frac{a+b}{2}, \quad \text{Var}[X] = \frac{(a-b)^2}{12}. \tag{B.17}$$

Proposition B.1 *Let $X \sim \text{Unif}[0, \theta]$. Then $X/\theta \sim \text{Unif}[0, 1]$.*

Proof Exercise. □

B.8 THE EXPONENTIAL DISTRIBUTION

Definition B.3 A random variable X has the exponential distribution with parameter $\lambda > 0$ if its pdf is

$$f(x) = \lambda e^{-\lambda x}, \quad x \geq 0. \tag{B.18}$$

We will write $X \sim \text{Exp}(\lambda)$. ||

Another common parameterization is

$$f(x) = \frac{1}{\rho} e^{-x/\rho}, \quad x \geq 0, \tag{B.19}$$

with $\rho = 1/\lambda > 0$. Here, ρ is a *scale parameter*—doubling it corresponds to doubling all values—and is equal to the mean of the distribution.

Theorem B.8 *Let X have an exponential distribution with $\lambda > 0$. Then*

$$\text{E}[X] = \frac{1}{\lambda} = \rho, \quad \text{Var}[X] = \frac{1}{\lambda^2} = \rho^2. \tag{B.20}$$

The exponential distribution is used to model *waiting times*: that is, time between the occurrence of successive events, such as the time between groups of customers arriving at a fast food restaurant. It is notable for being the only distribution with the *memoryless property*, that is,

$$P(X \geq t + h | X \geq h) = P(X \geq t) \tag{B.21}$$

The proof is an exercise.

If X represents the time in minutes until an event occurs, then Equation B.21 says that the probability the event will not occur within the next t minutes does not depend on how long you have already waited. If the current time is h (and the event has not occurred yet), the probability that the event will wait an additional t minutes is the same as the probability back at time $h = 0$.

Note that the times between individuals (opposed to groups) arriving at a fast food restaurant would not be memoryless. If someone just walked in the door, there is a higher than normal chance that someone else will arrive within the next 10 s (namely a friend or family member). The parameter λ is a *rate parameter*—it represents how quickly events occur. Doubling λ cuts the expected time until the next arrival in half.

For some calculations with exponential variables, it is useful to *standardize* to rate 1 (and mean and variance 1).

Proposition B.2 *Let X be an exponential random variable with parameter λ. Then $Y = \lambda X$ is exponential with parameter 1.*

Proof Let $X \sim \text{Exp}(\lambda)$ with pdf $f_X(x) = \lambda e^{-\lambda x}$. Then, the cdf is $F_X(x) = 1 - e^{-\lambda x}$.

Let $F_Y(x)$ and $f_Y(x)$ denote the cdf and pdf, respectively, of $Y = \lambda X$. Then

$$F_Y(x) = P(Y \leq x) = P(\lambda X \leq x) = P\left(X \leq \frac{x}{\lambda}\right) = F_X\left(\frac{x}{\lambda}\right).$$

Thus,

$$F_Y'(x) = f_Y(x) = F_X'\left(\frac{x}{\lambda}\right)\frac{1}{\lambda} = e^{-x},$$

so $Y \sim \text{Exp}(1)$. □

B.9 THE GAMMA DISTRIBUTION

Definition B.4 The gamma function is defined, for $r > 0$, by

$$\Gamma(r) = \int_0^\infty x^{r-1} e^{-x}\, dx \tag{B.22}$$

Theorem B.9

1. For $r > 1$,

$$\Gamma(r) = (r-1)\Gamma(r-1). \tag{B.23}$$

2. *For positive integer n,*

$$\Gamma(n) = (n-1)! \tag{B.24}$$

3.

$$\Gamma(\frac{1}{2}) = \sqrt{\pi}. \tag{B.25}$$

The proof is an exercise.

Definition B.5 A continuous random variable X has the gamma distribution with parameters $r > 0$, $\lambda > 0$ if its pdf is given by

$$f(x) = \frac{\lambda^r}{\Gamma(r)} x^{r-1} e^{-\lambda x}, \quad x \geq 0. \tag{B.26}$$

We denote this by $X \sim \text{Gamma}(r, \lambda)$.
Some examples are shown in Figure B.1. ||

The r and λ are called *shape* and *rate* parameters respectively. Another common parameterization is to let $\rho = 1/\lambda$

$$f(x) = \frac{1}{\rho^r \Gamma(r)} x^{r-1} e^{-x/\rho}, \quad x \geq 0. \tag{B.27}$$

Some books will use this and call ρ the *scale* parameter.
Note that the exponential distribution is a special case of the gamma distribution. That is, $X \sim \text{Exp}(\lambda)$ is the same as $X \sim \text{Gamma}(1, \lambda)$.

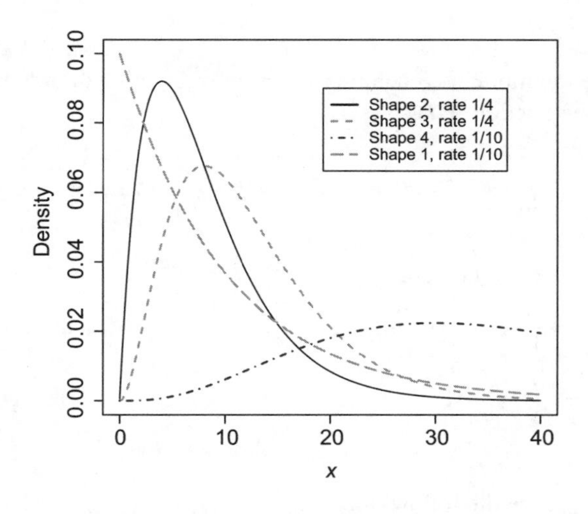

FIGURE B.1 Densities of gamma distribution.

By using the substitution $u = \beta x$, we find

$$\int_0^\infty x^{r-1} e^{-\lambda x}\, dx = \frac{\Gamma(r)}{\lambda^r} \tag{B.28}$$

so that Equation B.26 integrates to 1.

Theorem B.10 *Let X be a gamma random variable with parameters r, λ. Then*

$$\mathrm{E}[X] = \frac{r}{\lambda}, \qquad \mathrm{Var}[X] = \frac{r}{\lambda^2}. \tag{B.29}$$

Proof

$$\begin{aligned}
\mathrm{E}[X] &= \int_0^\infty x \frac{1}{\Gamma(r)} \lambda^r x^{r-1} e^{-\lambda x} \\
&= \frac{\lambda^r}{\Gamma(r)} \int_0^\infty x^r e^{-\lambda x}\, dx \\
&= \frac{\lambda^r}{\Gamma(r)} \frac{\Gamma(r+1)}{\lambda^{r+1}} && \text{by Equation B.28} \\
&= \frac{r}{\lambda}. && \text{by Theorem B.9.}
\end{aligned}$$

We leave the proof of the variance as an exercise. □

The proof to the following proposition is an exercise (Exercise 8).

Proposition B.3 *Let X be a gamma random variable with parameters r, λ and suppose $c > 0$ is a constant. Then cX is also a gamma random variable with parameters r, λ/c.*

We compute the moment generating function of a gamma random variable $X \sim \text{Gamma}(r, \lambda)$ (see Section A.7).

$$\begin{aligned}
M(t) &= \int_0^\infty e^{tx} f(x)\, dx \\
&= \frac{\lambda^r}{\Gamma(r)} \int_0^\infty x^{r-1} e^{-(\lambda - t)x}\, dx \\
&= \frac{\lambda^r}{\Gamma(r)} \cdot \frac{\Gamma(r)}{(\lambda - t)^r} && \text{by Equation B.28} \\
&= \left(\frac{\lambda}{\lambda - t}\right)^r, && \text{(B.30)}
\end{aligned}$$

for $t < \lambda$.

Thus, we can prove the following.

Theorem B.11 *Let $X_1, X_2, \ldots, X_n$ be independent random variables with $X_i \sim$ Gamma(r_i, λ),
$i = 1, 2, \ldots, n$. Then the sum $X = \sum_{i=1}^{n} X_i$ is also a gamma random variable with parameters $r_1 + r_2 + \cdots + r_n$ and λ.*

Proof Let $M_i(t)$ denote the mgf of X_i, $i = 1, 2, \ldots, n$, and $M(t)$ the mgf of the sum X. Then,

$$M(t) = \prod_{i=1}^{n} M_i(t) = \prod_{i=1}^{n} \left(\frac{\lambda}{\lambda - t}\right)^{r_i} = \left(\frac{\lambda}{\lambda - t}\right)^{r_1+r_2+\cdots+r_n}$$

for $t < \lambda$ (Theorem A.14). We see that $M(t)$ is the mgf of a gamma distribution with parameters $r_1 + r_2 + \cdots + r_n$ and λ. Hence, $X = X_1 + X_2 + \cdots + X_n \sim$ Gamma$(r_1 + r_2 + \cdots + r_n, \lambda)$. □

B.10 THE CHI-SQUARE DISTRIBUTION

Definition B.6 A continuous random variable X has chi-square distribution with m degrees of freedom if its pdf is

$$f(x) = \frac{x^{m/2-1}e^{-x/2}}{2^{m/2}\Gamma(m/2)}, \quad \text{for } 0 \leq x < \infty. \tag{B.31}$$

We write $X \sim \chi^2_m$.

Some examples are shown in Figure B.2. ||

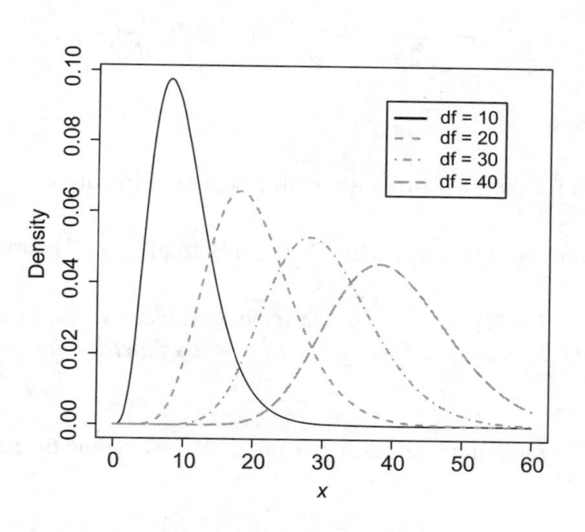

FIGURE B.2 Densities for the chi-square distribution.

Theorem B.12 *Let X be a random variable following a chi-square distribution with m degrees of freedom. Then*

$$\mathrm{E}[X] = m, \qquad \mathrm{Var}[X] = 2m. \tag{B.32}$$

The chi-square distribution with m degrees of freedom is equal to the gamma distribution with $r = m/2$ and $\lambda = 1/2$. Thus, it follows from Theorem B.11 that

Theorem B.13 *Let $X_1, X_2, \ldots, X_k$ be independent chi-square random variables with degrees of freedom $m_1, m_2, \ldots, m_k$, respectively. Then $X = X_1 + X_2 + \cdots + X_k$ is a chi-square random variable with $m_1 + m_2 + \cdots + m_k$ degrees of freedom.*

Theorem B.14 *Let Z be a random variable from a standard normal distribution. Then Z^2 has a chi-square distribution with 1 degree of freedom.*

Proof Let f_{Z^2} denote the pdf for Z^2.

$$\begin{aligned}
\int_{-\infty}^{x} f_{Z^2}(y)dy &= P(Z^2 \leq x) \\
&= P(-\sqrt{x} \leq Z \leq \sqrt{x}) \\
&= P(Z \leq \sqrt{x}) - P(Z \leq -\sqrt{x}) \\
&= \frac{1}{\sqrt{2\pi}} \int_{-\sqrt{x}}^{\sqrt{x}} e^{-y^2/2} dy.
\end{aligned}$$

Differentiating both sides with respect to x, we obtain

$$\begin{aligned}
f_{Z^2}(x) &= \frac{1}{\sqrt{2\pi}} \left(e^{-x/2} \frac{d}{dx}\sqrt{x} - e^{-x/2} \frac{d}{dx}(-\sqrt{x})\right) \\
&= \frac{1}{\sqrt{2\pi}\sqrt{x}} e^{-x/2},
\end{aligned}$$

which is the pdf for the χ^2 distribution with 1 degree of freedom. □

Thus, Theorem B.14 together with Theorem B.13 gives us Theorem B.15.

Theorem B.15 *Let $Z_1, Z_2, \ldots, Z_k$ be an independent random sample from the standard normal distribution. Then $\sum_{i=1}^{k} Z_i^2$ has a χ^2 distribution with k degrees of freedom.*

Recall that for random variables $X_1, X_2, \ldots, X_n$, we define the sample variance

$$S^2 = \frac{1}{n-1} \sum_{i=1}^{n} (X_i - \bar{X})^2.$$

Theorem B.16 *Let $X_1, X_2, \ldots, X_n$ be a random sample from $N(\mu, \sigma^2)$. Then*

1. *$\bar{X}$ and S^2 are independent random variables.*
2. *$(n-1)S^2/\sigma^2$ has a χ^2 distribution with $n-1$ degrees of freedom.*

Proof We present only the proof of (2). We have $(X_i - \mu)/\sigma \sim N(0, 1)$, $i = 1, 2, \ldots, n$, so by Theorem B.15, we have

$$\frac{1}{\sigma^2}\sum_{i=1}^{n}(X_i - \mu)^2 = \sum_{i=1}^{n}\left(\frac{X_i - \mu}{\sigma}\right)^2 \sim \chi_n^2.$$

Now, recalling that $\sum_{i=1}^{n}(X_i - \bar{X}) = 0$,

$$\begin{aligned}\frac{1}{\sigma^2}\sum_{i=1}^{n}(X_i - \mu)^2 &= \frac{1}{\sigma^2}\sum_{i=1}^{n}\left[(X_i - \bar{X}_i) + (\bar{X}_i - \mu)\right]^2 \\ &= \frac{1}{\sigma^2}\sum_{i=1}^{n}(X_i - \bar{X})^2 + \left(\frac{\bar{X} - \mu}{\sigma/\sqrt{n}}\right)^2.\end{aligned}$$

Let $U = (1/\sigma^2)\sum_{i=1}^{n}(X_i - \mu)^2$, $V = (1/\sigma^2)\sum_{i=1}^{n}(X_i - \bar{X})^2$, $W = (\bar{X} - \mu)/(\sigma/\sqrt{n})$. Then by Theorem B.15 and Theorem B.14, U and W are chi-square random variables. It is a fact that their mgf's are $M_U(t) = 1/(1-2t)^{n/2}$ and $M_W(t) = 1/(1-2t)^{1/2}$, respectively.

Now, $U = V + W$, and since V and W are independent, by Theorem A.14 we have $M_U(t) = M_V(t)M_W(t)$, or

$$\begin{aligned}M_V(t) &= \frac{M_U(t)}{M_W(t)} \\ &= \frac{(1-2t)^{-n/2}}{(1-2t)^{-1/2}} \\ &= (1-2t)^{-(n-1)/2},\end{aligned}$$

which is the mgf of a chi-square distribution with $n-1$ degrees of freedom.

Since

$$\begin{aligned}V &= \frac{1}{\sigma^2}\sum_{i=1}^{n}(X_i - \bar{X})^2 \\ &= \frac{n-1}{n-1}\cdot\frac{1}{\sigma^2}\sum_{i=1}^{n}(X_i - \bar{X})^2 \\ &= \frac{n-1}{\sigma^2}\cdot\frac{1}{(n-1)}\sum_{i=1}^{n}(X_i - \bar{X})^2 \\ &= \frac{n-1}{\sigma^2}S^2,\end{aligned}$$

and this completes the proof. □

B.11 THE STUDENT'S t DISTRIBUTION

Definition B.7 A random variable has a t distribution with k degrees of freedom ($k > 0$) if its pdf is

$$f(x) = \frac{\Gamma((k+1)/2)}{\Gamma(k/2)\sqrt{k\pi}} \left(1 + \frac{x^2}{k}\right)^{-(k+1)/2}, \quad \text{for } -\infty < x < \infty.$$

||

Theorem B.17 *Let Z be a standard normal random variable and let W denote a chi-square distribution with k degrees of freedom. Then the random variable defined by*

$$T = \frac{Z}{\sqrt{W/k}}$$

has a t distribution with k degrees of freedom.

Proof Let $f_T(t)$, $f_Z(z)$, $f_W(w)$ denote the pdf's of T, Z, and W, respectively. If $f(z, w)$ denotes the joint pdf of Z and W, then by independence, $f(z, w) = f_Z(z) f_W(w)$. Thus,

$$P(T \leq t) = P\left(\frac{Z}{\sqrt{W/k}} \leq t\right) = P\left(Z \leq t\sqrt{\frac{W}{k}}\right)$$

$$= \int_0^\infty \int_{-\infty}^{t\sqrt{w/k}} f_Z(z) f_W(w)\, dz\, dw.$$

By the fundamental theorem of calculus, $(d/dt)P(T \leq t) = f_T(t)$, so

$$f_T(t) = \int_0^\infty \frac{d}{dt} \int_{-\infty}^{t\sqrt{w/k}} f_Z(z) f_W(w)\, dz\, dw$$

$$= \int_0^\infty f\left(t\sqrt{\frac{w}{k}}\right) \sqrt{\frac{w}{k}}\, f_W(w)\, dw$$

$$= \int_0^\infty \frac{\sqrt{\frac{w}{k}}\, e^{-\frac{t^2 w}{2k}}}{\sqrt{2\pi}} \frac{w^{\frac{k}{2}-1} e^{-\frac{w}{2}}}{2^{\frac{k}{2}}\Gamma(k/2)}\, dw$$

$$= \frac{1}{\sqrt{2\pi}\, 2^{k/2}\Gamma(k/2)\sqrt{k}} \int_0^\infty w^{\frac{k}{2}-\frac{1}{2}} e^{-w\left(\frac{1}{2}+\frac{t^2}{2k}\right)}\, dw.$$

Now, make a u-substitution, $u = w(\frac{1}{2} + t^2/(2k))$ to get

$$\begin{aligned} f_T(t) &= \frac{1}{\sqrt{2\pi}\, 2^{k/2}\Gamma(k/2)\sqrt{k}} \int_0^\infty \frac{u^{\frac{k}{2}-\frac{1}{2}} e^{-u}}{(1/2 + t^2/(2k))^{\frac{k}{2}-\frac{1}{2}}(1/2 + t^2/(2k))} du \\ &= \frac{1}{\sqrt{2\pi}\, 2^{k/2}\Gamma(k/2)\sqrt{k}(1/2 + t^2/(2k))^{(k+1)/2}} \int_0^\infty u^{\frac{k+1}{2}-1} e^{-u} du \\ &= \frac{\Gamma((k+1)/2)}{\sqrt{2\pi}\, 2^{k/2}\Gamma(k/2)\sqrt{k}(1/2 + t^2/(2k))^{(k+1)/2}} \\ &= \frac{\Gamma((k+1)/2)}{\sqrt{\pi k}\, \Gamma(k/2)(1 + t^2/k)^{(k+1)/2}} \end{aligned}$$

which is the desired pdf. □

The pdf for a t distribution is bell shaped and symmetric about 0 with heavier tails than the standard normal. As k tends toward infinity, the density of the t distribution tends toward the density of the standard normal. Some examples are shown in Figure B.3.

Theorem B.18 *Let T denote a random variable with a t distribution with k degrees of freedom. Then*

$$\mathrm{E}[T] = 0, \qquad \mathrm{Var}[T] = k/(k-2),\ k > 2. \tag{B.33}$$

Theorem B.19 *Let $X_1, X_2, \ldots, X_n \overset{i.i.d.}{\sim} N(\mu, \sigma^2)$. Then the random variable $T = (\bar{X} - \mu)/(S/\sqrt{n})$ has a t distribution with $n-1$ degrees of freedom.*

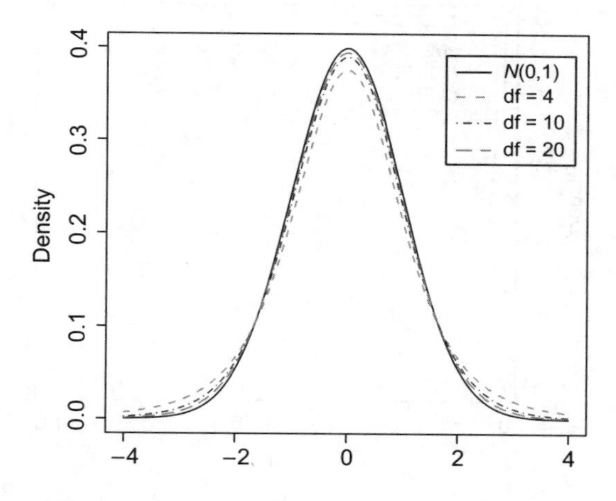

FIGURE B.3 Densities for the t distribution.

Proof Let $X_1, X_2, \ldots, X_n \sim N(\mu, \sigma^2)$. Then $\bar{X} \sim N(\mu, \sigma^2/n)$, so $Z = (\bar{X} - \mu)/(\sigma/\sqrt{n}) = \sqrt{n}((\bar{X} - \mu)/\sigma)$ is a standard normal random variable. From Theorem B.16, we know that $W = (n-1)S^2/\sigma^2$ is a chi-square random variable with $n-1$ degrees of freedom and that Z and W are independent.

Thus, by Theorem B.17,

$$T = \frac{Z}{\sqrt{W/(n-1)}} = \frac{\sqrt{n}(\bar{X} - \mu)/\sigma}{\sqrt{((n-1)S^2/\sigma^2)/(n-1)}} = \frac{\bar{X} - \mu}{S/\sqrt{n}}$$

has a t distribution with $n-1$ degrees of freedom. □

B.12 THE BETA DISTRIBUTION

Definition B.8 A random variable X has a beta distribution with parameters $\alpha > 0$ and $\beta > 0$ if its pdf is

$$f(x) = \frac{\Gamma(\alpha + \beta)}{\Gamma(\alpha)\Gamma(\beta)} x^{\alpha-1}(1-x)^{\beta-1}, \quad \text{for } 0 \le x \le 1. \tag{B.34}$$

We will denote this by $X \sim \text{Beta}(\alpha, \beta)$.

Some examples are shown in Figure B.4. ||

Note that

$$\begin{aligned}\Gamma(\alpha)\Gamma(\beta) &= \int_0^\infty u^{\alpha-1} e^{-u}\, du \int_0^\infty v^{\beta-1} e^{-v}\, dv \\ &= \int_0^\infty \int_0^\infty u^{\alpha-1} v^{\beta-1} e^{-(u+v)}\, du\, dv.\end{aligned}$$

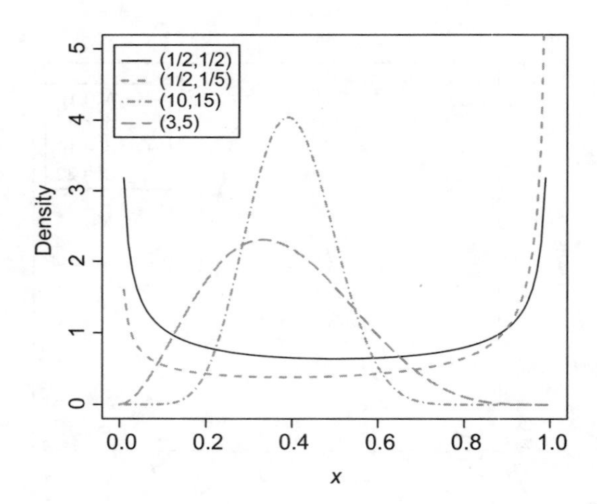

FIGURE B.4 Densities for the Beta distribution (α,β).

Now, make the substitution $x = u/(u+v)$ and $y = u + v$. Then $u = xy$ and $v = (1-x)y$. Computing the Jacobian for this transformation and making the appropriate changes in the limits of integration, we find

$$\begin{aligned}\Gamma(\alpha)\Gamma(\beta) &= \int_0^1 \int_0^\infty x^{\alpha-1}(1-x)^{\beta-1} y^{\alpha+\beta-1} e^{-y}\, dy\, dx \\ &= \Gamma(\alpha+\beta) \int_0^1 x^{\alpha-1}(1-x)^{\beta-1}\, dx.\end{aligned}$$

Thus, the function in Equation B.34 integrates to 1.

Theorem B.20 *Let X have a beta distribution, $X \sim \text{Beta}(\alpha, \beta)$. Then*

$$\mathrm{E}[X] = \frac{\alpha}{\alpha+\beta}, \qquad \mathrm{Var}[X] = \frac{\alpha\beta}{(\alpha+\beta)^2(\alpha+\beta+1)}. \tag{B.35}$$

The proof is left as an exercise.

B.13 THE *F* DISTRIBUTION

Definition B.9 Let Y be a chi-square random variable with m degrees of freedom and W be a chi-square random variable with n degrees of freedom with Y and W independent. The random variable

$$X = \frac{Y/m}{W/n} \tag{B.36}$$

is called an F distribution with m and n degrees of freedom. We denote this by $X \sim F_{m,n}$.

Some examples are shown in Figure B.5. ||

Theorem B.21 *The pdf of an F distribution with m, n degrees of freedom is*

$$f(x) = \frac{\Gamma((m+n)/2) m^{m/2} n^{n/2}}{\Gamma(m/2)\Gamma(n/2)} \cdot \frac{x^{m/2-1}}{(mx+n)^{(m+n)/2}}, \tag{B.37}$$

for $x > 0$.

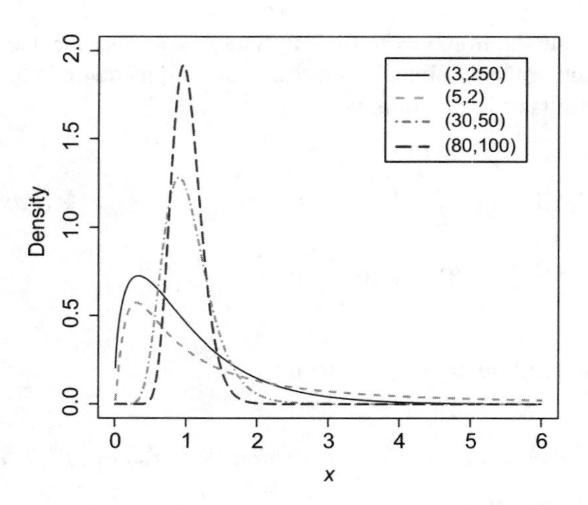

FIGURE B.5 F distribution with (m, n) degrees of freedom.

Proof Let $f_Y(y) = \frac{1}{2^{m/2}\Gamma(m/2)} y^{m/2-1} e^{-y/2}$ and $f_W(w) = \frac{1}{2^{n/2}\Gamma(n/2)} y^{n/2-1} e^{-w/2}$ denote the pdf's for Y and W, respectively. Then by independence,

$$\begin{aligned} F_X(x) &= P\left(\frac{Y/m}{W/n} \leq x\right) = P(Y \leq \frac{m}{n} xW) \\ &= \int_0^\infty \int_0^{mwx/n} \frac{1}{2^{m/2}\Gamma(m/2)2^{n/2}\Gamma(n/2)} w^{n/2-1} e^{-w/2} y^{m/2-1} e^{-y/2} \, dy \, dx \\ &= \frac{1}{2^{m/2+n/2}\Gamma(m/2)\Gamma(n/2)} \int_0^\infty \int_0^{mwx/n} w^{n/2-1} y^{m/2-1} y e^{-w/2-y/2} \, dy \, dw. \end{aligned}$$

Now differentiate with respect to x and then set $z = (mx/(2n) + 1/2)w$:

$$\begin{aligned} f_X(x) &= \frac{1}{2^{m/2+n/2}\Gamma(m/2)\Gamma(n/2)} \int_0^\infty w^{n/2-1} (mwx/n)^{m/2-1} e^{mwx/(2n)-w/2} \frac{mw}{n} \, dw \\ &= \frac{1}{2^{n/2+m/2}\Gamma(m/2)\Gamma(n/2)} \left(\frac{m}{n}\right)^{m/2} x^{m/2-1} \int_0^\infty \frac{z^{m/2+n/2-1}}{(mx/(2n) + 1/2)^{m/2+n/2-1}} e^{-z} \\ &\quad \times \frac{dz}{(mx/(2n) + 1/2)} \\ &= \frac{1}{2^{m/2+n/2}\Gamma(m/2)\Gamma(n/2)} \left(\frac{m}{n}\right)^{m/2} x^{m/2-1} \frac{2^{m/2+n/2} n^{m/2+n/2}}{(mx+n)^{m/2+n/2}} \\ &\quad \times \int_0^\infty z^{m/2+n/2-1} e^{-z} \, dz \\ &= \frac{m^{m/2} n^{n/2}}{\Gamma(m/2)\Gamma(n/2)} \frac{x^{m/2-1}}{(mx+n)^{(m+n)/2}} \Gamma\left(\frac{m+n}{2}\right). \end{aligned}$$

□

Theorem B.22 *Let X be a random variable from an F distribution with m and n degrees of freedom, $X \sim F_{m,n}$. Then*

$$\mathrm{E}[X] = \frac{m}{n-2}, \quad n > 2, \tag{B.38}$$

$$\mathrm{Var}[X] = \frac{2n^2(m+n-2)}{m(n-2)(n-4)}, \quad n > 4. \tag{B.39}$$

We leave the proof of the expectation as an exercise (Exercise 14).

B.14 EXERCISES

1. A box contains 30 red balls and 20 blue balls.
   (a) If you draw 15 balls at random without replacement, what is the probability that you draw 8 red balls and 7 blue balls?
   (b) If you draw 15 balls at random without replacement, how many red balls do you expect?
2. You draw 100 values at random from the exponential distribution with $\lambda = 1$. Find the probability that 30 of these values fall in the interval $[0, 0.25]$, 30 fall in $(0.25, 0.75]$, 22 fall in $(0.75, 1.25]$, and the rest fall in $(1.25, \infty)$.
3. According to `http://www.aabb.org`, the distribution of blood types in the United States includes 48% of type O, 37% of type A, 11% of type B, and 4% of type AB. If you select 45 people at random,
   (a) what is the probability that the distribution of their blood types will be 20 of type O, 15 of type A, 6 of type B, and 4 of type AB?
   (b) what is the expected number of people in your sample with type A blood?
4. Prove Proposition B.1.
5. Prove the memoryless property for the exponential distribution (Equation B.21).
6. Prove Theorem B.9.
7. Find the variance of the gamma distribution with parameters r and λ.
8. Prove Proposition B.3
9. Verify Theorem B.13.
10. Suppose $X_1 \sim \chi^2_5$, $X_2 \sim \chi^2_4$, and $X_3 \sim \chi^2_8$. Let $W = X_1 + X_2 + X_3$. Find $P(W \leq 10)$.
11. Let $X \sim \mathrm{Gamma}(2, \lambda)$. Show that $Y = 2\lambda X$ has a chi-square distribution with 4 degrees of freedom.
12. Let $X_1, X_2, \ldots, X_n \sim N(\mu, \sigma^2)$. Define $S^2 = 1/(n-1)\sum_{i=1}^{n}(X_i - \bar{X})^2$. Prove that

$$\mathrm{Var}[S^2] = \frac{2\sigma^4}{n-1}.$$

13. Prove Theorem B.20.
14. Prove that the expected value of $X \sim F_{m,n}$ is $m/(n-2)$ for $n > 2$. *Hint*: Suppose $W \sim \chi_n^2$. Show that $\mathrm{E}\left[1/W\right] = 1/(n-2)$, $n > 2$.
15. Let $X \sim F_{m,n}$. Show that $1/X \sim F_{n,m}$.
16. Let $X \sim F_{m,n}$. Define $Y = mX/(mX+n)$. Show that Y has a beta distribution with $\alpha = m/2$ and $\beta = n/2$.

APPENDIX C

DISTRIBUTIONS QUICK REFERENCE

Mathematical Statistics with Resampling and R, First Edition. By Laura Chihara and Tim Hesterberg.

Discrete

Probability Mass Function	Mean and Variance	Moment Generating Function $M_X(t)$
Bernoulli		
$f(x;p) = p^x(1-p)^{1-x}$	$\mu = p$	$(1-p) + pe^t$
$x = 0, 1, 0 \le p \le 1$	$\sigma^2 = p(1-p)$	
Binomial		
$f(x;n,p) = \binom{n}{x} p^x(1-p)^{1-x}$	$\mu = np$	$\left[pe^t + (1-p)\right]^n$
$x = 0, 1, 2, \ldots, n, 0 \le p \le 1, n \in \mathbf{Z}^+$	$\sigma^2 = np(1-p)$	
Geometric		
$f(x;p) = (1-p)^{1-x}p$	$\mu = \frac{1}{p}$	$\frac{pe^t}{1-(1-p)e^t}$
$x = 1, 2, 3, \ldots, 0 < p \le 1$	$\sigma^2 = \frac{1-p}{p^2}$	$t < -\ln(1-p)$
Hypergeometric		
$H(x;n,M,N) = \frac{\binom{M}{x}\binom{N}{n-x}}{\binom{M+N}{n}}$	$\mu = \frac{nM}{M+N}$	Omitted
$x = 0, 1, \ldots, n, n, M, N \in \mathbf{Z}^+$	$\sigma^2 = \frac{nMN(M+N-n)}{(M+N)^2(M+N-1)}$	
$\max\{0, n-N\} \le x \le \min\{M, n\}$		
Negative Binomial		
$f(x;r,p) = \binom{x-1}{r-1} p^r(1-p)^{x-r}$	$\mu = \frac{r}{p}$	$\left[\frac{pe^t}{1-(1-p)e^t}\right]^r$
$x = r, r+1, r+2, \ldots, 0 < p \le 1, r \in \mathbf{Z}^+$	$\sigma^2 = \frac{r(1-p)}{p^2}$	$t < -\ln(1-p)$
Poisson		
$f(x;\lambda) = \frac{\lambda^x e^{-\lambda}}{x!}$	$\mu = \lambda$	$e^{\lambda(e^t-1)}$
$x = 0, 1, 2, \ldots, \lambda > 0$	$\sigma^2 = \lambda$	

Probability Density Function	Mean and Variance	Moment Generating Function
Beta		
$f(x;\alpha,\beta) = \dfrac{\Gamma(\alpha+\beta)}{\Gamma(\alpha)\Gamma(\beta)} x^{\alpha-1}(1-x)^{\beta-1}$	$\mu = \dfrac{\alpha}{\alpha+\beta}$	$1 + \sum_{k=1}^{\infty}\left(\prod_{s=0}^{k} \dfrac{\alpha+s}{\alpha+\beta+s}\right)\dfrac{t^k}{k!}$
$0 < x < 1, \alpha > 0, \beta > 0$	$\sigma^2 = \dfrac{\alpha\beta}{(\alpha+\beta)^2(\alpha+\beta+1)}$	
Cauchy		
$f(x;\alpha,\beta) = \dfrac{\beta/\pi}{(x-\alpha)^2+\beta^2}$	μ does not exist	Does not exist
$-\infty < x < \infty, -\infty < \alpha < \infty, \beta > 0$	σ^2 does not exist	
Chi-Square		
$f(x;m) = \dfrac{1}{2^{m/2}\Gamma(m/2)} x^{(m-2)/2} e^{-x/2}$	$\mu = m$	$(1-2t)^{-p/2}$
$x \geq 0, m > 0$	$\sigma^2 = 2m$	$t < 1/2$
Exponential		
$f(x;\lambda) = \lambda e^{-\lambda x}$	$\mu = \dfrac{1}{\lambda}$	$(1-\lambda t)^{-1}$
$x \geq 0, \lambda > 0$	$\sigma^2 = \dfrac{1}{\lambda^2}$	$t < 1/\lambda$
F		
$f(x;m,n) = \dfrac{\Gamma(\frac{m+n}{2}) m^{m/2} n^{n/2}}{\Gamma(\frac{m}{2})\Gamma(\frac{n}{2})} \dfrac{x^{m/2-1}}{(mx+n)^{(m+n)/2}}$	$\mu = \dfrac{n}{n-2}$	Does not exist
$x > 0, m > 0, n > 0$	$\sigma^2 = \dfrac{2n^2(m+n-2)}{m(n-2)(n-4)}$	

(Continued)

Probability Density Function	Mean and Variance	Moment Generating Function $M_X(t)$
Gamma		
$f(x;r,\lambda) = \frac{1}{\Gamma(r)}\lambda^r x^{r-1} e^{-\lambda x}$	$\mu = r/\lambda$	$(1-\lambda t)^{-r}$
$x \geq 0, r > 0, \lambda > 0$	$\sigma^2 = r/\lambda^2$	$t < 1\lambda$
Normal		
$f(x;\mu,\sigma) = \frac{1}{\sigma\sqrt{2\pi}} e^{-(x-\mu)^2/(2\sigma^2)}$	$\mu = \mu$	$e^{\mu t + \sigma^2 t^2/2}$
$-\infty < x < \infty, \mu > 0, \sigma > 0$	$\sigma^2 = \sigma^2$	
Student's t		
$f(x;k) = \frac{\Gamma(\frac{k+1}{2})}{\sqrt{\pi k}\Gamma(\frac{k}{2})}\left(1 + x^2/k\right)^{-(k+1)/2}$	$\mu = 0$	Does not exist
$-\infty < x < \infty, k \in \mathbf{Z}^+$	$\sigma^2 = \frac{k}{k-2}$	
Uniform		
$f(x;a,b) = \frac{1}{b-a}$	$\mu = \frac{a+b}{2}$	$\frac{e^{bt} - e^{at}}{(b-a)^t}$
$a < x < b, -\infty < a < b < \infty$	$\sigma^2 = (b-a)^2/12$	
Weibull		
$f(x;k,\lambda) = \frac{k \cdot x^{k-1}}{\lambda^k} e^{-(x/\lambda)^k}$	$\mu = \frac{\Gamma(1+1/k)}{\lambda^{1/k}}$	$\sum_{n=0}^{\infty} \Gamma\left(1+\frac{n}{k}\right)\frac{t^n \lambda^n}{n!}$
$x > 0, k > 0, \lambda > 0$	$\sigma^2 = \frac{\Gamma(1+2/k) - \left(\Gamma(1+1/k)\right)^2}{\lambda^{2/k}}$	

SOLUTIONS TO ODD-NUMBERED EXERCISES

Chapter 1

1. (a) Population: high school students. Sample: 2000 high school students. Statistic 47%. (c) Population: people living in the United States Parameter: 13.9%.
3. (a) Observational study. (b) No, we cannot establish causation. The participants were not randomized into the two groups. (c) No, we cannot generalize to all urban American adolescents.
5. n/N.

Chapter 2

1. $\bar{x} = 12$, $m = 11.5$. $\tilde{x} = 2.26$, $\tilde{m} = 2.39$. No, not the same.
3. (a) No. (b) No. (c) f is linear. (d) f is an increasing (or decreasing) function and n is odd.
5. (a) Favor: 1885; Oppose: 930. (b) The `summary` command indicates that there were many nonresponses. The `table` command does not give any indication that there were nonresponses.

Mathematical Statistics with Resampling and R, First Edition. By Laura Chihara and Tim Hesterberg.

(c)

	Death Penalty	
Own Gun?	Favor	Oppose
No	375	199
Refused	7	2
Yes	243	59

(d) 80.4% of gun owners favor the death penalty, and 65.3% of non-gun owners favor the death penalty.

7. (a) True.

$$\frac{w_1 + w_2 + \cdots + w_n}{n} = \frac{n\bar{x} + n\bar{y}}{n} = \bar{x} + \bar{y}.$$

(b) True.

9. *Hint:* Show that the function is symmetric about θ: $f(\theta - x) = f(\theta + x)$.

11. $F^{-1}(y) = a\sqrt{y}$, so $a\sqrt{\alpha/2}$ and $a\sqrt{1-\alpha/2}$ are the desired quantiles.

13. No, the discrete nature of this random variable results in $P(X \leq 1) = 0.007$ and $P(X \leq 2) = 0.035$.

15. (a) 3. (b) 4. (c) 6.

Chapter 3

1. (a) $\bar{X}_t - \bar{X}_c = 11 - 7 = 4$. (b) $t = 4 =$ observed statistic; P-value $= 4/10 = 0.2$. (c) 0.2.

3. (a) Observed test statistic 5.886; one simulation gives a P-value $= 0.0003$, so we conclude that the difference in mean delay times between the two carriers is statistically significant. (b) Observed test statistic -5.663; one simulation gives P-value $= 0.0001$, so the difference in mean delay times between May and June is statistically significant.

5. Code provided at `https://sites.google.com/site/ChiharaHesterberg`.

7. (a) The difference in proportions is statistically significant.

9. (b) Mean number of strikeouts in away games: 7.31; mean number of strikeouts in home games: 6.95. (c) Observed test statistic is 0.358. P-value $= 0.21$.

11. $c = 6.6814$, df$= 2$, P-value $= 0.0354$. Therefore, we conclude there is an association between age and support of marijuana for medicinal purposes.

13. (a) This is a test of homogeneity. Let π_{ij} denote the proportion of fish from region i ($i = 1, 2, 3$) and j rays ($j = \geq 36, 35, 34, 33, 32, \leq 31$). Then H_0: $\pi_{1j} = \pi_{2j} = \pi_{3j}$ versus H_a: at least one pair of proportions not the same. (b) $c = 12.803$, df $= 10$, P-value $= 0.23$. We conclude there is not

enough evidence to support the hypothesis that fin ray counts differ across the regions.

15. $c = 13.4621$, df $= 1$, P-value $= 0.0002$. The difference between the two airlines is statistically significant.
17. P-value=0.005. Gender and happiness are not independent.
19. (a) Marginal probabilities do not change; degree of freedom will not change either.
21. $c = 9.0978$, df $= 5 - 1 = 4$, P-value is 0.0587. Yes, it is plausible that these numbers are drawn from a distribution with pdf $f(x) = 2/x^3, x \geq 1$.
23. (a) Using the intervals (0,17], (17,22], (22,28], (28,33], and (33,44], $c = 28.388$, df $= 5 - 0 - 1 = 4$, and P-value $= 0$. Test statistics may vary depending on the cutoffs used for the intervals. Conclude that the data do not come from $N(24, 10^2)$.
25. $c = 2.26$, df $= 4$, P-value=0.69. The numbers appear to be drawn randomly.
27. (a) Expected counts: 3.43,11.57, 4.57, 15.42. Two are less than 5. (b) $\binom{20}{2}\binom{15}{6}/\binom{35}{8} = 0.0404$. (c) $\sum_{k=0}^{2}\binom{20}{k}\binom{15}{8-k}/\binom{35}{k} = 0.0461$. (d) If the selection had been completely random, then the chance of obtaining a committee consisting of two or fewer seniors is only 4.6%, which gives (mild) evidence that there is cause for suspicion.

Chapter 4

1. There are 20 possible sets of size 3. The mean of the medians is 5.7. The median of the population is 5.5.
3. (a) {6, 8, 8, 9, 10, 10, 10, 11, 12, 13, 14, 14}. (b) No. (c) $\mathrm{E}[X] + \mathrm{E}[Y] = 10.25$.
5. 0.001.
7. 0.332.
9. 0.022.
11. 8.
13. (a) $W = \bar{X} + \bar{Y} \sim N(36, 3.11^2)$. (c) 0.901.
15. See Theorem B.15.
17. (c) 0.506.
19. (b) 10, 5. (c) 0.5.
21. (a) 13.86.
23. (a) $f_{\max}(x) = 8(1/x^2 - 1/x^3)$. (b) 1.545.
25. $120e^{-12x}(1 - e^{-12x})^9$.
27. $P(X = k) = (30^k e^{-30})/k!, k = 0, 1, \ldots$.
29. (a) Yes. (c) 4.471.

Chapter 5

1. Answers will vary.
3. (a) 3^3, order matters. (b) 10 distinct. (c) 3. (d) No.
5. (a) $\binom{9}{5}$. (b) $\binom{2n-1}{n}$.

7. (a) $N(36, 8^2/200)$. (c) Bootstrap distribution using sample in (b): mean 35.91, SE 0.53. (d)

	Mean	Standard Deviation
Population	36	8
Sampling distribution of $\bar{X}$	36	$8/\sqrt{200} = 0.566$
Sample	35.01	7.386
Bootstrap distribution	35.91	0.53

9. For odd n, median will be one of the sample points. Thus, for small n, there will be only n possible values for the median, so the sampling distribution is much more “granular" than when n is even. As n increases, this becomes less apparent.
11. About (13.68, 23.23). Bias is about 25%.
13. (a) Bell shaped, mean at about 5.196, standard deviation at about 1.43. (b) (2.33, 8). (c) For the bootstrap distribution, we sample with replacement from the original sample: we sample from the men and then from the women. With the permutation distribution, we sample assuming that there is no difference in the means; in particular, we sample without replacement from the pooled data.
15. (3.49, 11.49).
17. (c) 16.3, SE 0.319. (d) (1.17, 2.22).

Chapter 6

1. $\hat{p} = X/n$.
3. $\hat{\theta} = 4/(\sqrt{5} + 3 + 3 + \sqrt{10}) = 0.351$.
5. (a) $\hat{\mu} = \bar{X}$. $\hat{\sigma} = \sqrt{(1/n)\sum_{i=1}^{n}(X_i - \mu)^2}$.
7. Does not exist.
9. $\hat{N} = 9$.
11. $\hat{\lambda} = (n+m)/(\sum_{i=1}^{n} X_i + 2\sum_{j=1}^{m} Y_j)$.
13. (a) $\hat{\alpha} = n/\sum_{i=1}^{n} x_i^{\beta}$.
    (b)

$$n/\alpha = \sum_{i=1}^{n} x_i^{\beta},$$

$$\alpha \sum_{i=1}^{n} X_i^{\beta} \ln(x_i) - n/\beta = n \sum (\ln(x_i)).$$

15. $\hat{r} = 22.88$, $\hat{\lambda} = 3.138$.
17. Shape=0.917, scale= 17.344, c=0.8095, so the evidence supports the theory that times between successive earthquakes follow a Weibull distribution.
19. (a) $\bar{X}/(\bar{X} - 2)$.
21. p.
23. $\theta + 1$, so $\bar{X}$ is biased.
25. $\sum_{i=1}^{n} a_i = 1$.
27. (a) $-\sigma^2/n$. (b) $\sigma^4/n^2 \sum_{i=1}^{n}(X_i - \bar{X})^2$. (c) $2(n-1)\sigma^4/n^2 + (\sigma^2/n)^2$.

29. $\text{Var}[\hat{\theta_1}] = \theta^2$, $\text{Var}[\hat{\theta_2}] = (1/4)(2\theta^2) = (1/2)\theta^2$ and $\text{Var}[\hat{\theta_3}] = (1/9)(\theta^2 + 2^2\theta^2) = (5/9)\theta^2$, so $\hat{\theta_2}$ is the most efficient and $\hat{\theta_1}$ is the least efficient.
31. The two curves approach each other.
33. (a) $2/(3\theta)$. (b) $-17/270$. (c) $27T/10$, where $T = X_1/9 + X_2/9 + X_3/3$.
35. (a) $f_{\max}(x) = (n\alpha x^{\alpha n-1})/(\beta^{\alpha n})$. (b) $n\alpha\beta/(\alpha n + 1)$. (c) Bias: $-\beta/(\alpha n + 1)$. (d) MSE: $2\beta^2/((\alpha n + 1)(\alpha n + 2))$.
37. (d) $(\pi - 4)/(4\lambda)$.

Chapter 7

1. (a) Population mean here is not random. The probability that it is contained in the interval is 0 or 1. (c) Not 95% of the time, but 95% of the confidence intervals generated by each sample will contain true mean. (e) Each sample gives rise to a different confidence interval and 95% of these intervals will contain the true mean.
3. (a) (201.8, 218.2). (b) 97. (c) 166.
5. $4n$.
7. 117.624.
9. (9.19, 13.39) cm.
11. (b) (−533.61, −83.294).
13. (b) $(0.772, \infty)$.
15. (b) (59.451, 70.266). (c) Yes, (59.1, 66.8). The upper endpoint decreased 3 in.
17. Unbalanced sample sizes, pooled CI's are not capturing true mean difference at 95%. Balanced case, pooled CI does better.
19. $[5.14, \infty)$.
21. (a) 1021. (b) 968.
23. (a) (0.0691, 0.1343). (c) Yes, the intervals overlap; thus, we cannot make the firm conclusion that taking the drug makes a difference.
25. (b) (52.87, 103.29). (c) (54.78, 105.01). Bootstrap t: (56.84, 112.15). Bootstrap t. Intervals will vary slightly.
27. (a) 1497 nonsmokers, 90 smokers. (c) (−27.68, 190.69). This interval contains 0. Percentile: (−24.79, 187.13). Bootstrap t: (−27.46, 190.27). They are all roughly the same. Report formula t or bootstrap t.
29. (95.76, 155.51).

Chapter 8

1. P-value = 0.237, so we conclude that mean calcium levels are the same.
3. P-value = 0.007, so we conclude that, on average, body temperatures are higher for children in Sodor.
5. P-value = 0.0002, so we conclude that mean pH levels are different at the two locations.

7. P-value $= 0.001$; with 95% confidence, we conclude that, on average, ales have at least 7.1 more calories than lagers.
9. P-value $= 0.0187$. Evidence supports the hypothesis that the proportion of sex workers in Bamako who are HIV positive is greater than 0.391.
11. P-value $= 0.01$. Evidence supports the claim that the proportions in the two regions are not the same.
13. *Hint:* Use the `rbinom` command to draw random samples. For example, `rbinom(1000, 10, .5)` will result in 1000 samples of size 10 with $p = 0.5$.
15. (a) Type I error: we conclude that arsenic levels in the community are higher than 10 ppb even though it is, in fact, not higher than or equal to 10 ppb. The community may take unnecessary and expensive measures to lower the arsenic levels. Type II error: we conclude that arsenic levels in the community are less than or equal to 10 ppb even though in reality they are actually higher than 10 ppb. Community is drinking unsafe water and not taking any action.
(b) Not necessarily, since the t distribution has fatter tails.
17. 8.
19. $n \geq 15$.
21. $X \leq 22$ or $X \geq 38$.
23. (a) 0.04. (b) 0.527.
25. (a) 0.0497. (b) 0.548.
27. $1 - (3/4)^{\theta+1}$.
29. $1 - \beta = P\left(Z < q_1 + \dfrac{\mu_0 - \mu_1}{\sigma/\sqrt{n}}\right) + P\left(Z < q_2 - \dfrac{\mu_1 - \mu_0}{\sigma/\sqrt{n}}\right)$
31. 0.0003.
33. (b) 0.224.

Chapter 9

1. $-1/100$.
3. 118.
5. 1.
7. (a) 0.4996. (c) 0.992.
9. When solving for a and b that minimizes $g(a, b)$, we take the partial derivative: $\partial g/\partial a = 2\sum_{i=1}^{n}(y_i - (a + bx_i))(-1) = 0$ Thus, $2\sum_{i=1}^{n}(y_i - (\hat{a} + \hat{b}x_i))(-1) = 0$. But $\hat{a} + \hat{b}x_i = \hat{y}_i$. So $\sum_{i=1}^{n}(y_i - \hat{y}_i) = 0$.
11. (a) `weight` $= 20.078 + 1.607$`height`. (b) 116.498. (c) $R^2 = 0.5625$; 56% of the variability in weight is explained by the regression line.
13. (a) Increasing variance. (c) Indicates that a linear model is appropriate.
15. (b) `Kills` $= 1.736 + 0.947$`Assts`. For every additional assist per set, there is associated nearly one additional kill. About 93.7% of the variability in kills per set is explained by this model. (c) Yes, a straight line model is appropriate.
17. (a) 0.349. (b) `Weight` $= -2379.69 + 148.995 \times$ `Gestation`. (c) $R^2 = 0.122$. (d) The problem is that σ is not constant for different gestation lengths (in part because there are few data points at 42 weeks).
19. (0.6994, 1.0345).
21. (b) `Level` $= -3280 + 1.83$`Year`. (c) No, there appears to be serial correlation.

23. 0.4397; (0.206, 0.663).

25. Proof omitted. 27. Proof omitted. 29. Proof omitted.

31. (a) $\ln(\hat{p}/(1-\hat{p})) = 2.905 - 0.037\texttt{Temperature}$. (b) $e^{-0.037\times(-5)} = 1.203$. A 5°F decrease in temperature increases the odds of an O-ring incident by a (multiplicative) factor of 1.203. Or, there is a 20% increase in the odds of an O-ring incident for every 5°F drop in temperature. (d) The range of temperatures in the data set is 53–81°F. We are extrapolating quite a bit at temperature 33°F.

33. (a) $\ln(\hat{p}/(1-\hat{p})) = -2.23 + 0.29\texttt{Hits}$. (b) $e^{0.29} = 1.03$. Each additional hit increases the odds of winning by 3%. (d) (0.86, 0.98). Answers will vary.

Chapter 10

1. (a) Posterior column: 0.954, 0.5828, 0.3221. (b) After observing four wins out of five, you now believe that there is a 32.2% chance that your long-term probability of winning is 0.6. (c) 0.5, 0.5227.

3. (b) After observing four wins out of seven, you now believe that there is a 2.3% chance that your probability of winning is 0.2. (c) 0.5, 0.55.

5. (a) (0.2048, 0.2424). (b) (0.228, 0.261). (c) 0.0693.

7. 0.55, 0.0012. 9. $\mu\,|\,\mathbf{x} \sim N(0.829, 0.0323^2)$.

11. (a) $\mu\,|\,\mathbf{x} = 19.53$, $\sigma\,|\,\mathbf{x} = 1.4797$, precision 0.457. (c) $\mu\,|\,\mathbf{x} = 19.14$, $\sigma\,|\,\mathbf{x} = 1.531$, precision 0.4267.

13. First, it is not conjugate, so computations are difficult and second it is nonzero for $\theta < 0$ and $\theta > 1$, values that are impossible.

15. (a) $f(\theta) = 1/\theta^n$. (b) Pareto distribution with parameters $\alpha + n, \beta$. (c) 0.0089.

17. (a) $\theta^n e^{-\theta\sum_{i=1}^{n} X_i}$. (c) Gamma with parameters 14, 17.

Chapter 11

1. (a) The female distribution is centered about 9.3 and male distribution about 14.5, the means of the original data. The male distribution has a larger standard deviation, 1.16 versus 0.92; these numbers match $s/\sqrt(n)$ for each data set. Both smoothed bootstrap distributions are very close to normal. (d) Both are very close to normal, with the same mean −5.2 (for female minus male); the smoothed bootstrap standard error is larger, 1.48 compared to 1.44; for comparison, the usual formula standard error is 1.48.

3. The distribution is close to normal, with slight positive skewness. The mean is 7.30 and standard deviation 6.9. A 95% bootstrap percentile interval is (6.01, 8.70).

5. (a) $r/(n\lambda^2)$. (b) $f(x) = x^2$, so $f'(\mu) = 2\mu$. $\text{Var}[\bar{X}^2] \approx (2\mu)^2 r/(n\lambda^2)$. (c) $f(x) = \log(x)$, so $f'(\mu) = 1/\mu$. $\text{Var}[\bar{X}^2] \approx (1/\mu)^2 r/(n\lambda^2)$. (d) $f(x) = \sqrt(x)$, so $f'(\mu) = 1/(2\sqrt{\mu})$. $\text{Var}[\bar{X}^2] \approx (1/(4\mu))r/(n\lambda^2)$. (e) Pick sample size n and parameters r and λ. Generate N samples, say $N = 10^4$, for each compute $\bar{x}$, $\bar{x}^2$, $\log \bar{x}$, and $\sqrt{\bar{x}}$. Compute the variance of each set of N statistics and compare. Double-check by repeating with other r, λ, and n.
7. (a) 0.9046; a confidence interval based on 10^6 replications; would be $0.9046 \pm 2 \times 0.000125$. (b) 0.9045242 with absolute error < 1e−14.
9. If $f(x) = \exp(-x)$ for $x > 0$ and $h(x) = \sin(x)$, then compare `integrate(function(x) sin(x) * exp(-x), 0, Inf)` and `N <- 1000000; x <- rexp(N); mean(sin(x))`.
11. Using $N = 10^6$ gives approximately (a) $0.597 \pm 2 \cdot 0.00037$ and (b) $0.597 \pm 2 \times 0.00025$. (d) The second one gives more accurate estimates. The corresponding graph is flatter, the h values do not vary as much.
13. Proof omitted.

Appendix B

1. (a) 0.201. (b) 9.
3. (a) 0.00138. (b) 16.65.

BIBLIOGRAPHY

Agresti, A. (1992). A survey of exact inference for contingency tables. *Statistical Science 7*(2), 131–177.

Agresti, A. and B. Caffo (2000). Simple and effective confidence intervals for proportions and differences of proportions result from adding two successes and two failures. *The American Statistician 54*(4), 280–288.

Agresti, A. and B. Coull (1998). Approximate is better than "exact" for interval estimation of binomial proportions. *The American Statistician 52*(2), 119–126.

Barnsley, R., A. Thompson, and P. Legault (1992). Family planning: Football style. The relative age effect in football. *International Review for the Sociology of Sport 27*, 77–87.

Bellman, R. (1966). *Adaptive Control Processes: A Guided Tour*. Princeton, NJ: Princeton University Press.

Box, G. (1953). Non-normality and tests on variances. *Biometrika 40*, 318–335.

Brashares, J., P. Arcese, M. Sam, P. Coppolillo, A. Sinclair, and A. Balmford (2004). Bushmeat hunting, wildlife declines, and fish supply in West Africa. *Science 306*(5699), 1180–1183.

Brown, L., T. Cai, and A. DasGupta (2001). Interval estimation for a binomial proportion. *Statistical Science 16*(2), 101–133.

Camill, P., L. Chihara, B. Adams, C. Andreassi, A. Barry, S. Kalim, J. Limmer, M. Mandell, and G. Rafert (2010). Early life history transitions and recruitment of *Picea mariana* in thawed boreal permafrost peatlands. *Ecology 2*, 448–459.

Carlin, B. and T. Louis (2009). *Bayesian Methods for Data Analysis* (third ed.). Boca Raton, FL: Chapman & Hall/CRC.

Casella, G. and R. Berger (2001). *Statistical Inference*. Belmont, CA: Duxbury Press.

Mathematical Statistics with Resampling and R, First Edition. By Laura Chihara and Tim Hesterberg.

Catchpole, N. (2004). Beer and hot wings data. Private communication.

Chan, K. (2008). Chinese children's perceptions of advertising and brands: An urban rural comparison. *Journal of Consumer Marketing 25*(2), 74–84.

Chance, B. and A. Rossman (2005). *Investigating Statistical Concepts, Applications, and Methods* (first ed.). Belmont, CA: Duxbury Press.

Chernick, M. (1999). *Bootstrap Methods: A Practitioner's Guide*. New York: Wiley.

Cobb, G. (2007). The introductory statistics course: A ptolemaic curriculum? *Technology Innovations in Statistics Education 1*. `http://repositories.cdlib.org/uclastat/cts/tise/vol1/iss1/art1`.

Cochran, W. (1954). Some methods for strengthening the common χ^2 tests. *Biometrics 10*, 417–451.

Collett, D. (2003). *Modelling Binary Data* (second ed.). Boca Raton, FL: Chapman & Hall/CRC.

Craig, A. (1932). On the distributions of certain statistics. *American Journal of Mathematics 54*, 353–366.

DASL. Data and stories library. Carnegie Mellon University Statistics Department. `http://lib.stat.cmu.edu/DASL/DataArchive.html`.

Davison, A. and D. Hinkley (1997). *Bootstrap Methods and Their Application*. Cambridge: Cambridge University Press.

Davy, S., B. Benes, and J. Driskell (2006). Sex differences in dieting trends, eating habits, and nutrition beliefs of a group of Midwestern college students. *Journal of the American Dietetic Association 106*, 1673–1677.

DeGroot, M. and M. Schervish (2002). *Probability and Statistics* (third ed.). New York: Pearson.

Dobson, A. (2002). *An Introduction to Generalized Linear Models* (first ed.). Boca Raton, FL: Chapman & Hall/CRC.

Draper, N. and H. Smith (1998). *Applied Regression Analysis* (third ed.). New York: Wiley.

Efron, B. and R. Tibshirani (1993). *An Introduction to the Bootstrap*. New York: Chapman & Hall.

Ernst, M. (2004). Permutation methods: A basis for exact inference. *Statistical Science 19*(4), 676–685.

Evans, J. (2008). Starcraft data set. Private communication.

Fisher, R. A. (1922). On the interpretation of chi-square from contingency tables, and the calculation of *P*. *Journal of the Royal Statistical Society 85*, 87–94.

Fisher, R. A. (1924). The conditions under which chi-square measures the discrepancy between observation and hypothesis. *Journal of the Royal Statistical Society 87*, 442–450.

Gelman, A., J. Carlin, H. Stern, and R. Rubin (2010). *Bayesian Data Analysis* (second ed.). Boca Raton, FL: Chapman & Hall/CRC.

Ghahramani, S. (2004). *Fundamentals of Probability with Stochastic Processes* (third ed.). Upper Saddle River, NJ: Prentice Hall.

Gilks, W., S. Richardson, and D. Spiegelhalter (1996). *Markov Chain Monte Carlo in Practice* (first ed.). London: Chapman & Hall/CRC.

Goodman, L. (1952). Serial number analysis. *Journal of the American Statistical Association 47*, 622–634.

Gregory, P. and R. Tasto (1976). Results of the jack mackerel subpopulation discrimination feasibility study. Technical report, California Department of Fish and Game. http://aquacomm.fcla.edu/84/1/Marine_Resources_Administrative_Report_No[1]._76-2.pdf.

Greif, R., O. Akca, E.-P. Horn, A. Kurz, and I. Sessler (2000). Supplemental perioperative oxygen to reduce the incidence of surgical-wound infection. *The New England Journal of Medicine 342*, 161–167.

Hammersley, J. and D. Handscomb (1965). *Monte Carlo Methods*. London: Fletcher and Son Ltd.

Hasumi, T., T. Akimoto, and Y. Aizawa (2009). The Weibull–log Weibull distribution for interoccurrence times of earthquakes. *Physics A: Statistical Mechanics and Its Applications 388*, 491–498.

Haynor, B., D. Lojovich, B. Starr, and S. Sultan (2010). Service times data. Private communication.

Hesterberg, T. (1988). Advances in importance sampling. Ph.D. thesis, Statistics Department, Stanford University.

Hesterberg, T. (1991). Importance sampling for Bayesian estimation. In A. Buja and P. Tukey (Eds.), *Computing and Graphics in Statistics*, Volume 36 of *Volumes in Mathematics and Its Applications*, pp. 63–75. New York: Springer.

Hesterberg, T. (1995). Weighted average importance sampling and defensive mixture distributions. *Technometrics 37*(2), 185–194.

Hesterberg, T. (1998). Simulation and bootstrapping for teaching statistics. In *Proceedings of the Section on Statistical Education*, pp. 44–52. American Statistical Association.

Hesterberg, T. (2006). Bootstrapping students' understanding of statistical concepts. In G. F. Burrill and P. C. Elliot (Eds.), *Thinking and Reasoning with Data and Chance: NCTM Yearbook*, pp. 391–416. Reston, VA: National Council of Teachers of Mathematics.

Hesterberg, T. (2008). It's time to retire the $n \geq 30$ rule. In *Proceedings of the American Statistical Association, Statistical Computing Section.*

Hesterberg, T., D. Moore, S. Monaghan, A. Clipson, and R. Epstein (2007). Bootstrap methods and permutation tests. In D. Moore and G. McCabe (Eds.), *Introduction to the Practice of Statistics*. New York: W. H. Freeman.

Hien, R. and S. Baker (2010). Book prices data. Private communication.

Justus, C. G., W. R. Hargraves, A. Mikhail, and D. Graber (1978). Methods for estimating wind speed frequency distributions. *Journal of Applied Meteorology 17*, 350–353.

Kutner, M., C. Nachtsheim, J. Neter, and W. Li (2005). *Applied Linear Statistical Models* (fifth ed.). New York: McGraw-Hill.

Larsen, J. and M. Marx (2006). *An Introduction to Mathematical Statistics and Its Applications* (fourth ed.). Upper Saddle River, NJ: Pearson Prentice Hall.

Latter, O. (1902). An enquiry into the dimensions of the cuckoo's egg and the relation of the variations to the size of the eggs of the foster-parent, with notes on coloration. *Biometrika 1*(2), 164–176.

Liu, J. (2001). *Monte Carlo Strategies in Scientific Computing*. New York: Springer.

Lohr, S. (1991). *Sampling: Design and Analysis* (second ed.). Belmont, CA: Duxbury Press.

Ludbrook, J. and H. Dudley (1998). Why permutation tests are superior to t and F tests in biomedical research. *The American Statistician 52*, 127–132.

McCullagh, P. and J. Nelder (1989). *Generalized Linear Models* (second ed.). Boca Raton, FL: Chapman & Hall/CRC.

McCullough, B. D. (2000). Is it safe to assume that software is accurate. *International Journal of Forecasting 16*, 349–357.

Miao, W. and P. Chiou (2008). Confidence intervals for the difference between two means. *Computational Statistics and Data Analysis 52*, 2238–2248.

Monson, B. (2010). Walleye data. Minnesota Pollution Control Agency, private communication.

Moore, D., G. McCabe, and B. Craig (2009). *Introduction to the Practice of Statistics* (sixth ed.). New York: W. H. Freeman.

Moser, B. and G. Stevens (1992). Homogeneity of variance in the two-sample means test. *The American Statistician 46*, 19–21.

Moser, B., G. Stevens, and C. Watts (1989). The two-sample t test versus Satterthwaite's approximate F test. *Communication in Statistics: Theory and Methods 18*, 3963–3075.

Mukamal, K., L. Kuller, A. Fitzpatrick, W. Longstreth, M. Mittleman, and D. Siscovich (2003). Prospective study of alcohol consumption and risk of dementia in older adults. *Journal of the American Medical Association 289*, 1405–1413.

Newcombe, R. (1998a). Interval estimation for the difference between independent proportions: Comparison of eleven methods. *Statistics in Medicine 17*, 873–890.

Newcombe, R. (1998b). Two-sided confidence intervals for the single proportion: Comparison of seven methods. *Statistics in Medicine 17*, 857–872.

Nolan, D. and T. Speed (2009a). *Stat Labs: Mathematical Statistics Through Applications*. New York: Springer.

Nolan, D. and T. Speed (2009b). Teaching statistics theory through applications. *The American Statistician 53*, 370–375.

O'Hagan, A. (2009). *Kendall's Advanced Theory of Statistics*, Volume 2B. New York: Wiley.

Pan, W. (2009). Approximate confidence intervals for one proportion and difference of two proportions. *Computational Statistics and Data Analysis 40*, 143–157.

Paulson, E. (1940). A note on the distribution of the median. *National Mathematics Magazine 14*, 379–382.

Pearson, K. (1900). On a criterion that a given system of deviations from the probable in the case of a correlated system of variables is such that it can be reasonably supposed to have arisen from random sampling. *Philosophical Magazine Series 5 50*, 157–175.

Pearson, K. (1922). On the χ^2 goodness of fit. *Biometrika 14*, 186–191.

Pearson, K. (1923). Further note on the χ^2 goodness of fit. *Biometrika 14*, 418.

Pitman, J. (1993). *Probability*. New York: Springer.

Primack, B., E. Douglas, and K. Kraemer (2010). Exposure to cannabis in popular music and cannabis use among adolescents. *Addiction 105*, 515–523.

Rencher, A. (2000). *Linear Models in Statistics*. New York: Wiley.

Rice, J. (2006). *Mathematical Statistics and Data Analysis* (third ed.). Belmont, CA: Duxbury Press.

Ricker, W. (1973). Linear regressions in fishery research. *Bulletin of Fisheries Research Board of Canada 30*, 409–433.

Ricker, W. (1975). Computation and interpretation of biological statistics of fish populations. *Bulletin of Fisheries Research Board of Canada 191*, 382.

Rider, P. (1960). Variance of the median of samples from a Cauchy distribution. *Journal of the American Statistical Association 55*, 322–323.

Robert, C. and G. Casella (2004). *Monte Carlo Statistical Methods* (second ed.). New York: Springer.

Robert, C. and G. Casella (2010). *Introducing Monte Carlo Methods with R*. New York: Springer.

Rodgers, B. and T. Robinson (2004). Do basic television channels have more commercials than extended cable channels? `http://apstats.4t.com`.

Ross, S. (2009). *A First Course in Probability* (eighth ed.). Upper Saddle River, NJ: Prentice Hall.

Rubinstein, B. (1981). *Simulation and the Monte Carlo Method*. New York: Wiley.

Samaha, F. F., N. Iqbal, P. Seshadri, et al. (2009). A low-carbohydrate as compared with a low-fat diet in severe obesity. *New England Journal of Medicine 348*(21), 2074–2081.

Sandler, R. S., S. Halabi, J. A. Baron, B. S. Bandinger, et al. (2003). A randomized trial of aspirin to prevent colorectal adenomas in patients with previous colorectal cancer. *New England Journal of Medicine 348*(10), 883–890.

Scheaffer, R. and L. Young (2010). *Introduction to Probability and Its Applications* (third ed.). Boston, MA: Brooks/Cole, Cengage Learning.

Scheffe, H. (1970). Practical solutions of the Behrens–Fisher problem. *Journal of the American Statistical Association 65*, 1501–1508.

Schenker, N. and J. Gentleman (2001). On judging the significance of differences by examining the overlap between confidence intervals. *The American Statistician 55*, 182–186.

Seguro, J. V. and T. W. Lambert (2000). Modern estimation of the parameters of the Weibull wind speed distribution for wind energy analysis. *Journal of Wind Engineering and Industrial Aerodynamics 85*, 75–84.

Stuart, A. and K. Ord (2009). *Kendall's Advanced Theory of Statistics*, Volume 1: *Distribution Theory* (sixth ed.). New York: Wiley.

Stuart, A., K. Ord, and S. Arnold (2009). *Kendall's Advanced Theory of Statistics*, Volume 2A (sixth ed.). New York: Wiley.

Suess, E. and B. Trumbo (2010). *Introduction to Probability Simulation and Gibbs Sampling with R*. New York: Springer.

Tiampo, K., D. Weatherley, and S. Weinstein (2008). *Earthquakes: Simulations, Sources and Tsunamis*. Pageoph Topical Volumes. Basel: Birkhauser.

Tippett, L. H. C. (1952). *The Methods of Statistics* (fourth ed.). New York: Wiley.

Tukey, J. (1977). *Exploratory Data Analysis* (first ed.). Reading, MA: Addison Wesley.

Verzani, J. (2010). UsingR: Data sets for the text "Using R for Introductory Statistics". R package version 0.1–12.

Voss, L. and J. Mulligan (2000). Bullying in school: Are short pupils at risk? Questionnaire study in a cohort. *British Medical Journal 320*, 612–613.

Wackerly, D., W. I. Mendenhall, and R. Scheaffer (2002). *Mathematical Statistics with Applications* (sixth ed.). Pacific Grove, CA: Duxbury Press.

Waldrop, R. (1995). *Statistics: Learning in the Presence of Variation* (first ed.). Dubuque, IA: William C. Brown Publishers.

Wang, C., C. Schmid, P. Hibberd, R. Kalish, R. Roubenoff, R. Rones, and T. McAlindon (2009). Tai Chi is effective in treating knee osteoarthritis: A randomized controlled trial. *Arthritis Care and Research 61*, 1545–1553.

Weisberg, S. (2005). *Applied Linear Regression* (third ed.). New York: Wiley.

Weisser, D. (2003). A wind energy analysis of Grenada: An estimation using the 'Weibull' density function. *Renewable Energy 28*, 1803–1812.

Westman, E., W. Yancy, J. Mavropoulos, M. Marquart, and J. McDuffie (2008). The effect of a low-carbohydrate, ketogenic diet versus a low-glycemic index diet on glycemic control in type 2 diabetes mellitus. *Nutrition and Metabolism 5*. BioMed Central, The Open Access Publisher, doi:10.1186/1743-7075-5-36.

Wilson, E. B. (1927). Probable inference, the law of succession, and statistical inference. *Journal of the American Statistical Association 22*, 209–212.

Zhou, J., E. Erdem, G. Li, and J. Shi (2010). Comprehensive evaluation of wind speed distribution models: A case study for North Dakota sites. *Energy Conversion and Management 51*, 1449–1458.

INDEX

α (Significance level), 37, 223
Agresti-Coull interval, 191–194
Alternative hypothesis, 36
Asymptotic bias, 151

β (Type II error), 226
Barchart, 14
Bayes, 301–322
  Bayes Theorem, 302
  computational issues, 340–341
  importance sampling, 355–359
Bellman, Richard, 341
Bernoulli distribution, 373
Beta distribution, 390–391
  as prior for binomial data, 310
Bias, 122–124, 148
Binomial distribution, 373–374
Bonferroni correction, 233
Bootstrap, 99–129
  accuracy, 125
  bias, 122
  bias for parametric bootstrap, 335
  bootstrap distributions and sampling distributions, 104
  bootstrap idea, the, 100
  bootstrap sample, 100
  center, bias, spread, shape, 103–104
  distribution too narrow, 330–331
  estimate cdf with ecdf, 107
  how many resamples?, 129
  Johnson's t interval, 207
  parametric, 331–335
  percentile interval, 113–114
  vs. permutation test, 116
  ratio of means, 120
  regression, 279
  relative risk, 123–124
  sample rows of the data, 279
  single population, 101
  smoothed, 327–331
  standard error, 102
  stratified, 340
  t confidence interval, 195–200
  two populations, 114
  variation of bootstrap distribution, 125, 128
  Z interval, 208
Boxplot, 19

Mathematical Statistics with Resampling and R, First Edition. By Laura Chihara and Tim Hesterberg.

χ^2_m, 57, 385
Case studies
  Beer and hot wings, 8, 26–27, 38–42, 47–48, 295, 359
  Birth weights of babies, 2, 20, 99–102, 190, 198, 215–217, 297
  Black spruce seedlings, 8, 31, 70, 202, 240, 248, 255,261–262
  Bushmeat, 283–286
  Flight delays, 1, 13–15, 20, 30, 69, 72, 133, 203
  General Social Survey (GSS), 5–6, 31, 52–56, 72, 192, 195, 205, 218, 219 244,324
  Verizon repair times, 3, 44–47, 48–51, 117–118, 120–122, 199
  Wind energy, 144–146, 154, 331–334
Categorical variable, *see* Variable, categorical, 406
Cauchy distribution, 142, 159
Cause and effect, 9
cdf, *see* Cumulative distribution function
Census, 6
Center, 14, 17
Central Limit Theorem, 84–92
  accuracy, 90
  binomial data, 87–90
  finite population, 91
Chebyshev's Inequality, 158, 368
Chi-square distribution, 57, 385–388
  sum of chi-square random variables, 386
Chi-square test, *see* Independence, 48
Confidence interval
  Agresti-Coull, 193–194
  bootstrap t vs. formula t, 200
  for difference of two proportions, 194–195
  for difference of two means, 178–183
  in general, 183–189
  Johnson's t, 207
  for one mean (σ known), 167–172
  for one mean (σ unknown), 172–178
  for one proportion, 191–194
  one-sided, 189–191
  Score, 192
  t confidence interval, 174, 197, 199
  z, 167–171
Conjugate family, 312
Consistency, 157–159
Contingency table, 14, 52–58
Continuity correction, 89–90, 220
Convergence in probability, 157
Correlation, 251–254
  bootstrap, 279–283
  coefficient, 251
  permutation test for independence, 282
  sample, 253
Covariance, 247–251
Cramer-Rao Inequality, 154
Credible interval, 313
Critical region, 224
Critical value, 224
Cumulative distribution function, 364
Curse of dimensionality, 341

Data sets
  Alelager, 241, 298
  Bangladesh, 110–112, 131, 196, 205, 359
  BookPrices, 133
  Case studies, *see* Case studies
  Cereals, 71
  Challenger, 298
  Fatalities, 286–293
  FishMercury, 132
  Girls2004, 132, 202
  IceCream, 132
  Illiteracy, 295, 298
  Lottery, 73
  Maunaloa, 297
  MnGroundwater, 206
  Phillies2009, 66–67, 70, 73, 299
  Quakes, 162, 359
  Service, 162
  Skating2010, 266–267, 275, 276, 283
  Titanic, 299
  TV, 114–117
  TXBirths2004, 206
  Vollebyall2009, 296
  Walleye, 297
  Watertable, 299
Degrees of freedom
  Welch's approximation, 179, 208–209
Delta method, 335–339
Density estimate
  kernel, 328